Calcium and Cellular Secretion

Calcium and Cellular Secretion

RONALD P. RUBIN

Medical College of Virginia
Virginia Commonwealth University
Richmond, Virginia

PLENUM PRESS • NEW YORK AND LONDON

Library of Congress Cataloging in Publication Data

Main entry under title:

Calcium and cellular secretion.

Bibliography: p.
Includes index.
1. Secretion. 2.Calcium—Physiological effect. I. Rubin, Ronald P. I. Title: Cellular
secretion.
QP190.C34   574.87'6   82-7489
ISBN 0-306-40978-X   AACR2

QP
190
.R8
1982

© 1982 Plenum Press, New York
A Division of Plenum Publishing Corporation
233 Spring Street, New York, N.Y. 10013

To my many colleagues,
whose forays into the arcane land of
secretory phenomena provided the impetus
to undertake and complete this mission.

Preface

Since the publication of my previous monograph in 1974, important progress has been made in the broad area of calcium research, particularly as it pertains to secretory phenomena. The significant advance in methodology, while widening the scope of our knowledge, has caused research in this particular area to become more and more specialized. It has, therefore, become increasingly difficult for researchers to consider and evaluate work outside their own areas of specialization and to comprehend the field from a broad perspective. While many valuable reviews on the importance of calcium in cellular function are being produced, they have not attempted to pull together all of the existing knowledge into a more general narrative. This volume brings together experimental data and theories from diverse sources, and attempts to synthesize into a broad conceptual framework the massive amount of specialized information that presently exists in the general area of calcium metabolism and the secretory process.

It is impossible in a book of this length to discuss specific references in detail and still maintain readability, so generalizations must be made. With this approach, one must be selective, and I apologize to those individuals whose contributions have been omitted. In many instances, I have cited recent references to provide for the reader a current bibliography in a particular area. However, anyone who attempts an undertaking of this magnitude has an intellectual responsibility to inform himself of all of the available information, in order to curtail personal bias. A sincere attempt has been made to delve deeply enough to limit this liability.

Although this narrative will describe the present state of the art regarding the control mechanisms regulating secretory phenomena, a paramount aim is not only to inform the reader but to encourage development of a greater interest in one or another of the areas discussed. This will enable us to continue to build knowledge onto knowledge, which is vital for a true understanding of the fundamental role of calcium in the functioning of the living cell.

Finally, I would like to acknowledge the valuable contributions of those who provided indispensable assistance in completing this prodigious task. A number of my colleagues were kind enough to read sections of the monograph relative to their own areas of expertise and to make helpful suggestions. For this I wish to thank: Per-Olof Berggren, Michael Berridge, Paul Churchill, Cary Cooper, Jacques Dumont, Maurice Feinstein, John Foreman, Adron Harris, S.M. Kirpekar, Suzanne Laychock, Paul Munson, Ole Petersen, Harvey Pollard, James Putney, Michael Schrey, Ramadan Sha'afi, and George Weiss. To these individuals I owe a great deal, not only for their valid criticisms, but also for their much needed encouragement. Reading of the manuscript by Stephen Halenda aided in clarifying the presentation. I would also like to thank Jim Putney and Suzanne Laychock, who gave so unselfishly of their time to engage in helpful discussions that aided in clarifying my thoughts on so many of the diverse subjects that had to be covered.

Special acknowledgment is also given to Karen Soderquist who organized the bibliography and to Rebecca Harris and Pat Nowell for patiently typing and retyping the manuscript and bibliography.

In conclusion, I would like to express my sincere appreciation to The Commonwealth Fund for a Book Program Award that supported this enterprise.

Ronald P. Rubin

Contents

General Introduction

A. BIOLOGICAL IMPORTANCE OF CALCIUM

All living cells are bathed in extracellular fluid that differs in composition from the intracellular milieu. Activation of cells is characterized by alterations in cell membrane permeability and the resultant transmembrane movement of ions. The crucial membrane components are the various "channels" or "gates" that define the permeability or conductances of various ions. Sodium, potassium, magnesium, and calcium all play significant roles in biological activity, but calcium, because of its unique chemical properties, is the most crucial in terms of possessing a wide diversity of functions. The importance of this cation is underscored by the fact that no electrolyte is under more vigorous control in the body than calcium.

The complex functions of the cells that have evolved require messengers to mediate and coordinate cell responses, and inorganic ions are a likely choice as mediators, because they have relatively high diffusion coefficients and are therefore readily diffusible. Sodium might be considered a likely candidate as an intracellular messenger since it is present in appreciable concentrations in the extracellular fluid and can readily traverse cells down an electrochemical gradient. However, sodium only weakly binds to macromolecules, thus limiting its biological activity. Moreover, a low cellular calcium concentration is maintained by a complex system of sequestration and expulsion, which will be discussed in more detail later. Since the sodium concentration as free ion inside the axon is greater by a factor of 10^6 than the free calcium concentration, even with a 1000-fold smaller calcium influx during activity, appreciable changes in the intracellular free calcium concentration may be brought about more rapidly than corresponding changes in sodium (Mullins, 1978). Thus, because the concentration of free calcium in the cytosol is very low compared to that of other physiological cations, calcium is singularly suited to play a diverse and crucial role in physiological function.

However, the fact that rapid increases in cell calcium can be achieved does not alone endow this cation with its diverse functions, for calcium possesses special chemical properties that enable this cation to achieve such diversity of action (R. J. P. Williams, 1970). Calcium, uniquely among physiological cations, manifests versatility in accepting complex ligands, particularly large anions, and this property may account for its ability to be absorbed onto specific sites on the cell membrane (Urry, 1978). On the cell surface, calcium stabilizes the ordered state of phospholipids by decreasing lipid fluidity. X-ray diffraction studies show that calcium forms tightly packed, highly ordered structures with phospholipids (Newton et al., 1978), and, because of its specific charge-to-size ratio, the resultant effect of calcium binding is a segregation of membrane phospholipids into discrete domains, and the resulting conversion of the membrane to a more ordered state (Lee, 1975; Papahadjopoulos et al., 1978). Other calcium binding sites of importance are localized to peptides and include hydroxyl, carbonyl, and carboxyl groups (Williams, 1977). Calcium binding to proteins, if proper positioning occurs sterically, allows cross-linking between proteins, producing a more compact membrane, with resultant effects on membrane activity. The diverse actions of calcium within the cell are also attributed to its ability to bind to nonmembranous ligands such as adenine nucleotides, citric acid, phosphate, and calcium-binding proteins (Rubin, 1974a).

Although the concentration of magnesium in the cell is several orders of magnitude higher than that of calcium, magnesium is much more hydrated and, for this reason, is restricted in its binding potentialities (R. J. P. Williams, 1970). Magnesium–ligand interactions contain more water molecules and fewer ligand groups. Hence, magnesium is not able to cause the segregation of membrane phospholipids and bring about a "more-ordered" membrane. Similarly, monovalent cations possess a much lower affinity for specific sites on the surface of the cell membrane, and, in fact, appear to penetrate into the interior of the membrane, while increasing membrane fluidity. All of these factors help, at least in part, to explain in chemical terms some of the singular properties possessed by calcium that enable it to exert the myriad of biological effects cited below.

For purposes of classification and exposition, four principal roles of calcium have been defined (Ashley and Campbell, 1979): (1) structural, as a constituent of bone and biomembranes; (2) co-factor for enzymatic reactions, particularly those localized extracellularly (conversion of prothrombin to thrombin) or on the cell surface (adenylate cyclase, phospholipase A_2); (3) electrical, to stabilize membranes and act as a current carrier; and (4) intracellular regulator of such fundamental processes as secretion, contraction, cation fluxes, and hormone action (glycogenolysis). Such a classification, while providing a template for describing the actions of calcium, does not enable us to fathom the fundamental and diverse aspects of its biological activity. The role of calcium in the constitution of biological systems extends from the provision of rigidity to the organism through its presence in bone; to the control of not only communication between cells, but also the structure and integrity of individual cells; and, of course, to the regulation of many fundamental cellular processes within cells, such as motility, cellular transport, muscle contraction and secretion. Even the most basic of all biological processes, namely fertilization, cellular proliferation, and cell death are calcium-mediated events.

About a volume devoted to so specialized a domain of calcium action as secretion, the reader may legitimately ask why the full range of calcium's protean actions are considered. At the very least, the effects of calcium on other biological processes can, and in fact do, indirectly influence the secretory process and therefore can hardly be neglected. In a broader context, knowledge of other actions of calcium, particularly when in-depth information is available, may help us to solve the mystery of the molecular mechanisms governing the secretory process. For example, all cellular movements require a force-generating reaction and a regulatory mechanism that appears to be controlled by calcium. Understanding the control mechanisms of such a basic process affords us the flexibility of applying this knowledge to other calcium-dependent systems, and, with the necessary modifications of methodology and interpretative reasoning, enables us to dissect out the underlying mechanisms for these other related processes, including secretion. It is, therefore, entirely appropriate that we consider the varied actions of calcium, particularly those that may relate directly or even remotely to the secretory apparatus.

1. Cellular Functions of Calcium

Many cellular responses to stimuli are wholly or partly mediated by changes in the distribution of calcium. At the lowest level of the biological spectrum, this cation not only performs functions akin to those that it performs in mammalian cells but also enables the organism to perform the most basic of functions. In fact, fertilization, the forerunner of all other biological processes in higher forms of life, requires calcium (Gwatkin, 1977). The acrosome reaction of sea urchin sperm is involved in sperm–egg attachment, and the induction of this reaction by egg jelly involves an increase in calcium permeability and a discharge of acrosomal granules by fusion with the plasma membrane (Collins and Epel, 1977). The acrosome reaction thus bears a marked resemblance to the release of secretory product by exocytosis. Free calcium also causes egg cleavage and induces the granule release reaction by the ovum. Thus, an increase of free cytoplasmic calcium is a general feature of the mechanism coupling membrane excitation to the process of fertilization. Calcium ions also positively modulate the proliferation of epithelial and mesenchymal cells by regulating DNA synthesis and mitotic activity in concert with other cellular mediators (Whitfield et al., 1979). Calcium also plays an important role in delivering nutrients to the cell, being a positive modulator of endocytosis and axoplasmic flow (Prusch and Hannafin, 1979; Hammerschlag, 1980).

A general theme that encompasses these various actions of calcium is best expressed in the following manner. Eukaryotic cells show various forms of cell movement, including contraction, protoplasmic streaming, and amoeboid and ciliary movement, as well as active movements of cell organelles, including secretory vesicles. These expressions of what has been generally termed cell motility all have features in common (Porter, 1976), with calcium being the common thread linking all of them (Allison and Davies, 1974; Weber, 1976). A wide variety of eukaryotic cells contain filamentous structures that have been characterized as actin– and myosinlike in nature (Hatano and Oosawa, 1978). The interaction of these proteins with calcium provides the molecular basis for cell motility and the movement of cytoplasmic components within cells. Since all cellular movements require a force-generating reaction, interaction of myosin-ATP and actin may generate the contractile force through a mechanism controlled by calcium (Clarke and Spudich, 1977; Hitchcock, 1977).

Calcium not only controls internal cellular events but also modulates cation and water permeability of cells. Calcium exerts significant stabilizing effects on the structural and functional properties of biomembranes. Simply stated, membranes are less leaky in the presence of calcium. The early biologists, well aware of this important function of calcium, observed that when red blood cells were placed in a hypotonic solution of such nonelectrolytes as dextrose or lactose they rapidly hemolyzed, and calcium was very effective in delaying or preventing hemolysis induced by osmotic or mechanical means (Lucke and McCutcheon, 1932).

There is general agreement that the channels responsible for excitation in biomembranes are discrete structures capable of allowing selective ionic flux, and the presence of calcium in the extracellular fluid is essential for the proper functioning of these channels. Since cell function, at least in excitable cells, appears related to membrane potential, the influence of calcium is expressed by altering the potential-dependent characteristics of the channels; this action of calcium is thought to be exerted by the screening of fixed negative charges on the membrane.

Calcium ions are also employed as charge carriers (Reuter, 1973; Hagiwara and Byerly, 1981), so that calcium channels are subjected to the same general regulatory influences as sodium and potassium channels. However, the effects of calcium on sodium and potassium conductances are quite complex (Putney, 1978). Not only does external calcium inhibit sodium influx, but cytoplasmic calcium enhances sodium influx by altering activity in the sodium channel. Moreover, cellular calcium increases potassium permeability in a wide variety of cell systems, perhaps by an action on the inside surface of the cell membrane (Meech, 1978). So an important role for calcium as a second messenger may be exerted through its effects on sodium and potassium conductances. Calcium also seems to stabilize a particular conformation of membranes to ensure an optimal response to membrane active agents (Triggle, 1980).

Calcium also exerts an effect at the next higher level of development, cell-to-cell communication (Gilula and Epstein, 1976; Loewenstein and Rose, 1978). Direct intracellular communication involving exchange of ions and molecules is mediated by gap junctions. Gap junctions, which are found in a wide variety of tissues, including those specialized for secretion, appear to be the agency through which cells adjust their activity to meet specific needs by information transfer. Calcium exerts a key role here as well, as demonstrated by the fact that calcium deprivation causes an uncoupling of previously coupled cells (Meda et al., 1980; Peracchia and Peracchia, 1980). The basis of this phenomenon may rest with the basic importance of calcium in cellular adhesion and aggregation, for it is well known that, in calcium-deficient media, epithelial cells lose their integrity and become detached from one another. Peculiarly enough, high cell calcium is also associated with loss of cell coupling (Petersen and Iwatsuki, 1979), and, in fact, uncoupling of electrical activity in adjacent cells is one parameter by which electrophysiologists monitor increases in free intracellular calcium.

But, just as important as calcium is in initiating and maintaining life by its actions on fertilization and cell growth and development, it also is a mediator of cell injury and death. Damage to plasma membranes disrupts permeability barriers with a consequent influx of calcium ions; impaired function is a common sequela, and even cell death may be a specific consequence of a disturbance in intracellular calcium homeostasis. The ability of calcium-buffering systems such as the mitochondria to accumulate calcium is readily compromised during ischemic injury, a fact that emphasizes the importance of calcium as a mediator of cell injury and death (Schanne *et al.*, 1979). It is therefore mandatory that there exist inactivation mechanisms in the plasma membrane to regulate calcium permeability and cellular buffering mechanisms to curtail the potential for calcium ions to accumulate to toxic levels within the cell.

B. METHODS FOR ANALYSIS AND TISSUE CONCENTRATIONS OF CALCIUM

For the organism to carry out its very basic functions and even to survive, the calcium concentration in biological fluids must be maintained within narrow limits. Plasma calcium concentration (10 mg/100 ml), which normally fluctuates only by $\pm 3\%$, is maintained by a number of homeostatic mechanisms. The principal calcium-regulating hormones are parathyroid hormone, vitamin D_3, and calcitonin, although the latter's role in man is still controversial. Many other hormones influence calcium metabolism; they include thyroid hormone, prolactin, growth hormone, insulin, the somatomedins, and adrenal and gonadal steroids. These agents regulate calcium concentration by controlling (1) net absorption from the gut, (2) net loss through the urine, and (3) net deposition in bone. Total body content of calcium is 1–1.5 kg, and the skeleton contains more than 98% of this calcium. The exchangeable pool of calcium in body fluids and cells, which is critical for regulating basic physiological functions, represents only 1–2% of the total body calcium. Only 50% of the plasma calcium concentration is present in diffusible form, ionized or bound to citrate and phosphate; the remainder is nondiffusible and bound to proteins (albumin).

Earlier methods employed to measure total tissue calcium concentrations included conventional spectrophotometry, fluorescent spectrophotometry, as well as atomic absorption and dual-wavelength spectrophotometry (for references see Rubin, 1974a). Radiotracer analysis is also employed to analyze the kinetics of calcium binding and transport in cells. These methods do not distinguish between free and bound calcium, but they do provide useful information concerning calcium distribution and kinetics when utilized in subcellular fractions. The calcium concentrations of most tissues tend to cluster around the value of 1 μmole/g wet weight. This is true for muscle, peripheral nerve, brain, and endocrine tissues. Certain tissues, such as kidney and salivary glands, have specialized mechanisms for sequestering calcium and have higher calcium concentrations, whereas erythrocytes, which lack these mechanisms, have lower concentrations (Rubin, 1974a).

1. Subcellular Distribution

Because the role of calcium in physiological and biochemical processes is so diverse and complex, buffer compartments are required to protect the cells from the vagaries of the environment. These compartments not only provide buffer systems for calcium, but are also a source of intracellular calcium to be released to the cytoplasm during exposure to appropriate stimuli. The compartmentalization of calcium, which varies with the morphological makeup of the cell, can be arbitrarily divided into (1) membrane fractions, which include plasma membrane and endoplasmic reticulum—the latter including specialized membrane systems such as the sarcoplasmic reticulum in muscle, and (2) other cellular organelles, which include the mitochondria, secretory granules, and lysosomes. Binding to membranes accounts for some of the calcium sequestration in cells, although another important source of sequestered calcium in certain systems appears to be the mitochondria (Fig. 1).

The physiologically significant form of calcium bears little relevance to total amount of cellular calcium. This is most strikingly illustrated by the fact that of a total of 400 μmoles calcium in squid axoplasm, only 10 μmoles is ionized (Baker, 1976). Hence, the overwhelming proportion of the total calcium of a resting cell is physiologically inactive. But to

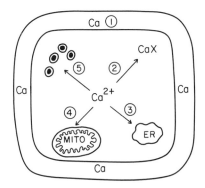

FIGURE 1. Mechanisms involved in the cellular sequestration of calcium. (1) Calcium bound to the plasma membrane complexed with ATP and phospholipid. (2) Calcium buffering by macromolecular components and small molecules of the cytoplasm. (3) Calcium uptake into endoplasmic reticulum (ER). (4) Calcium uptake into mitochondria. (5) Calcium sequestration by secretory organelles.

correlate calcium fluxes with cellular activity such as secretion, one must be able to utilize ionic fluxes as a direct index of changes in calcium concentration occurring within the cell during activation. None of the aforementioned techniques possesses the critical feature of being able to detect small, rapid changes in the free calcium concentration of the cell and thus directly link the rise in the free cellular calcium concentration to the physiological response.

The advent of three analytical tools—calcium-activated photoproteins, metallochromic indicator dyes, and calcium-selective electrodes—is helping to address this problem. These calcium probes possess one or another of the following properties to make them sufficiently useful for monitoring the concentrations of free intracellular calcium: sufficient sensitivity to detect micromolar concentrations of free cellular calcium, low toxicity, rapid reaction kinetics to detect changes in free calcium concentration in rapidly responding cells, and freedom from interference by other cations, especially magnesium. Although no one single method is ideal, all three of these techniques are useful in determining the free calcium concentration within cells. Moreover, the time course of the changes in the free cell calcium can be correlated with the response times of cell activation, a procedure that allows a more direct approach to elucidating the physiological role of calcium in cell activation. Finally, these methods can also be used to monitor the calcium-buffering capacity of cells during the increase of free calcium concentration that follows stimulation.

a. Photoproteins

Two coelenterate photoproteins (aequorin and obelin) are employed as calcium indicators because of their favorable kinetic properties with respect to affinity and speed of reaction with calcium, as well as their accessibility and stability (Blinks *et al.*, 1976; Blinks, 1978; Carbone, 1979). In aqueous solution, photoproteins emit a flash of blue luminescence in response to free calcium, which is the only commonly occurring cation that will stimulate photoprotein luminescence, although other biologically important cations, such as potassium, sodium, and magnesium, inhibit the reaction by competing with calcium for binding sites on the protein. Calcium concentrations as low as 10 μM can be easily detected. However, the cell must be sufficiently large to incorporate the injected protein. Once calcium gains access to the cell, not only can the free cellular calcium concentration be measured, but also, with the aid of an image intensifier, analysis of the spatial distribution of the free calcium concentration within cells can be carried out. Although aequorin is not ideally suited for measurement of extracellular calcium, an aequorin signal that increases with increasing external calcium and decreases or disappears in low external calcium indicates that calcium has entered the cell from the external medium, rather than being released from internal stores.

b. Metallochromic Indicators

Metallochromic indicators (murexide, arsenazo III, and antipyrlazo III) undergo color changes when the free calcium concentration of the solution is altered, so that kinetic changes in the cellular calcium concentration can be detected by spectrophotometry (Scarpa et al., 1978; Scarpa, 1979). These indicators are hydrophilic and, like aequorin, do not bind to membranes. They have faster response times than photoproteins, but may be less sensitive and selective. Another group of calcium indicators, the tetracyclines, readily penetrate into cell membranes and emit a fluorescence response that requires the simultaneous presence of both calcium and membrane. Thus, the tetracyclines are useful in analyzing calcium mobilization from internal stores, a condition which is monitored by a decrease in fluorescence (Chandler and Williams, 1978; Feinstein, 1980). However, these fluorescence indicators lack specificity for particular divalent cations, and it is difficult to identify quantitatively the cellular locus of the signal.

c. Calcium-Selective Electrodes

Calcium-sensitive microelectrodes now have a sufficiently high degree of sensitivity and selectivity to calcium, as well as a rapid enough response time, to follow biologically-induced changes in calcium distribution (Simon et al., 1978; O'Doherty et al., 1980). Through the development of ion exchangers that selectively extract calcium from aequous solutions into an organic membrane phase, liquid-membrane electrode systems have been obtained that generate a potential difference that is directly proportional to the log of the calcium concentration between 10^{-3} and 10^{-7} M. The response time of calcium-selective electrodes is approximately 1 sec, as compared to the response times of photoprotein and murexide of less than 1 msec. This is a very convenient method and estimates of free calcium in cells have confirmed the values of approximately 10^{-7} M that were determined in nerve and salivary glands. But improvement in sensitivity and response time still is needed to expand the utility of this technique. While researchers now have cause for optimism about the prospect of significant advances in our knowledge of calcium activity in cells during rest and stimulation, one of the difficulties in making accurate assessments of calcium fluxes is the recognized ability of cells to buffer changes in the calcium concentration. And, in fact, these techniques for measuring calcium concentrations are not rapid enough to cope with the speed of buffering.

Morphological methods have also been employed to detect calcium in cells and to ascertain changes in its distribution. One relatively popular approach involves the use of anions such as oxalate or pyroantimonate (Herman *et al.*, 1973; Wooding and Morgan, 1978; Henkart, 1980). These agents penetrate to sites of calcium sequestration and precipitate the cation; the complex appears as an opaque particle under electron microscopy. Rapid and sophisticated freezing techniques have been employed in an attempt to minimize calcium loss and redistribution during fixation, since it is apparent that the conventional methods of tissue fixation are much slower than the biological movements that they are presumed to demonstrate. Electron-probe microanalysis offers a sophisticated and specialized approach to localizing calcium within cells (McGraw *et al.*, 1980; Ornberg and Reese, 1980). Tissues are fixed with or without the inclusion of the calcium-precipitating agent and microanalysis is performed under an electron microscope fitted with an X-ray spectrometer. This technique, which utilizes the emission by atoms of X-rays of characteristic energy when the atoms are bombarded with radiation of a sufficiently high energy, measures mass of total calcium (free and bound) and can detect as little as 10^{-18} to 10^{-19} g calcium.

C. REDISTRIBUTION OF CALCIUM DURING STIMULATION

1. Influx of Extracellular Calcium

Although the resting permeability of cells to calcium is low, electrical or chemical stimulation of many types of tissues is associated with an enhanced movement of this cation from the extracellular fluid across the plasma membrane into the cell. According to the ionic theory of excitation as originally described by Hodgkin and Huxley (1952), action potentials of excitable cells are produced as a result of an increase in membrane permeability to sodium ions. The increase in sodium permeability is followed by an increase in potassium permeability that brings the membrane potential back to its resting value. These workers did not consider the contribution of calcium as a charge carrier during the action potential of the squid axon, although in 1957 Hodgkin and Keynes, using radiotracer analysis, showed that a small amount of calcium entered the axon during the action potential. Small electrical currents carried by calcium ions moving across the membrane are also detectable by voltage clamp techniques, if the squid axon is immersed in a solution free of sodium to block the sodium currents (Meves and Vogel, 1973).

Nerve terminals and brain synaptosomal preparations (pinched-off resealed nerve endings) also manifest enhanced calcium permeability following depolarization (Blaustein, 1975). While sodium is the primary current carrier in these systems, calcium ions act as the principal charge carrier in skeletal muscle fibers of crustacea and most invertebrates, as well as in smooth muscle fibers of certain vertebrates (Reuter, 1973; Hagiwara and Byerly, 1981). Despite the relatively low extracellular concentration of calcium, a large driving force for inward calcium movement is fostered by a large electrochemical calcium gradient, which is maintained by the relative impermeability of the plasma membrane and by active extrusion and buffering mechanisms. The cell calcium concentration can therefore be increased severalfold despite the small magnitude of the calcium current and the rapid inactivating mechanisms that are known to exist. However, since cell survival appears to depend on keeping cellular calcium levels low, it is not surprising that the calcium current is usually very small and can be rapidly inactivated.

Not only is there known to be an increase in ^{45}Ca uptake into nerve and muscle during stimulation, but also it is now well established that, in many secretory organs, cell activation is accompanied by an increase in ^{45}Ca uptake into the tissue. Electrophysiological studies have also been instrumental in elucidating the ionic mechanisms associated with secretion (Williams, 1981). A prominent feature is a graded electrical response to depolarization produced by an inward current carried by calcium or sodium. Douglas and his associates first demonstrated that in isolated adrenal chromaffin cells acetylcholine-induced depolarization possesses a calcium component. The blockade of this inward calcium current by the local anesthetic tetracaine depresses the secretory response (Douglas, 1975). Glucose depolarizes β cells of the endocrine pancreas and produces spiking activity when present in high concentrations (Matthews, 1977). The amplitude of these spikes varies directly with the external calcium concentration and not with that of sodium. Thus, the β-cell membrane appears to possess a regulatory system such that glucose-induced depolarization leads to calcium influx.

The properties of calcium permeability at presynaptic nerve terminals appear to be similar to those observed in certain endocrine glands in that, when the sodium and potassium conductances are depressed by appropriate pharmacological means, the membrane becomes capable of producing an all-or-none calcium spike (Katz and Miledi, 1969). The release of transmitter becomes a function of calcium entry that depends on the presynaptic membrane potential. The action of calcium as a current carrier may be directly linked to the coupling mechanism that is responsible for triggering secretion. Alternatively, the effects of the small influx of calcium during membrane depolarization may be amplified by the subsequent release of a larger amount of intracellular calcium as in mammalian cardiac muscle (Fabiato and Fabiato, 1978), or calcium entry may activate sodium conductance as observed in exocrine pancreas or potassium conductance as observed in salivary glands (Putney, 1978).

2. Fast and Slow Channels

Studies utilizing radiotracer analysis that describe an increase in ^{45}Ca uptake do not unequivocally connote a primary increase in calcium permeability; the increase may, for example, be secondary to changes in cellular calcium distribution. A much more precise method for following the voltage-sensitive changes in calcium entry is to use the calcium-sensitive photoprotein, aequorin, and measure the increase in light output in response to calcium entry that can be seen following even a single action potential (Hallett and Carbone, 1972). By comparing the calcium entry associated with trains of depolarizing pulses of different durations, Baker and his associates (Baker, 1972) were able to identify two phases of calcium inward movement in squid axons: an early phase, roughly coincident with the increase in sodium permeability in which the channel is approximately 100 times more accessible to sodium than to calcium, and a second phase, roughly parallel with the increase in potassium permeability. The two phases of calcium entry can be distinguished not only temporally but also pharmacologically. Hille (1970) presented a view of separate channels for independent ion movements based upon the pharmacological dissection of these channels. The sodium channel could be selectively blocked by tetrodotoxin, the poison of the Japanese puffer fish, while the potassium channel could be selectively blocked by tetraethylammonium. The early phase of calcium entry is also blocked by tetrodotoxin, a fact that seems to reflect the entry of calcium into the axon through the sodium channels. The second phase of calcium entry is not blocked by tetrodotoxin or tetraethylammonium, but is impaired by magnesium, manganese, cobalt, and organic calcium antagonists such as methoxy-verapamil (D-600) (Table 1).

The properties of the late phase reveal a marked similarity to calcium channels that normally control transmitter release at nerve terminals. In neuroendocrine and endocrine glands such as adrenal chromaffin cells,

TABLE 1.   Comparison of the Effects of Various Pharmacological Agents on the Early and Late Components of Depolarization-Induced Calcium Entry in Squid Axons with the Effects of These Same Agents on Transmitter Release

Agent(s)	Early calcium entry (via the sodium channel)	Late calcium entry (via the late calcium channel)	Transmitter release
Tetrodotoxin	Blocks	no effect	no effect
Tetraethylammonium ions	No effect	no effect	no effect or enhanced
Mg^{2+}, Mn^{2+}, Co^{2+}, La^{3+}	Some reduction	blocked	blocked
D-600, verapamil	Some reduction	blocked	blocked

[a] Modified from Baker, 1976.

neurohypophysis, adenohypophysis, and pancreatic β cells, which have been designated as excitable by virtue of their ability to respond to high potassium, voltage-sensitive calcium channels can also be identified by neurophysiological and pharmacological means (Williams, 1981). These channels operate to control directly the calcium-gating mechanisms responsible for enhanced calcium entry into the cell during activation of the secretory process.

3.   Depolarization as a Prelude to Calcium Entry

In excitable cells, voltage-dependent sodium and calcium channels operate to regulate the conductance of these cations during depolarization, but a question which we have not yet addressed is whether depolarization is a *sine qua non* for stimulating calcium entry, or, in other words, whether activation of the sodium channel is an obligatory prelude to activation of the calcium channel. Insight into this problem was first gleaned by Hutter and Kostial (1955), who elicited a discharge of acetylcholine from the cat superior cervical ganglion in response to high potassium in a medium devoid of sodium. Early experiments on perfused cat adrenal glands also demonstrated the expendability of sodium in the secretory process, as shown by the ability of acetylcholine to evoke secretion in the absence of external sodium and by the relative ineffectiveness of tetrodotoxin in inhibiting the secretory response to acetylcholine (Douglas, 1975). Moreover, acetylcholine was able to stimulate catecholamine release from an adrenal gland perfused with a sodium-deprived medium containing an osmotic equivalent of potassium that completely depolarized the chromaffin cell membrane.

Such early findings indicated that the critical permeability change produced by depolarization and associated with the activation of secretion did not directly involve sodium ions. More direct evidence favoring this concept was provided by Katz and Miledi (1969), who found that nerve terminals in squid axon were large enough to permit impalement by intracellular electrodes for measuring the membrane potential, for passing current, and for injecting drugs. When the sodium current was blocked with tetrodotoxin and the delayed outward potassium current blocked by injected tetraethylammonium, local regenerative responses capable of eliciting postsynaptic responses were obtained during depolarizing current pulses; the amplitude of these pulses was found to be calcium-dependent. On the basis of such evidence it seems apparent that sodium entry *per se* is not required for transmitter release, but is only a prelude to calcium entry.

This is not to say that the activation of the voltage-dependent sodium channel cannot regulate secretion; in fact, veratridine is a pharmacologic agent that activates the voltage-dependent sodium channels and is able to trigger a secretory response that is blocked by tetrotodoxin (Nordmann and Dyball, 1978; Kirpekar and Prat, 1979). Thus, activation of sodium channels is able to enhance the calcium current required for secretion, although it is not a mandatory prelude to calcium entry. The concept that gross electrical changes are not required for the initiation of the secretory process—but rather only specific effects on calcium distribution brought about by stimulation—implies that electrical activity can be uncoupled from secretion. As a matter of fact, chemical or electrical activation of the secretory process can occur in the absence of any discernible sodium current or when the cell is completely depolarized by excess potassium, as long as permeability changes permit calcium to traverse the cell membrane.

A related issue concerns the existence of receptor-operated channels in secretory cells that are activated in the absence of any discernible change in electrical activity. While this type of channel is characteristic of exocrine glands (Putney, 1978), the ability of acetylcholine, for example, to elicit adrenomedullary catecholamine secretion when the gland is completely depolarized by high potassium (Douglas, 1975) connotes the existence of receptor-operated channels in excitable secretory cells as well.

4. Release of Intracellular Calcium

Alteration in plasma membrane transport is not the only way to modulate functionally important calcium in the cytoplasm. If external calcium cannot reach the cellular site of action in sufficient concentrations rapidly enough to elicit a secretory response, an alternate mode of calcium mobilization must be employed. Thus, skeletal muscle can contract and relax by mechanisms of calcium control that are exclusively intra-cellular (Endo, 1976). Since the endoplasmic reticulum, isolated as mi-crosomes, accumulates and stores calcium in the cell, it is reasonable to consider that this system will also release calcium upon appropriate stimulation. Microsomal preparations from muscle tissue have been studied more extensively. However, energy-dependent calcium uptake acitvity has been found in microsomal fractions from a variety of non-muscle tissues, including brain synaptosomes, adrenal glands, blood platelets, and the endocrine and exocrine pancreas; these fractions rep-resent reservoirs of cellular calcium potentially available for release.

The question whether the endoplasmic reticulum can release cal-cium, as well as bind it, has not been satisfactorily answered, although there is some experimental evidence, accumulated using chlorotetra-cycline as a calcium probe, that the release of calcium observed during the blood-platelet-release reaction provoked by thrombin emanates from the endoplasmic reticulum (Feinstein, 1980). While sarcoplasmic retic-ulum of muscle represents a well-characterized and well-defined mem-brane system, in nonmuscle tissue the membranes are more heteroge-neous and include not only endoplasmic reticulum but also plasma membranes; the latter—relatively rich in calcium—may also provide the source of activator calcium, since cell activity is generally initiated by chemical or electrical stimulation through events occurring on the surface of the cell.

Calcium is also sequestered by active transport in mitochondria, which may provide an additional reservoir for calcium mobilization (Carafoli and Crompton, 1978a; Lehninger et al., 1978). In addition, cal-cium finds its way into the secretory granules of various tissues to such an extent that these organelles may contain a relatively high calcium concentration as compared to other subcellular fractions (Rubin, 1974a; Clemente and Meldolesi, 1975; Ravazzola et al., 1976). However, not only have the specific cellular sources eluded identification, but likewise the mechanism by which calcium is released from intracellular stores is not understood. We do know that the endoplasmic reticulum in squid axon forms junctions with the surface membrane similar to those found in skeletal muscle, so the endoplasmic reticulum may release calcium in response to surface membrane stimulation by the spread of activating currents through these membrane systems. Extending the analogy with contractile systems, external calcium may enter the cell to trigger the release of calcium from intracellular stores. On the other hand, influx of sodium into the secretory cell may also promote calcium release from intracellular stores (Lowe et al., 1976).

The locus of calcium release during activation may depend on the type of stimulation as well as the type of tissue involved. The notion that different stimulating agents acting on the same system initiate activation by affecting different calcium fractions is favored by the consistently reported finding of a differential sensitivity of certain systems to calcium deprivation or calcium antagonists upon exposure to a variety of stimulating agents. For example, in the adrenal medulla, after the onset of calcium-free perfusion, the secretory response to excess potassium is lost long before the response to acetylcholine or to sympathomimetic amines (Rubin, 1974a). The release of amylase from the rat salivary (parotid) gland induced by α-adrenergic or cholinergic receptor activation is dependent upon the presence of extracellular calcium, whereas the secretory response to β-adrenergic receptor activation is unimpaired by the lack of extracellular calcium (Butcher and Putney, 1980). A similar phenomenon has been reported in various smooth muscle preparations where high- and low-affinity calcium-binding sites can be identified by the more rapid loss of contractile responses to excess potassium than to norepinephrine after the introduction of a calcium-free medium (Weiss, 1981). However, whether the critical calcium fractions are extracellular or cellular, translocation of this cation is a fundamental requirement for many physiological processes, including secretion and contraction.

D. REGULATION OF CELL CALCIUM

Since a rapid change in the free cytoplasmic calcium concentration is the main mechanism by which calcium mediates its effects, efficient calcium-buffering mechanisms exist to terminate these effects. On the one hand, buffering systems cannot be infinitely efficient since this would prevent the increase of intracellular free calcium concentration; on the other hand, ineffective buffering systems would lead to calcium concentrations that are toxic to the cell. Obviously then, complex and coordinated systems must exist to allow a transient and limited increase in the free calcium concentration in the cell. Moreover, these buffers must maintain a fine structural compartmentalization of calcium in the vicinity of responsive receptors, for if the augmented levels of cell calcium were uniformly distributed, then all cellular functions controlled by calcium would be linked.

While calcium entry is a process of passive diffusion down an electrochemical gradient, the cellular calcium pumps that operate to maintain the intracellular calcium concentration at a very low level can be divided into two groups: (1) those that compartmentalize calcium within intracellular organelles, and (2) those that pump calcium out of the cell. The importance of these intracellular mechanisms for calcium regulation is reinforced by the notion that free calcium concentrations can be increased not only by increasing calcium entry but also by inhibiting the active mechanisms for calcium efflux and binding in cells.

1. Calcium Binding by Mitochondria

It has long been known that calcium uptake is a general property of mitochondria (Chance, 1965). Calcium ions are accumulated by mitochondria against a large electrochemical gradient, and the driving force appears to be the membrane potential of the mitochondria, which, being negative inside, creates a large electromotive force. The influx is energized by electron transport-driven hydrogen ion ejection, while a simultaneous entry of anions such as phosphate is also thought to occur. The avidity with which calcium is transported into mitochondria is illustrated by studies with liver mitochondria showing that calcium uptake takes primacy over oxidative phosphorylation, but, inexplicably, this is not seen in heart mitochondria (Jacobus *et al.*, 1975). It thus appears that the importance of mitochondria in calcium binding varies from tissue to tissue. For example, in salivary glands mitochondria can act as efficient calcium buffers when the intracellular free calcium concentration is suddenly increased (Rose and Lowenstein, 1975). In squid axon, inhibition of the calcium mitochondrial pump by inhibitors of oxidative phosphorylation increases calcium efflux. However, in this system the endoplasmic reticulum also plays an important role as a calcium buffer (Blaustein *et al.*, 1980).

For mitochondria to be physiologically important as a calcium buffer, the rate of calcium uptake must be sufficiently high to remove the free intracellular calcium, and the uptake capacity must be able to meet the requirements for the particular cell. Also, a physiological mechanism for releasing mitochondrial calcium should exist to prevent an excessive accumulation of this cation. But whether cyclic AMP or sodium serves as a modulator of mitochondrial calcium release remains a physiological conundrum (Borle, 1974; Mela, 1977; Carafoli and Crompton, 1978b). Despite these unanswered questions, mitochondrial calcium transport must be important for regulating calcium homeostasis in secretory cells, since the free cytosolic calcium concentration is, at least in part, regulated by the rate of energy-linked cycling of calcium into and out of mitochondria.

2. Calcium Binding by Microsomes

Although studies of intracellular calcium uptake systems in nonmuscle tissue have generally focused on mitochondria, the energy-dependent calcium uptake activity that exists in other cellular organelles, particularly the endoplasmic reticulum, may provide another important source of calcium buffering in the cell. The microsomal fractions obtained from homogenates of tissues specialized for secretion, including brain synaptosomes, adrenal cortex, endocrine and exocrine pancreas, and blood platelets, manifest ATP-dependent calcium-accumulating activity, which may function in the intact cell to sequester calcium. In blood platelets the activity is derived from the dense tubule system, which may function in a manner analogous to the sarcoplasmic reticulum of muscle (White, 1972).

To press the analogy, morphological studies in squid axon and vertebrate nerve reveal that the endoplasmic reticulum—which sequesters intraneuronal calcium—forms junctions with the surface membrane similar to those seen in muscle (Henkart, 1980), thus enabling this type of system not only to sequester free calcium but also to release it during stimulation. In brain synaptosomes, nonmitochondrial storage sites have a higher affinity for calcium than do mitochondria and appear important as a calcium buffer when the calcium load is relatively small (Blaustein et al., 1978). This finding, taken together with the close proximity between the endoplasmic reticulum and mitochondria, suggests that the endoplasmic reticulum protects the mitochondria from exposure to high calcium concentrations. This would free the mitochondria to use energy from oxidative metabolism to synthesize ATP rather than to pump calcium. The physiological importance of these calcium buffers is exemplified by the fact that phenomena related to neurotransmitter release such as facilitation and posttetanic potentiation (Rahamimoff et al., 1978) may result from the temporary retention of excess calcium within the terminals following generation of action potentials.

While attention has focused on the mitochondria and endoplasmic reticulum as the prime calcium-buffering systems in the cell, the secretory organelles themselves also provide another means by which the free calcium ion concentration can be lowered in the cell. Calcium uptake also occurs into isolated secretory vesicles (Russell and Thorn, 1975; Michaelson et al., 1980; Häusler et al., 1981), from which the cation is subsequently released into the extracellular medium together with the primary secretory product. The rapidity with which calcium is accumulated by isolated secretory vesicles is consistent with a physiological role for these organelles in calcium sequestration in the intact cell. Finally, other cellular organelles termed "microvesicles," found in neural tissue that contains ATPase, have the capacity to accumulate calcium by an ATP-dependent process (Shaw and Morris, 1980; Torp-Pedersen et al., 1980). The organelles, which do not appear to represent retrieved granule membranes following exocytosis, may function in concert with the other systems to sequester calcium, and, like the secretory vesicles, may rid the cell of excess calcium by discharging it into the extracellular space.

It is apparent then that these are diverse systems for carrying out the important function of calcium sequestration. It would be foolhardy to attempt to offer any generalizations regarding the relative importance of each of these systems in light of our limited knowledge and because the activity of each system, in all probability, varies significantly from one cell type to another. Nevertheless, the close relationship between calcium sequestration and the action of calcium on the secretory process mandates that inroads must be continually made into this very significant facet of calcium metabolism.

3. Calcium Extrusion Mechanisms

In addition to the compartmentalization of calcium within intracellular organelles, there are mechanisms for extruding calcium from the cell (Fig. 2). An outwardly directed calcium pump localized at the cell surface provides another mechanism for buffering cellular calcium. An ATP-driven calcium pump has been described in human erythrocyte plasma membrane (Schatzmann and Burgin, 1978). It is linked to a calcium-stimulated, magnesium-dependent ATPase, which can be distinguished from the sodium–potassium ATPase by its cation specificity and by being insensitive to inhibition by cardiac glycosides. In addition to the ATP-driven calcium extrusion mechanism described in erythrocyte ghosts, a similar mechanism operates in cardiac sarcolemma (Sulakhe and St. Louis, 1976) kidney (Moore et al., 1975), and certain types of

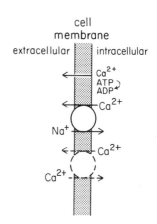

FIGURE 2.   Components of calcium efflux from the cell. Calcium may be extruded by an energy-requiring pump, or by an ion exchange mechanism involving either extracellular sodium or calcium.

cultured cells (Borle, 1969) and may also possibly operate in nerve (DiPolo, 1978; Gill et al., 1981). But, because of the morphological complexity of other cell types compared to erythrocytes, it is more difficult to define the precise location of the membranes involved in calcium transport, although in certain systems mentioned above, the calcium-dependent ATPase has been localized to the plasma membrane (for example see Gill et al., 1981). However, the assumption must be made that these plasma membrane fractions have been completely separated from the microsomal fractions derived from the endoplasmic reticulum.

While it seems likely that all cells share the ability to extrude calcium against a large electrochemical gradient, this critical process can be brought about by diverse mechanisms. In the erythrocyte, net calcium transport against an electrochemical gradient is effected by direct utilization of energy derived from ATP hydrolysis. This calcium efflux, which reflects the activity of calcium-ATPase, is not coupled to the movement of other ions. In excitable cells, an important component of calcium efflux is dependent on the external sodium (Na) concentration, i.e., Na–Ca exchange (Blaustein, 1974). The existence of a countertransport mechanism of sodium–calcium exchange in the squid axon was first established by Baker and his associates (Baker, 1972), and extended to cardiac muscle fibers (Reuter, 1973) and adrenal chromaffin cells (Rink, 1977). It is not unreasonable to speculate that all excitable prokaryotic cells may utilize this calcium–sodium cotransport system. Calcium–sodium exchange mechanisms also appear to exist in prokaryotic cells as well as in eukaryotic cells, although calcium–hydrogen antiporters appear to predominate in prokaryotic cells (Stroobant et al., 1980).

The sodium-dependent calcium efflux is also under metabolic control and stimulated by intracellular ATP (Brinley, 1978; Mullins, 1978, 1979). However, the energy is not immediately derived from the hydrolysis of ATP but from the sodium gradient established by the sodium–potassium ATPase. The effect of ATP on the sodium-dependent calcium efflux is thus a catalytic one, rather than that of a substrate of an ion pump. One important property of the sodium–calcium exchange is that it is decreased by membrane depolarization, suggesting that the exchange is associated with a net inward movement of positive charges, i.e., that the pump is electrogenic. The sodium–calcium exchange is also reversible in the sense that calcium can move in either direction across the membrane; the direction calcium is driven depends on the sodium electrochemical gradient. If the sodium electrochemical gradient is greater than the calcium gradient, calcium will be extruded by the cell.

Conversely, if the sodium electrochemical gradient is less than the calcium gradient, calcium will move into the cell in exchange for sodium.

Apart from the ATP-dependent calcium pump and sodium–calcium exchange, there is also a component of calcium efflux in certain tissues that is sensitive to external calcium (Fig. 2). This calcium efflux—which reflects a cellular exchange mechanism not involving a pump—is most prominent in systems that utilize extracellular calcium and where stimulation is accompanied by a large increase in calcium influx (Rink, 1977; Baker and McNaughton, 1978). Calcium–calcium exchange obviously does little to bring about a lowering of intracellular calcium, and at present its physiological significance is uncertain.

We have attempted in this brief discussion to establish that the low level of free calcium in the cell makes this cation an excellent candidate for producing a transient increase in activity, even with only modest increases in the free cytoplasmic calcium concentration. This increase may be brought about by either calcium entry or by calcium mobilization from internal stores. To prevent calcium from attaining toxic levels within the cell, inactivating mechanisms exist to terminate the entry of cations. During recovery the excess calcium may be sequestered by internal stores, particularly the endoplasmic reticulum, mitochondria, and the secretory granules, where it is slowly released and extruded from the cell by carrier-mediated transport systems, thus restoring the original steady-state conditions. The maintenance of the free intracellular calcium concentration is thus a delicate balance between the activities of membrane channels for allowing passage of calcium into the cell, those of cellular organelles for sequestering calcium, and those of pumps for extruding calcium from the cell.

E. CALCIUM AND THE REGULATION OF MEMBRANE PERMEABILITY

1. Membrane Stabilization

We have seen that the plasma membrane is a barrier between an extracellular calcium concentration of 2 mM and an intracellular concentration of less than 1 μM and that the membrane plays an important role in maintaining this differential by controlling ion permeability. The crucial membrane components in the control of membrane permeability are various "channels" or "gates" that define the permeabilities or conductances of various ions. Although it is not clear just what a channel or gate is, since they have not been isolated or identified, the presence of calcium in the extracellular fluid is essential for the functioning of membrane ionic channels.

The early biologists, working with less sophisticated biological systems, noted the importance of calcium (and other divalent cations) in maintaining the normal permeability of cells by protecting against lysis by osmotic, mechanical, and pH effects (Lucke and McCutcheon, 1932). Hence the action of calcium on cell membranes may be viewed as a stabilizing action. As the external calcium concentration is increased, nerve becomes less responsive to stimulation by electric current or by high potassium, although the resting potential undergoes little change (Shanes, 1958). Other divalent cations such as magnesium and strontium produce the same phenomenon. Thus, in the gated sodium channel of nerve, increasing the extracellular calcium increases the critical depolarization necessary to reach the firing threshold. The stabilizing effect of calcium is not restricted to excitable tissue, as demonstrated by the ability of calcium to inhibit sodium transport in epithelial cells such as the kidney (Taub and Saier, 1979).

Two general hypotheses have been presented to explain calcium's stabilizing action (Hille, 1976). The first suggests that calcium acts as a "plug" to the sodium channel. But a "plug" would not be expected to carry currents through both the sodium and calcium channels, which is clearly the case for calcium. A second theory proposes the presence of fixed anionic sites on the surface of the membrane which induce a negative surface potential (zeta potential). This potential difference, which cannot be recorded with microelectrodes since it is confined to the membrane, acts on ion channels to alter current–voltage relations. Calcium is thought to influence the electric behavior of the membrane by causing it to sense an increase in the potential difference between its inner and outer surface, by "screening" these fixed negative charges. While an increase in external calcium produces a stabilization of membranes, calcium deprivation results in an increase in the excitability of nerve and muscle (Brink, 1954). The threshold for inducing electrical activity is markedly reduced in low-calcium media and spontaneous firing is observed, although the release of neurotransmitter is depressed. A decrease in the resting membrane potential is demonstrable in many excitable tissues after calcium deprivation (Shanes, 1958); and Frankenhauser and Hodgkin (1957) have estimated that in the squid axon a fivefold decrease in calcium concentration is similar to depolarization of 10 to 15 mV. One may thus generalize that calcium excess and lack may be analogous to anodal and cathodal stimulation, respectively.

The role of calcium in regulating membrane permeability is in all probability related, at least in part, to its characteristic property of interacting with membrane phospholipids. The plasma membrane is viewed as a two-dimensional mosaic of integral membrane proteins embedded in a fluid lipid layer, with peripheral proteins bound loosely to either surface. Both rigid and nonrigid (fluid) lipid regions exist in membranes, and there is abundant evidence that calcium opposes membrane expansion and decreases lipid fluidity or motional freedom by complexing with head groups of acidic phospholipids (Papahadjopoulos *et al.*, 1978). Electron spin resonance (spin label) techniques show that the density of packing of individual phospholipids in lipid bilayers (model membrane systems) increases when calcium is present. All of these effects of calcium are consistent with its role as a stabilizer of membranes.

Since calcium plays a critical role in regulating the permeability of cells in the resting state, it therefore follows that the permeability changes that follow membrane stimulation also involve calcium. Calcium binding to the receptor area stabilizes receptor conformation and thereby regulates receptor-mediated conductance changes (Triggle, 1980). The fact that calcium is critical for maintaining the resting state might lead one to suspect that cell activation is associated with a mobilization of membrane calcium. In fact, the key process during cell activation is in all probability a displacement of calcium from negatively charged membrane sites, which causes macromolecules to undergo a conformational change leading to an increase in permeability (Tobias, 1964; Somlyo and Somlyo, 1968; Douglas, 1975). Monovalent cations may compete with calcium for these negatively charged membrane sites and thereby depress the rate of calcium entry. This competition may be the basis of the well-known ability of sodium and potassium to antagonize the effects of calcium in a variety of plant and animal cells (Höber, 1945).

Although cell membranes have pores or channels through which ions enter and leave, it is difficult to explain ion antagonisms simply in terms of competition for entry. Calcium currents are demonstrable during the entire sequence of conductance changes of sodium and potassium, but the pharmacological dissection of these ionic currents clearly shows that the critical calcium current for triggering secretion does not enter the neuron through either the sodium or the potassium channels (Baker, 1972). In adrenal chromaffin cells, which have the same embryological ancestry as neuronal tissue, the calcium current required for catecholamine release also enters through channels functionally distinct from the inward sodium current (Rubin, 1981). Moreover, lowering the sodium concentration of the medium bathing isolated neurohypophyses augments secretion but does not enhance calcium entry (Douglas and

Poisner, 1964a). Therefore, the relationship between ionic events cannot be simply viewed as a competition for membrane or channel binding sites prior to entry into the cell. Moreover, the laboratory demonstration of cations competing for the same binding sites does not signify that this competition is physiologically relevant. The question whether a decrease in cell sodium or potassium may occur to the extent of causing an increase in the amount of calcium bound to a membrane *in vivo* cannot now be answered.

Hence, it is crucial for understanding the nuances of ion antagonisms to recognize other controlling factors. Calcium efflux involves the exchange of outside sodium for inside calcium, so in a sodium-free solution calcium efflux will decrease, making it difficult to ascertain under sodium-deprived conditions whether the increase in cell calcium is a result of an increased uptake or a decreased efflux. Moreover, an increase in cell calcium is known to inhibit the sodium–potassium pump, which would bring about an increase in the sodium concentration in the cell, and in this way affect the sodium–calcium exchange mechanism by reducing the electrochemical gradient for sodium.

2. Activation of Membrane Permeability

We have seen that external calcium regulates excitability either by entering the nerve fiber during the action potential or by controlling the permeability of the membrane to sodium and potassium, and to calcium as well, by its stabilizing action on the plasma membrane. Calcium also functions as a chemical messenger to regulate sodium and potassium permeability. The presence of a calcium-sensitive potassium channel was first established in the erythrocyte, where, as expected, the passive permeability to potassium was high in low calcium, but unexpectedly, potassium permeability was also elevated when the intracellular calcium was high (Lew and Ferreira, 1978) (Fig. 3). The added finding that injected calcium increased potassium permeability in central neurons (Meech, 1978) was also surprising, since, as we have seen, an increase in the extracellular calcium causes an increase in membrane resistance.

However, in many types of cells, including secretory cells, calcium entry causes changes in membrane potassium conductance (Putney, 1978). This calcium-sensitive potassium current is distinct from the other major late potassium current, the delayed rectifying current, which is activated in direct response to depolarization. Activators (propranolol, tetracaine) and inhibitors (quinine and quinidine) of the calcium-sensitive potassium channel show it is different from the tetraethylammonium-sensitive potassium channel found in excitable membranes (Lew

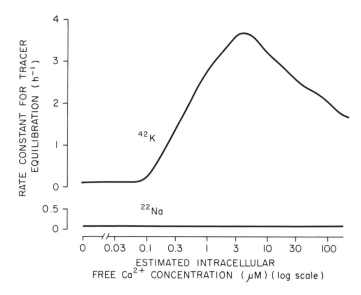

FIGURE 3. The stimulatory effect of intracellular calcium on the permeability of red cell membranes to potassium. Ghosts were prepared after lysing red cells in a calcium-buffered solution containing ^{42}K and/or ^{22}Na. The increased potassium permeability is proportional to the intracellular calcium concentration in the range of 0.1–3 μM, but there is no change in sodium permeability. The intracellular free calcium concentration is calculated on the assumption that the contents of the ghosts have the same composition as the lysing solution (redrawn from Simons, 1975).

and Ferreira, 1978). The calcium-sensitive potassium channel is not found in all cells and its physiological function has not been elucidated, but the importance of this mechanism is most clearly seen in the salivary gland, where an increase in cell calcium activity is associated with the secretion of water and ions, particularly potassium. Calcium binding to the potassium gate may regulate the opening and closing of the potassium channel. On the other hand, this mechanism may provide a feedback to make an excitable system self-limiting, in the sense that calcium entry will cause cell activation, but, at the same time, by producing an increase in potassium efflux will tend to restore the membrane potential to normal and therefore limit further calcium entry.

Electrophysiological studies of the salivary glands have demonstrated that receptor agonists may cause a hyperpolarization of the membrane that is thought to be the result of an enhanced potassium permeability caused by calcium entry (Petersen, 1976; Putney, 1978). However, under certain conditions the changes in the membrane potential can be biphasic with a depolarization preceding the hyperpolarization. Analysis of these electrophysiological changes reveals that an increase in sodium conductance precedes the increase in potassium conductance, and these changes in ionic permeability are mimicked by the intracellular injection of calcium. Salivary secretagogues such as carbachol and phenylephrine increase sodium uptake into cells of the salivary glands, and this process requires the presence of extracellular calcium. Not only does an increase in cytoplasmic calcium activate sodium and potassium conductances, but in the blowfly (insect) salivary gland 5-hydroxytryptamine stimulates calcium influx, which is responsible for activation of chloride channels (Berridge, 1975).

It is thus apparent that in nonexcitable tissues calcium modulates ion permeabilities by a mechanism that Putney (1978) has referred to as stimulus–permeability coupling. This aspect of calcium action has opened new vistas in the consideration of ion interactions. Activation of receptor-operated calcium channels in nonexcitable cells mediates an increase in the sodium and potassium conductances. Conversely, in excitable cells a decrease in extracellular calcium may lead to an accumulation of intracellular sodium by promoting sodium uptake or by impeding the calcium–sodium exchange mechanism. Ultimately, then, any analysis of calcium metabolism must always be considered in light of potential interactions with other physiologically relevant cations.

3. Biochemical Description of the Calcium Channel

Although ion channels have not been isolated and identified, the primary proof of their existence has been the demonstration by the electrophysiologist of ionic conductances too large to be accounted for by carrier mechanisms. Our conception of ionic channels has also been greatly clarified by the physiologist's and pharmacologist's use of drugs and toxins to define the membrane currents in cells. Thus, studies demonstrating the selectivity and permeability of ion channels (Hille, 1970) and the measurement of gating currents extended our understanding beyond a formal description of the events associated with channel activation. Biochemical studies also aided our understanding of the nature of the calcium channel. This approach began developing some years ago when Hokin (1968) showed that phosphatidylinositol (PI) turnover as assessed by ^{32}P incorporation could be activated by pancreatic secretagogues; however, the fact that this increase in PI turnover was unaffected by calcium deprivation left open the relation of this event to secretion.

Robert Michell (1975), extending this work, demonstrated that PI metabolism is stimulated by all receptors that utilize calcium as their second messenger—particularly α-adrenergic and muscarinic cholinergic receptors—and not by agents that do not activate the receptors. However, high potassium also stimulates PI turnover in smooth muscle, suggesting some similarity between the receptor-sensitive and potential-sensitive gates. The lack of calcium dependency for the PI breakdown and turnover prompted Michell to propose that the PI effect is not a consequence of, but rather a prelude to, calcium mobilization. So within the framework of this concept, PI breakdown is envisioned as a coupling reaction essential to the mechanism by which receptor activation is linked to calcium mobilization. The idea that PI breakdown is an essential coupling step in calcium mobilization receives support from experiments carried out in insect salivary glands in which 5-hydroxytryptamine-desensitized glands depleted of PI stores regained responsiveness when allowed to resynthesize PI (Fain and Berridge, 1979).

Since PI turnover can only be demonstrated in intact cell systems, it has not been possible to investigate the molecular mechanisms of the functional link between PI breakdown and calcium mobilization, so there is as yet no information on how hydrolysis of PI is related to calcium flux. The fact that the dose–response curves for agonists exhibiting the PI response (as measured by an increase in ^{32}P turnover) are quite different from the dose–response curves of the physiological responses such as secretion or contraction has made it necessary to evoke the concept of "spare receptors" to explain this disparity. That is, the occupation of only a small proportion of available receptors resulting in a relatively small PI response is necessary to produce a maximal change in the physiological response. Moreover, the demonstration in the rabbit neutrophil that the PI effect can be abolished by removing the external calcium and can be mimicked by the divalent cationophore A23187 casts doubt on the universality of the Michell concept (Cockcroft et al., 1980).

However, when exceptions to a theory periodically arise they do not necessarily invalidate the whole concept. This usually requires only that the theory be modified so as to include these exceptions. Also, it often happens that two or more theories can explain the same phenomenon, and then we must search for some fundamental concept common to the contending theories, for this will likely be a better approximation to the true situation. Indeed, we are now in an era where other theories involving phospholipid turnover are being put forward to explain calcium mobilization and cell activation. In particular, existing evidence supports the idea that calcium-activated phospholipase A_2 may play a pivotal role in cell surface phenomena associated with secretion. The turnover of arachidonic acid in PI and other phospholipids by way of

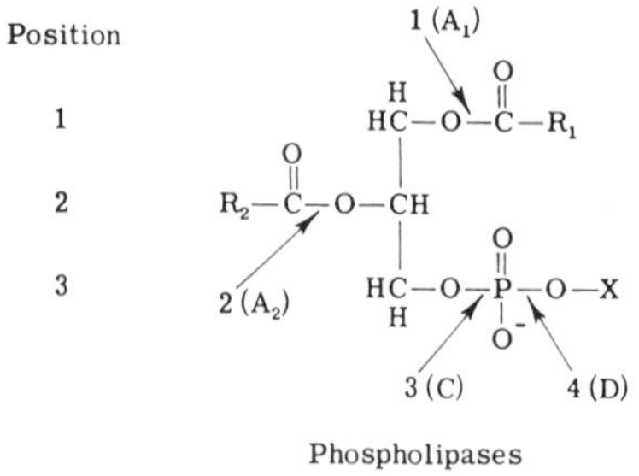

Position   1 (A$_1$)

1

2

3

FIGURE 4. Sites of phospholipid hydrolysis by phospholipases. R_1, saturated fatty acid; R_2, unsaturated fatty acid; X, polar head group (e.g., inositol).

a deacylation–reacylation cycle involving position 2 of phospholipids may provide a means to alter membrane structure and function, since phospholipid acyltransferases possess a selectivity that promotes incorporation of unsaturated fatty acids such as arachidonic acid into specific membrane phospholipids. And membranes enriched in unsaturated fatty acids manifest enhanced fluidity, an effect expected to increase membrane permeability.

So, in addition to the reactions inherent in Michell's theory which implicate a phospholipase-C-mediated turnover of the polar head group of PI, theories involving phospholipase A$_2$ and arachidonic acid are now being elucidated (Fig. 4). The culmination or resolution of all such theories will be a clearer definition of the mechanisms regulating not only calcium fluxes but also the calcium activation of sodium and potassium movements.

F. CALCIUM IONOPHORES

An important event in excitable tissue that is activated with a depolarizing electrical current or high potassium, or in cells responding to receptor activation, is an increase in the free calcium concentration in the cell cytosol. To directly prove this concept necessitates that calcium be injected directly into the cell and the response measured. This technique of microinjection has been successfully employed in various tissues including large axons, mast cells, and salivary gland cells (Miledi, 1973; Kanno et al., 1973; Rose and Loewenstein, 1975). However, the microinjection technique is technically impossible in many cells owing to their relative smallness, so that other approaches must be employed. Antibiotics are being developed that, as the basis of their biological activity, facilitate the passive movement of ions across membranes. These agents, which have been given the name ionophores, have been used in different ways to investigate the biological phenomena (Pressman, 1976). They have been employed to perturb biological membranes by altering the ionic gradients and potentials that exist across the membranes; they have also been used as probes of membrane structure and function.

Ionophores facilitate the transfer of cations across cell membranes by functioning either as carriers of ions (Fig. 5) or pore-formers that remain relatively fixed while ions move through them (McLaughlin and Eisenberg, 1975). Moreover, ionophores have been developed with varying affinities for monovalent or divalent cations. For example, valinomycin is a neutral ionophore with a high selectivity for potassium; it transports potassium as a neutral carrier rather than by way of a pore as does gramicidin. Such agents, by affecting monovalent cation fluxes, will, of course, indirectly affect movement of calcium ions. However, with the advent of calcium-selective ionophores, an effective tool became available whose primary actions would be limited to promoting calcium movements (Reed and Lardy, 1972). The ability of these agents to complex with and transport calcium across biomembranes has been of inestimable importance and has led to a greater understanding of the direct role of calcium in various biological functions.

By far the most widely used of these so-called calcium-selective ionophores in biological systems up to now has been A23187. Two molecules of this ionophore form a lipophilic complex with a single calcium ion, which then migrates from the bulk phase across the cell membrane and releases calcium into the cell interior (Gomperts, 1976). As a result of this action, A23187 does not displace cell surface calcium, as is thought to occur with certain receptor activators, but there is a short circuiting of these events directly causing an increase in calcium in the cell. The formation constant of the ionophore–calcium complex must be high enough for complex formation on one side of the biomembrane and low enough for there to be calcium release on the other side of the membrane.

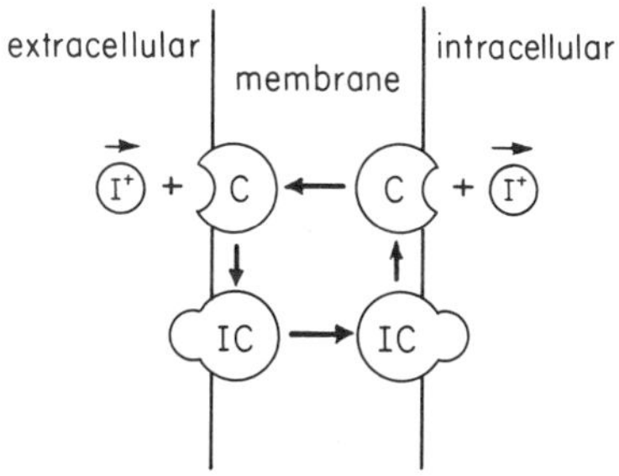

FIGURE 5.   Schematic representation of ionophore (carrier)-mediated ion translocation across the cell membrane. The ionophore (C) acts as a mobile carrier by binding the cation (I^+) at the external surface of the cell membrane forming a lipid soluble complex which transports the cation to the cytoplasmic side and releases it into the cell interior. The ionophore then returns to the external surface to repeat the cycle.

Thus, the use of A23187 allows us to circumvent certain of the steps at the membrane associated with receptor activation and more directly focus on the direct effects of calcium. The order of ability of A23187 to transport divalent cations is: calcium > magnesium = strontium > barium (Pfeiffer *et al.*, 1978). However, ionophore A23187 may also transport monovalent cations (Pfeiffer *et al.*, 1978), implying that the specificity of this agent for divalent metal ions is relative and not absolute. Moreover, we have also seen that primary changes in calcium distribution can, and do, alter the distribution of sodium and potassium. Nevertheless, the relative selectivity of A23187 for calcium makes it a valuable tool for studying the effects of calcium on a variety of cellular functions.

Cognizance of the role of calcium in a given biological function allows us, in large measure, to predict the actions of A23187. Thus, in every system where the existing evidence has implicated calcium as a mediator, the final response to A23187 has mimicked that of the primary signal. These effects extend from the enhancement of skeletal, cardiac, and smooth muscle contractility; to the acrosome reaction involved in the sperm–egg attachment; to activation of sea urchin eggs at fertilization; to blood platelet aggregation; and to the elevation in the rate of secretion from a wide diversity of cell types—which will be enumerated in the next chapter. Not only can the effects of A23187 be blocked or abolished by the removal of extracellular calcium in many cases, but they may be proportional to the amount of calcium in the extracellular medium. Moreover, the action of A23187 culminates in the release of secretory product by the normal physiological process, i.e., exocytosis (Kagayama and Douglas, 1974; Thoa *et al.*, 1974).

The fact that the actions of A23187 mimic those of microinjections of calcium directly into the cell—where this has proven possible—certainly confers strong support for the concept that this divalent cation-ophore is indeed exerting its actions by bringing about an increase in the free calcium concentration of the cell, and, in doing so, triggers the physiological response. Another calcium ionophore, ionomycin, which possesses an even greater selectivity for calcium than does A23187 (Kauffman *et al.*, 1980) and also stimulates calcium-dependent secretory systems, (Bennett *et al.*, 1979; Massini and Näf, 1980; Perlman *et al.*, 1980) may prove to be an even more useful research tool.

Despite the obvious utility of these agents in activating calcium-dependent systems by perturbing cation gradients, there are limitations to their use. As above, even A23187 and ionomycin are only selective, and not specific, for calcium, and, depending on relative affinities and concentrations, they can and do alter the gradients of other ions. This point is most clearly illustrated by the use of another cationophore, X-

537A, which in many cases exerts parallel actions to those of A23187, but is much less selective for calcium (Pressman, 1976). The actions of X-537A may be exerted through the transport of sodium ions, which in turn promotes calcium flux; alternatively, it may penetrate the cell to promote the release of intracellular calcium (Perlman *et al.*, 1980).

Another drawback is that cationophores, such as A23187 acting at the cell surface, do not partition calcium in cells as do the physiological mediators and thus may saturate intracellular calcium reservoirs. Furthermore, the action of A23187 is not confined to the surface of the cell, but penetrates intracellularly to enhance the release of calcium from mitochondria (Borle and Studer, 1978), thus inducing secretion independently of extracellular calcium (Nordmann and Currell, 1975; Ponnappa and Williams, 1980). The ability of A23187 to enter the cell also enhances the opportunity for it to compromise the functionality of cells. In addition to inhibiting protein synthesis and impairing uridine metabolism in cell cultures, it causes the release of soluble enzymes and loss of viability (Chandler and Williams, 1977). Thus, in seeking the optimal and even the desired effects, the duration of exposure and concentration of ionophore, as well as the extracellular calcium concentration, will all be important variables.

Although experiments utilizing the divalent cationophores indicate that an increase in the free intracellular calcium concentration of the cell is critical for mediating certain important biological functions, the question still remains open whether the entry of calcium into the cell is the sole primary event for allowing these responses to occur. There are a diverse number of mediators that are purported to act in concert with calcium to trigger many physiological functions: these include cyclic AMP, cyclic GMP, and the prostaglandins. It is important, in fact crucial, for this discussion to note that by bypassing the receptor events associated with cell surface activation, A23187 is unable to increase levels of cyclic AMP in most tissues, and, in fact, A23187 may decrease cyclic AMP levels in certain systems due to the abililty of calcium to depress the activity of adenylate cyclase, the enzyme responsible for synthesizing cyclic AMP. A23187 is also a known activator of cyclic GMP and prostaglandin synthesis; these actions of A23187 are ascribed to the calcium-dependent activation of guanylate cyclase and phospholipase A_2, respectively. The latter enzyme releases arachidonic acid from phospholipids, thereby providing the precursor for prostaglandin synthesis. While such knowledge does not, of course, arbitrarily discredit the candidacy of cyclic AMP and enhance that of cyclic GMP and prostaglandins as mediators of calcium-dependent cellular events, it does provide us with some insight into the nature of the complex interactions of these various mediators. We will return to a discussion of this particular problem in Chapter 4.

G. CALCIUM ANTAGONISTS

1. Verapamil, D-600, and Nifedipine

The physiological purist may take issue with the use of ionophores since these agents do not utilize the normal membrane channels, but merely render lipid membranes selectively permeable to cations. Ionophores cannot, therefore, provide us with complete information regarding the sequence of the ionic events associated with activation by the primary signal, whether it be chemical or electrical. For example, while A23187 exerts a positive inotropic action on the heart simply by increasing the cell calcium concentration, the mechanism underlying the action potential of the myocardium—a prelude to the mechanical events—is more complex, involving: (1) a rapid inward current carried by sodium that is responsible for the upstroke of the action potential, (2) a slow inward calcium current related to the plateau phase, and (3) an outward potassium current, related to repolarization of the membrane (Reuter, 1973).

The properties of these channels closely resemble those of sodium, potassium, and calcium channels of the squid axon. Calcium influx into the cardiac cell mediated by voltage-sensitive calcium channels is not blocked by tetrodotoxin, but is blocked by another group of agents designated organic calcium antagonists; verapamil is the prototype of this group of compounds (Fleckenstein, 1977). Other such drugs are methoxyverapamil (D-600), nifedipine, and prenylamine. The compounds of this group are structurally dissimilar, suggesting possible differences in their basic modes of action (Triggle, 1981). Nevertheless, all of these drugs antagonize calcium movements through an action on the slow influx channels in cardiac muscle cells and block myocardial contractions without abolishing transmembrane action potentials, i.e., they uncouple membrane excitation from cell contraction.

The ability of these organic calcium antagonists to block calcium influx is not confined to cardiac cells but can be demonstrated in tissues where voltage-dependent calcium channels are thought to exist. The resultant effects of these organic antagonists is to interfere with the action of calcium linking the electrical events at the cell surface with the biochemical reactions in the cell interior leading to the effector response. They do so by preferentially blocking the channel when it is activated by an adequate depolarizing stimulus (McDonald *et al.*, 1980). By contrast, the calcium antagonists do not substantially affect the outward potassium current which occurs similarly in time to the slow inward calcium current, thereby clearly distinguishing these two types of channels. These channel blockers have, therefore, proven valuable in not only providing indirect evidence that in certain systems the primary stimulus enhances calcium influx into the cell, but also demonstrating that the slow calcium channel is indeed a discrete entity.

The net effect of a calcium antagonist in a given tissue will relate to the overall importance of the voltage-sensitive calcium channels in that particular tissue. In the myocardium, the plateau phase will be mainly affected by these agents (Reuter, 1973), whereas, in crustacean muscle fibers, where electrical activity is mediated principally by calcium spikes, these agents will almost completely abolish electrical activity (Hagiwara and Byerly, 1981). These inhibitory effects will, of course, be reversed by raising the extracellular calcium concentration if the antagonism is competitive. The importance of voltage-dependent calcium channels in secretory cells is illustrated by the β cell of the endocrine pancreas, where electrical activity is inextricably linked to the secretory response. In this tissue, both electrical events and insulin secretion induced by the primary stimulus glucose are blocked by organic calcium antagonists, and not by tetrodotoxin, implying that the primary current is carried by calcium ions through the slow calcium channels (Matthews, 1977).

The use of these calcium antagonists has contributed greatly to our understanding of the physiological events associated with cell activation. We have seen that agents such as the calcium ionophores that enhance calcium permeability trigger activity, whereas the calcium antagonists depress activity. Such evidence provides support for the thesis that a rise in the free intracellular calcium concentration of the cell is the critical link in cell activation. However, again, one must be cognizant of the limitations in the use of these agents. While it is clear that D-600 and verapamil block the voltage-dependent calcium channels in excitable cells, they appear to be less potent on the presynaptic than on the postsynaptic components (Nachsen and Blaustein, 1979).

The question also arises as to the mechanism of their effects in nonexcitable cells, which are devoid of voltage-dependent channels but possess receptor-operated channels. D-600 and verapamil block the secretory response in such nonexcitable cells as the salivary gland and exocrine pancreas. But these agents generally are much less potent in nonexcitable tissues than in heart and their mechanism of action in such tissues remains uncertain (Putney, 1978). The problem is made even more complex by the fact that the so-called calcium channel blockers are also able to block sodium channels (McGee and Schneider, 1979). Actually, verapamil was first described as an adrenergic blocking agent, so it is not unlikely that certain effects ascribed to blockade of the calcium channels may indeed be the result of adrenergic blocking activity (Blackmore *et al.*, 1979). Alternatively, the effects of verapamil and D-600 may be the result of a "local anesthetic" effect, implying a nonspecificity of action. So not only should the use of these agents on nonexcitable secretory cells be discouraged, but perhaps there should even be some concern for indiscriminately utilizing them on excitable secretory cells, particularly when their relative lack of potency and selectivity is apparent.

2. Lanthanum and Other Polyvalent Cations

The organic calcium antagonists block calcium-dependent effects by an action on the voltage-dependent calcium channels. Certain polyvalent (di- and trivalent) cations are also well-known calcium antagonists, but their effects seem to be, in certain respects, more complex. As previously discussed, calcium has an affinity for anionic binding sites on the cell surface, and other cations compete with calcium for these binding sites. Lanthanum, which is the best known member of the rare earth or lanthanide series of elements, has been used as a valuable probe for elucidating the physiological significance of calcium-binding sites on the membrane (Weiss, 1974; Mikkelsen, 1976), since morphological evidence established that lanthanum's effects were selective for external membrane binding sites (Langer and Frank, 1972; Haksar *et al.*, 1976). Lanthanum, because of its size, high charge density, and flexible coordination geometry, has a much greater electrostatic attraction for anionic binding sites than calcium (Mikkelsen, 1976; Martin and Richardson, 1979). If these anionic sites also serve to transport calcium, lanthanum inhibits the actions of calcium by displacing calcium from these sites.

Consequently, lanthanum prevents the tetrodotoxin-insensitive regenerative calcium current in the presynaptic terminal of the giant squid synapses and blocks depolarization-evoked transmitter release (Miledi, 1971). Lanthanum is also a potent inhibitor of calcium spikes in barnacle muscle fibers, blocks calcium influx in smooth and skeletal muscle during potassium depolarization (Weiss, 1974), and is a potent inhibitor of calcium transport in mitochondria (Bygrave, 1977; Mela, 1977). Lanthanum also impairs the depolarization-induced calcium influx into squid axons (Baker, 1972) and into rat brain synaptosomes (Blaustein, 1975) and blocks the slow calcium-dependent hyperpolarizations in motoneurons that result from an increase in potassium permeability activated by calcium entry (Weiss, 1974; Mikkelsen, 1976).

The supposition that lanthanum is a specific antagonist for only a portion of the total cellular calcium implies that this trivalent cation should be very useful in delineating various calcium pools for cell activation. This approach has been used by Weiss (1981) to distinguish various calcium-binding sites in vascular smooth muscle and to establish that a more superficial calcium pool is utilized by certain agonists (high potassium) to a greater extent than by other agonists (norepinephrine) to induce contraction of smooth muscle. In mammalian cardiac muscle, lanthanum uncouples excitation from contraction, suggesting that calcium activation emanates from superficial cellular regions (Langer and Frank, 1972). In the insulin-secreting β cells of the endocrine pancreas, the utilization of lanthanum has helped to distinguish various calcium pools involved in secretion, including lanthanum-sensitive and lanthanum-resistant pools (Flatt *et al.*, 1980).

The premise that lanthanum's actions are restricted to external membrane binding sites provided the basis for the lanthanum wash technique developed by Van Breemen and his associates (1972) to distinguish intracellular from superficially bound calcium. Van Breemen's method, which was originally developed for smooth muscle, has been utilized in other systems and is based on certain assumptions: (1) lanthanum will displace and replace external calcium; (2) lanthanum will block calcium uptake and efflux; and (3) lanthanum will not enter the cell. Accepting these assumptions, we may conclude that lanthanum will displace noncellular and superficial calcium, and we may use radiotracer analysis to detect effects on the distribution of cellular calcium that are not obscured by large quantities of radioactive calcium localized to the extracellular fluid.

The fact that lanthanides have atomic radii very similar to calcium, but are also fluorescent and paramagnetic, allows them to be employed as probes for studying calcium interactions with proteins and lipids of cellular biomembranes and organelles (Mikkelsen, 1976). The interactions of lanthanum with intracellular organelles may appear paradoxical in light of the supposition that lanthanum does not penetrate into cells. However, lanthanides have a high electron-scattering ability, and, by staining tissues with these cations, cellular binding sites are detectable; so the use of lanthanides to reveal ultrastructure is a standard technique. Heuser and Miledi (1971) not only demonstrated that lanthanum can accumulate in nerve terminals following prolonged exposure to the polyvalent cation, but they showed that it increases the frequency of miniature endplate potentials even in the absence of extracellular calcium.

Lanthanum also stimulates spontaneous catecholamine release from adrenal medulla, (Borowitz, 1972), vasopressin release from the neurohypophysis (Thorn *et al.*, 1975), and histamine release from mast cells (Foreman and Mongar, 1973), while blocking evoked release. Such findings pose the question whether lanthanum directly mimics the effects of calcium when allowed to enter the cell or whether its effects are mediated by the release of calcium from the intracellular binding sites.

Before continuing with our discussion of the effects of polyvalent cations, the distinction between the actions of these agents and those of the organic calcium antagonists must be made clear. Whereas the organic calcium antagonists seem to act selectively at voltage-dependent calcium channels and do not affect the binding of calcium, the polyvalent cations exert effects both on calcium channels and on the anionic binding sites on the surface of the cell membrane; in addition to lanthanum (Weiss, 1974), other divalent cations including manganese, cobalt, and nickel, share this property (Langer *et al.*, 1975). We have seen that properties of augmented calcium permeability at presynaptic terminals are similar to those in squid axon in their insensitivity to tetrodotoxin and suppression by organic calcium antagonists. Calcium currents at these sites are also suppressed by lanthanum, manganese, cobalt, and nickel (Katz and Miledi, 1969; Baker *et al.*, 1973; Blaustein, 1975). But since a primary action of these polyvalent cations is a displacement of calcium from superficial membrane sites, these cations are more nonselective in their actions in that they block the secretory response of nonexcitable, as well as excitable, cells (Putney, 1978). By contrast, the organic calcium antagonists appear to exert their selective blocking action on voltage-dependent calcium channels in excitable cells.

Manganese, while blocking the actions of calcium in certain systems, also exerts effects in whole cells, as well as in mitochondria, that resemble those of calcium (Getz *et al.*, 1979). Manganese competes with calcium for anionic binding sites and thus protects cells against permeability changes usually induced by calcium deprivation (Impraim *et al.*, 1979). Its movements into and out of cells are also similar to calcium (Getz *et al.*, 1979), and it is accumulated by mitochondria like calcium (Mela and Chance, 1968). Like lanthanum, manganese, as well as cobalt, will block evoked release (Kidokoro and Ritchie, 1980); whereas spontaneous release is enhanced by these divalent cations (Balnave and Gage, 1973; Weakly, 1973). The ability of these divalent cations to block stimulus-evoked calcium uptake and neurotransmitter release, while increasing spontaneous secretion, may relate to their ability to displace calcium from critical superficial membrane binding sites, or to enter the cell and compete with calcium for critical receptor sites.

Finally, the cations most closely related chemically to calcium, the alkaline earths magnesium, strontium, and barium, must also be considered since they provide a clue to the nature of the secretory process. Although magnesium has a stabilizing action which qualitatively is similar to that of calcium, its chemical properties differ from those of calcium (R. J. P. Williams, 1970), and effects on isolated membrane models frequently do not resemble those of calcium (Papahadjopoulos *et al.*, 1978; Newton *et al.*, 1978). The magnesium antagonism of calcium actions may not be simply the result of a competition for membrane binding sites. Neither can the antagonism be readily explained on the basis of competition for entry through the slow channel, since magnesium appears to use alternate pathways (Baker and Crawford, 1972). Nevertheless, magnesium does mimic the stabilizing actions of calcium (Shanes, 1958); but magnesium, although more closely related to calcium chemically, is the weakest of the cations in this regard. Moreover, it is essentially free of the activating properties of calcium. But despite its relatively weak activity, the actions of magnesium cannot be ignored since the free magnesium concentration within the cell is several orders of magnitude higher than that of calcium.

In considering the role of calcium as a modulator of secretion, it is important to pose the question whether other divalent cations can also act in a manner similar to that of calcium? We have seen that the role of calcium is relatively unspecific in regard to its effects on permeability properties of tissue, and a number of divalent cations substitute for calcium in this action. On the other hand, evidence accumulated over the years generally bears out the basic conclusion of Mines (1911) (see Chapter 2, Section B) that only strontium and, at times, barium are effective substitutes for calcium in cell activation. This generalization applies in most instances to secretory systems (Rubin, 1974a; Nakazato and Onoda, 1980). The tetrodotoxin-insensitive calcium currents at presynaptic terminals and in barnacle muscle fibers, which are suppressed by manganese, are maintained by substituting strontium or barium for calcium (Hagiwara and Byerly, 1981). Strontium closely mimics the effect of calcium, although in certain systems it appears to be less potent than calcium (Silinsky, 1981). Strontium can also substitute for calcium in its stabilizing action, although in certain cells, most notably the mast cell, the resting permeability to strontium is quite high relative to calcium, and thus strontium is able to trigger the release response in the absence of any other stimulating agent (Foreman, 1977). In most cases, however, strontium is without a stimulant effect in the absence of the primary signal; like calcium, it basically links the membrane events associated with cell activation with the response.

Barium is also able to mimic the effects of calcium in a variety of biological systems although it differs from calcium, and for the most part strontium, by the fact that it can stimulate without prior activation by another primary stimulus (Rubin, 1974a). The stimulant effect of barium is mediated through depolarization and its activity is blocked by magnesium; however, the role of calcium in the stimulant action of barium is unusual in the sense that it inhibits barium-evoked secretion (Douglas and Rubin, 1964). Barium may act to alter cell permeability by displacing a critical calcium fraction from the plasma membrane, and, once inside the cell it mimics the action of calcium; but the effects of barium on membrane properties differ in some ways from those of calcium and more closely resemble the depolarizing effects of potassium. Barium also differs from calcium in another important respect, which may help to explain its actions to enhance activity and excitability of excitable cells. We have also seen that a calcium-sensitive increase in potassium efflux is an important factor in restoring cells to their resting state following activation. The presence of barium in the cell, in contrast to calcium, does not bring about an increase in potassium permeability but, in fact, impairs conductance (Eaton and Brodwick, 1980). Thus, the presence of barium, rather than calcium, will bring about cell activation in most cells, but will obstruct a major mechanism for returning the cell to its resting state.

While an introductory chapter is not the appropriate place to discuss the comparative effects of each of the divalent cations in the various secretory systems, it is important to emphasize that such an approach has been fruitful in gaining information as to the nature of the calcium-receptive system, particularly in secretory cells. Thus, in most secretory systems studied, strontium and barium have been found to substitute for calcium, whereas magnesium is inhibitory. The commonality of cation effects on the secretory process and on the contractile mechanism suggests that these two fundamental biological processes possess certain fundamental similarities. Such an extrapolation has been made to other systems. Thus, in the process of cell motility, the acrosome reaction, egg activation and fertilization, and endocytosis, the same general scenario is observed (cf. Collins and Epel, 1977). In fact, whenever a calcium-sensitive system is activated by strontium and barium and inhibited by magnesium, consideration must be given to the possibility that molecular events are taking place that are fundamentally similar to those observed in contractile systems and secretory cells.

Even exceptions to this general thesis provide us with a greater understanding of the nature of what appear to be fundamentally similar biological events. For example, a role for calcium has been established in the initiation of the fast axonal transport of proteins from the peri-karyal region of the neuron to distal sites of utilization. Proteins that undergo fast axonal transport are synthesized on the rough endoplasmic reticulum, migrate through the Golgi apparatus and subsequently reach the transport system—a process that is clearly dependent on calcium. Secretory proteins follow the same itinerary in exocrine cells. Along with other similarities, it was noted that, in contrast to endocrine secretion, strontium, but not barium, substitutes for calcium in both axonal trans-port and exocrine secretion, and this observation suggests that the same major elements are involved in the mechanisms by which exocrine cells transport and export proteins (Hammerschlag, 1980). The fact that cal-cium, strontium, and barium, but not magnesium, all bind to photo-proteins (aequorin) to elicit photoluminescence (Prendergast *et al.*, 1977) calls additional attention to the ubiquitous calcium-binding proteins that are being isolated and identified and may be involved in the activation of the secretory process (see Chapter 3). These proteins should provide a key to unraveling the mystery of the diverse actions of calcium within the cell.

3. Local Anesthetics

The foregoing considerations have established that in excitable cells the organic calcium antagonists such as D-600 and verapamil selectively block the voltage-dependent calcium channel; whereas the polyvalent cations, by virtue of their structural similarity to calcium, antagonize calcium's actions by displacing calcium from anionic membrane sites. Moreover, the ability of the inorganic cations to block calcium fluxes in several tissues is probably a reflection of their ability to utilize the calcium channel and in this way interfere with the permeation of calcium into the cell.

Another group of agents, the tertiary amine local anesthetics, due to their amphipathic structure, partition into and alter the structure and function of cell membranes. Local anesthetics were orginally viewed as agents that block nerve conduction. Voltage-clamp studies have clearly shown that local anesthetics do indeed block the sodium channels, and in higher concentrations the potassium channel as well (Narahashi *et al.*, 1976). The lack of selectivity of local anesthetics for monovalent cation channels contrasts with the selectivity of tetrodotoxin for the sodium channels (Moore *et al.*, 1967). The actions of local anesthetics to block the voltage-dependent sodium and potassium channels could, of course, account for their stabilizing actions on the membranes of excitable cells: that is, to increase the threshold for firing while not producing any radical change in the resting membrane potential (Shanes, 1958). This effect resembles that of high calcium and/or lanthanum.

However, the relative nonselectivity of local anesthetic action has been further extended to the voltage-dependent calcium channels, where it has been shown in smooth and cardiac muscle and in chromaffin cells of the adrenal medulla that local anesthetics are able to block the voltage-dependent calcium channels (Feinstein, 1966; Douglas and Kanno, 1967; Josephson and Sperelakis, 1976). These agents are also able to modify a variety of processes not associated with excitable tissue, including cellular fragility, adhesion and aggregation, and locomotion (Nicolson and Poste, 1976). These drugs also interfere with such fundamental biological processes as the acrosome reaction and other sperm–egg interactions, as well as with endocytosis and exocytosis. A cursory scrutiny of this list reveals two fundamental similarities: (1) all of these processes involve the interaction of membranes and (2) all are mediated by calcium. There are sufficient grounds for suspecting that local anesthetics inhibit these calcium-dependent processes not only because of their general stabilizing action to depress excitation but because of their ability to act as direct antagonists of calcium entry.

The pioneering efforts of Meyer, Overton, and Skou demonstrated that the primary event in the actions of an anesthetic involves an interaction with membrane lipids (Seeman, 1972). Cellular membranes, indeed, appear to be the site at which local anesthetics exert their pharmacological actions, since they can protect red cell membranes against osmotic hypotonic lysis (Seeman, 1972), inhibit ATP-dependent calcium uptake by microsomes (Kurebe, 1978), and block calcium uptake and efflux from isolated mitochondria (Mela, 1977). Indeed, local anesthetics have been shown to induce membrane expansion, a disordering of lipid bilayers, and to enhance the fluidity of phospholipids in membranes (Papahadjopoulos, 1972). These effects of local anesthetics on membranes are opposite from those of calcium (Papahadjopoulos *et al.*, 1975), which, as we have already seen, increases the ordering of lipid bilayers and the compactness of membranes.

Maurice Feinstein (1964) first showed that basic local anesthetics interact with acidic phospholipids by electrostatic attraction and that the more physiologically potent of these local anesthetics inhibit the extraction of calcium from an aqueous to an organic phase containing acidic phospholipids such as phosphatidylserine. The ability of certain local anesthetics to antagonize the transport of calcium into the organic phase correlates with data revealing that local anesthetics possess high affinity for calcium-binding sites and are able to displace calcium from cell membranes with an effectiveness that is related to the nerve blocking potency of a given local anesthetic (Seeman, 1972; Low *et al.*, 1979). Moreover, such data can be supported by studies showing that the inhibitory effects of local anesthetics can sometimes be reversed by high calcium (Rubin *et al.*, 1967; Rubin, 1974b), although there is controversy as to whether calcium ions alter the blocking effects of local anesthetics (Hille, 1977). But regardless of their specific site of interaction, local anesthetics, by competing with calcium for membrane binding sites, alter the structural properties of the membrane, perhaps by weakening lipid–protein interactions (Papahadjopoulos, 1972). These physiochemical effects of local anesthetics that alter the structural and functional properties of membranes impair calcium entry into the cell. Local anesthetics also depress active transport out of the cell by altering the properties of membrane-bound enzymes such as calcium-ATPase.

At another level, the ability of local anesthetics to alter the structure and function of membranes by displacing bound calcium appears to be responsible for their ability to induce changes in the cellular cytoskeletal (microfilamentous) systems associated with the plasma membrane (Nicolson and Poste, 1976). Changes in calcium binding to membranes perturb activity of surface membrane receptors that may be responsible for controlling the movement of intracellular organelles, including linkage of secretory granules to cytoskeletal elements. Alternatively, local anesthetics can penetrate intracellularly (Feinstein *et al.*, 1976) and exert direct effects on cellular membrane systems, as evidenced by their ability to stabilize or, in high concentrations, labilize lysosomal membranes, to inhibit binding of calcium to high affinity sites in mitochondrial membranes, and to inhibit calcium accumulation and stimulate calcium efflux from microsomal preparations. Such diverse effects of the local anesthetics curtail their usefulness in delineating specific ionic events, particularly with the advent of more selective blocking agents.

H.   SUMMARY AND CONCLUSIONS

In this chapter we have seen that calcium, due to its unique chemical properties and cellular distribution, plays a pivotal role in many biological processes, including secretion. The implication has been made, which we will return to later, that a similar fundamental mechanism may underlie these seemingly diverse calcium-dependent processes. In these systems, stimulation is associated with either an increase in membrane permeability and calcium entry through voltage-dependent or receptor-operated channels, or a mobilization of cellular calcium. The utilization by the cell of extracellular and intracellular calcium varies with the nature of the stimulus and consequently determines sensitivity to inhibition by drugs.

The recent development of methods to rapidly and accurately determine the free intracellular calcium concentration not only makes it possible to more closely correlate the intracellular calcium concentration and the physiological response but also to detect local changes in calcium concentration within the cell. This is of critical importance since convergent evidence suggests that the magnitude of cell activation in such fundamental processes as secretion and contraction is determined by the concentration of free intracellular calcium. Calcium is eminently suited to play such a pivotal role not only because of its distinctive chemical properties but also because the concentration of free cation in the cell is very low compared to that of other physiological cations such as sodium, potassium, or magnesium. This large gradient enables changes in calcium concentration in the cell to be brought about rapidly by calcium fluxes of a modest magnitude. The small increases in calcium flux prevent the intracellular calcium buffers (binding sites) from being overwhelmed. The importance of the cellular buffering systems cannot be overemphasized, because if they function too zealously, then they will not provide adequate time for the calcium ion to exert its action. On the other hand, if the buffering capacity of the cell is inefficient, then the free calcium concentration may rise to levels that are detrimental to the cell.

While we have seen that calcium enters the cell through discrete channels, this by no means implies that calcium exerts its effects in isolation from other important physiological cations such as sodium and potassium, for we have also considered that calcium controls the distribution of other organic ions either by inhibiting their flux as a result of stabilizing action on cell membranes, by cation exchange (calcium–sodium exchange), or by calcium-sensitive potassium efflux and sodium influx.

The secretory process—with which we will be primarily concerned—is viewed as consisting of two major events: membrane excitation induced by the primary signal and the cellular events directly associated with the release process. In subsequent chapters we will emphasize the mechanism whereby calcium couples the membrane events that lead to the secretory response. However, when we consider the action of calcium on the secretory process, its other varied effects must also be kept in mind. For superimposed on this simple concept of membrane activation, calcium entry or cellular mobilization, and secretory response are additional control mechanisms to consider. These mechanisms, while adding to the complexity of events, broaden our understanding of a biological system that has the capability of responding to a stimulus in a coordinated and self-limiting manner, thereby expediting a restoration of the resting state. In a more general perspective, the outcome of this narrative might, it is hoped, be to confirm our belief in Nature's organization, efficiency, and parsimony. However, as we will discover, such an outcome may not be inevitable.

Survey of Calcium Action in Stimulus–Secretion Coupling

A. GENERAL FEATURES OF THE SECRETORY PROCESS

1. The Secretory Granule

Before we survey the plethora of information concerning the regulatory role of calcium in the secretory process, it is necessary to review briefly the general features of this process in order to better appreciate how calcium acts in secretory systems. Under electron-microscopic analysis a typical feature of the secretory cell is the presence of numerous membrane-bound granules (100–400 nm in diameter) scattered throughout the cytoplasm that contain a variable quantity of electron-dense material. Small granular vesicles are also a uniform feature at chemical synapses, being situated close to the presynaptic nerve terminal.

But even before the advent of electron microscopy, histologists recognized that hormones were stored in granules. In 1918 Cramer reported that catecholamine granules in the adrenal medulla could be clearly distinguished from lipid droplets of the cortex when stained with osmic acid, and he noted that these granules, which gave the appearance of "fine coal dust scattered over the medulla," disappeared when the glands were stimulated. Direct evidence that these subcellular particles actually contained secretory product had to await the development of appropriate centrifugation techniques, and it was not until the early 1950s (Blaschko *et al.*, 1955) that the isolation of catecholamine-containing granules and their biochemical identification clearly established that these organelles contained hormone bound in a physiologically inactive form. About this time gonadotropic hormones were isolated from a particulate fraction of the pituitary gland by differential centrifugation (McShan and Meyer, 1952).

Since that time secretory product from almost every conceivable tissue has been isolated within subcellular particles, with steroids being the one notable exception. While the isolation of transmitters from peripheral nerve presented a somewhat greater logistical problem than the isolation of hormones, both the isolation of synaptic vesicles from cholinergic nerve endings and the demonstration that they contained acetylcholine were pioneered in the 1970s by the work of Whittaker and his associates (Whittaker, 1974; Zimmermann and Whittaker, 1974). At synapses where norepinephrine is the transmitter, a characteristic fluorescent staining of the vesicles denotes the presence of catecholamines (Corrodi and Jonsson, 1967). These vesicles, which typically have an electron-dense core, can be harvested from peripheral nerve endings, from brain, and from brain synaptosomes, and upon biochemical analysis can be shown to contain neurotransmitter. Evidence derived from histochemical, morphological, biochemical, and electrophysiological techniques supports the basic tenet that these membrane-limited organelles represent stored secretory product.

The classical studies of Jamieson and Palade (1971) elucidated the route and timetable of the intracellular transport of secretory product in exocrine pancreas. Following the synthesis of the proteins in the rough endoplasmic reticulum, packaging of secretory proteins occurs in the Golgi region. Stored products of endocrine cells also follow this basic route of synthesis on attached polysomes, segregation in the rough endoplasmic reticulum, and the concentration and packaging into granules in the Golgi complex (Farquhar, 1971). Since protein synthesis remains centralized to the perikaryal region of the neuron, the means of transfer of material and information known as axonal transport has evolved. Proteins being processed for fast axonal transport follow a route through the endoplasmic reticulum and Golgi region similar to that of other secretory proteins, and secretory granules formed in the region of the Golgi complex in the cell body are then transported distally down the axons (Hammerschlag, 1980).

Secretory organelles contain other constituents in addition to hormone and neurotransmitter. The chromaffin granules of the adrenal medulla for example, in addition to catecholamines, are constituted of macromolecules (glycoproteins and mucopolysaccharides) and smaller molecules including nucleotides and calcium. Included among the macromolecules is dopamine-β-hydroxylase, which is one of the enzymes involved in the biosynthesis of epinephrine and exists in both a membrane-bound and soluble form. An acidic protein called chromogranin is also localized in the chromaffin granule. The composition of synaptic vesicles in adrenergic and cholinergic nerves is similar to that of the catecholamine-containing granules of the adrenal medulla, in that, in addition to neurotransmitter they also sequester nucleotides, as well as several protein components.

2. Exocytotic versus Nonexocytotic Secretion

While it is agreed that the cytoplasmic granules are the reservoir for the preformed secretory product of the cell, the next question to be addressed is whether these structures are the primary source of secretory product. Different approaches have been utilized in an attempt to answer this question, and we will attempt to identify these various approaches and present our own conclusions based on the consensus of evidence.

Initially, the view of Katz and his associates, which explained the quantal character of the end-plate potential (Katz, 1969), focused on the synaptic vesicle at the neuromuscular junction as the primary site of transmitter release. Relying mainly upon electrophysiological evidence, these investigators showed that the release of transmitter substances in the absence of nerve impulses produces spontaneously occurring small depolarizations at the end plate that were termed "miniature potentials." These miniature potentials, which have been detected at a variety of cholinergic and adrenergic synapses, are discrete all-or-none events and therefore must represent the random escape of a packet (quantum) of transmitter from nerve ending. During nerve stimulation these quanta summate to evoke a larger postsynaptic potential. According to this thesis, the discharge of the vesicular content is envisioned as the result of a collision between the vesicular and axonal membranes. Release would occur only in the statistically improbable event of a vesicle membrane and the plasma membrane fusing and initiating a reaction which causes the colliding membrane barriers to burst open. The nerve impulse is perceived to change the axon membrane so that it presents temporarily many more reactive sites to the colliding vesicles, with the result that the statistical probability of release is increased many-fold.

Katz also recognized the parallels between the quantal theory and the ultrastructural evidence obtained for the secretion of zymogen granules and adenohypophysial hormones (Palade, 1959; Farquhar, 1971). In these latter systems morphological evidence by fusion of the granule membrane and the plasma membrane (termed exocytosis) was first demonstrated. The revelation by Whittaker (1974) that synaptic vesicles contain acetylcholine considerably strengthened the vesicular hypothesis but did not, however, provide direct evidence for the fusion of the vesicle with the plasma membrane. Nevertheless, the release of "quantal packets" of acetylcholine was certainly consistent with this concept. Moreover, morphological investigations attempting to correlate changes in the synaptic fine structure with evoked transmitter release reveal changes in synaptic vesicle number accompanying stimulation in peripheral nerve terminals, as well as in the central nervous system. For example, prolonged nerve stimulation by electric current or by drugs such as black widow spider venom toxin causes depletion of vesicles (Ceccarelli and Hurlbut, 1980a,b; Ceccarelli et al., 1972).

Conventional electron microscopy has provided morphological evidence favoring exocytosis in a wide variety of secretory cells, which has been confirmed by freeze-etching studies (Grynszpan-Winograd, 1975; Orci and Perrelet, 1978; Aunis et al., 1979). The range of systems where exocytosis operates includes neurons, neuroendocrine cells, and endocrine and exocrine cells, as well as other miscellaneous cell types, including the mast cell and blood platelet. Storage vesicles or granules are found close to or attached to the inner surface of the cell membrane, and images can be obtained of invaginations of the external surface of the plasma membrane that contain the dense core of the granule (Fig. 6). The fact that these granules, which originate from the fusion and opening of vesicle and plasma membranes, are devoid of a membranous envelope, rules out intact storage vesicle expulsion as a possible mechanism for secretion. While the morphological evidence for the existence of exocytotic images cannot be ignored, the relative paucity of these images observed with conventional electron-microscopic techniques might argue against it being the principal and/or exclusive mechanism of secretion. Such arguments can be countered by results obtained with freeze-fracture techniques, which, by producing a split through the plasma membrane, promote the incidence of exocytotic images and clearly correlate them temporally with enhanced secretion (Heuser et al., 1979).

While the morphological evidence alone might be sufficient to convince the skeptic of the importance of exocytosis, biochemical evidence also exists to strengthen the argument. For if the secretory granule or vesicle were to secrete its contents directly into the extracellular space, as would be expected during exocytosis, then all soluble components of the storage vesicle should appear in the venous outflow from the secretory organ, and vesicle membrane constituents should not be detected in the extracellular space unless the whole vesicle is exported during secretion. Moreover, substances free in the cell cytoplasm or

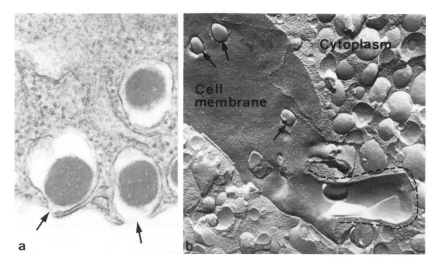

FIGURE 6. (a) Insulin-producing cell from an isolated rat islet stimulated with glucose. The arrows point to the opening of the inside of the granule in the extracellular space which occurs through fusion of the granule's limiting membrane with the cell (plasma) membrane. (b) Freeze-fracture replica of an islet cell stimulated with glucose and showing several exocytotic figures (arrows) on the P face of the cell (plasma) membrane. The exocytotic figures represent stages at which the granule core is being extruded after the fusion of the granule-limiting membrane with the plasma membrane. In the cross-fractured cytoplasm, one sees numerous profiles of secretory granule limiting membranes as well as a large concavity (outlined by a dotted line) merging with the cell membrane. The membrane limiting the concavity may represent the membrane of several secretory granules fused together in a so-called chain release. (a) × 50,000 (b) × 22,000 (courtesy of L. Orci, Geneva, Switzerland).

characteristic of other cell organelles should be retained in the cell. Thus the simultaneous release of catecholamines, ATP, chromogranin, and soluble dopamine-β-hydroxylase, and the retention of membranous dopamine-β-hydroxylase, have provided strong support for the concept that secretion from the adrenal medulla occurs by exocytosis. Experiments conducted under *in vivo* conditions support this concept (cf. Viveros, 1975). This biochemical approach has been successfully applied to the studies of secretion from the neurohypophysis and the parathyroid gland (Morris *et al.*, 1978; Habener and Potts, 1979). In these systems a granular protein contained within the same secretory granules as the hormone is secreted *pari passu* with the secretory product. On the other hand, cytoplasmic enzymes, such as lactate dehydrogenase, as well as mitochondrial enzymes, are retained within the cell during enhanced secretion.

Other more indirect arguments can also be invoked to support the concept of exocytosis. If secretion involved the diffusion of free hormone or transmitter across the cytoplasm, then it is difficult to discern how the secretory product would be protected from degradation by intracellular enzymes. Moreover, data demonstrating that nerve stimulation or high potassium cannot evoke calcium-dependent release of cytoplasmic transmitter that is artificially increased by pharmacological manipulation, support the conclusion that vesicular and not cytoplasmic transmitter constitutes the immediate source of releasable product during activity (Redburn et al., 1979).

While the secretory granules obviously represent an important element in the secretory process, the components of this apparatus are morphologically and functionally heterogeneous, as evidenced by the fact that a significant proportion of secretory product may be found in the cytoplasm. This nonparticulate material is not artifactually derived by damage to the particle during homogenization and centrifugation. Not only can granular and nongranular (free) secretory product be distinguished in many cases, but also various "pools" or compartments can be identified which participate in the release process. In their classsical study on the turnover and release of acetylcholine from the superior cervical ganglion, Birks and MacIntosh (1961) identified several "pools" of acetylcholine including (a) a readily releasable pool, (b) a depot or reserve pool, and (c) stationary (or nonmobilizable) pool. A "surplus" pool of acetylcholine was also detected when choline esterase was inactivated. This pool is localized to a compartment readily accessible to hydrolysis but does not contribute to the quantity of transmitter released. Similarly, early studies on norepinephrine release from sympathetic nerves demonstrated the presence of various "pools" of transmitter which could be selectively released by nerve stimulation and by indirectly acting sympathominetic amines, such as tyramine (Furchgott et al., 1963). The consistent findings that newly synthesized neurotransmitters and hormones are preferentially released over older storage forms support the concept of the functional heterogeneity of secretory systems (Collier, 1969; MacGregor et al., 1975; Walker and Farquhar, 1980).

Implicit in the original view of the quantal theory of release was the notion of the equal probability of release for each quantum within the nerve ending. This concept is no longer tenable in light of the heterogeneous nature of vesicular stores and most significantly the discovery of a small subpopulation of synaptic vesicles that appear after stimulation and incorporate and release newly synthesized transmitter at a higher rate than other subpopulations (Zimmermann, 1979). Similarily, a small vesicle population exists in adrenergic axons that may be derived from large synaptic vesicles following exocytosis, and these newly formed adrenergic vesicles are refilled with transmitter (DePotter and Chubb, 1977). Even in one of the best-studied exocytotic systems, the medullary chromaffin cell, the chromaffin granules do not form a single population of particles, and not only are heterogeneous in regard to size but also possess subpopulations that appear to be relatively devoid of ATP and dopamine-β-hydroxylase and that show certain different functional properties (Winkler and Smith, 1975).

The nonhomogeneity of the granular fraction is not at all inconsistent with the concept of exocytosis. Thus, the finding that the ratio of dopamine-β-hydroxylase to catecholamine in the perfusate may be only 10% of that in lysates of chromaffin granules (Dixon *et al.*, 1975) may not refute the concept of exocytosis as much as it reflects the presence of secretory organelles deficient in soluble protein. Obviously, it is necessary to ascertain the chemical makeup of the different granule fractions within a cell to accurately assess which substances are suitable to act as endogenous tracers for investigating exocytosis.

Despite the wealth of evidence for exocytosis as the primary mode of physiological discharge, we must acknowledge the contentious arguments of those who propose only a secondary role for the vesicle or granule. The discord stems from two major areas of investigation, acetylcholine release from peripheral nerves and from brain, and protein release from the exocrine pancreas. Evidence has been offered from isotopic labeling experiments that an extravesicular pool of acetylcholine—representing about 50% of the total tissue content—is functionally important in release (Marchbanks, 1976; Israel and Dunant, 1979; Tauc, 1979). A direct role for this soluble pool of acetylcholine is indicated by experiments that show an initial decrease of cytoplasmic acetylcholine during stimulation and only subsequently a depletion of the vesicular acetylcholine. These findings taken together with the well-known preferential release of newly synthesized acetylcholine have provided support for the notion that acetylcholine is released directly from the cytoplasmic compartment of the nerve ending. Even in the adrenal medulla, concerning which the best biochemical evidence supporting exocytosis is available, an apparent temporal dissociation between the release of catecholamines and other granule constituents could be explained on the basis of a release mechanism not involving exocytosis (Jacobs *et al.*, 1978). Alternatively, as aforementioned, the temporal dissociation could result from a change in the granule subpopulation participating in secretion as it proceeds.

Those eschewing mechanistic uniformity in secretory cells view exocytosis, as exemplified by hormonal secretory systems, as a relatively slow process, whereas the rapid transmission that characterizes the nerve ending necessitates the invoking of a "pore" mechanism. Within this view, a quantum is formed not by the fusion of the vesicle membrane with the plasma membrane, but by a depolarization-dependent opening of a pore for a constant time to allow the release of a constant amount of acetylcholine (Israel and Dunant, 1979). This hypothesis, while in keeping with the constancy of the amplitude of the miniature end plate potential, conveniently ignores the evidence that exocytosis, retrieval, and recycling of membranes have been clearly demonstrated in peripheral nerve and brain synaptosomes.

The general view of nonexocytotic secretion is shared by Rothman (1975), who proposes that a cytoplasmic secretory pool of secretory protein in equilibrium with other protein compartments in the pancreatic acinar cell represents the reservoir from which product is discharged. However, there is no direct evidence that exportable proteins are indeed released into the cytoplasm, and the nonparallel discharge of certain specific proteins can be readily explained by subpopulations of zymogen granules that contain varying mixtures of secretory protein.

We have now summarized what we feel are irrefutable arguments for exocytosis as the principal mode of physiological secretion. Even the contention that the wide variety of secretory products released by diverse cell types argues against a common mechanism of secretion can be countered by Douglas's rebuttal (1968) that the basic components of the secretory process, the plasma membrane and the vesicle membrane, are similar in different cells. Moreover, as we will see, the striking similarity of the requirements for secretion, particularly with regard to the role of calcium, reinforces a common mechanism of secretion. The critical requirement for exocytosis, then, is the obligatory fusion of the granule and cell membranes to establish a connection between the granule interior and the cell exterior.

Implicit in this discussion is the view that the vesicle or granule once having released its contents to the cell exterior is retrieved by the cell by a process akin to endocytosis and is reused to store and release secretory products (Meldolesi *et al.*, 1978; Morris *et al.*, 1978; Suszkiw *et al.*, 1980). Since the synthesis rate of the membrane proteins is lower than that of exportable proteins, the membranes of the secretory granule must be reused rather than incorporated into the plasma membrane. Added support for the postulate that endocytotic vesicles are derived from the plasma membrane has been obtained by the recovery within the vesicles of extracellular markers such as horseradish peroxidase, a macromolecule that is incapable of crossing intact membranes (Fried and

Blaustein, 1978). There is thus a coupling of exocytosis and endocytosis in the secretory cycle. On the other hand, the existing evidence while strongly favoring exocytosis does not preclude the possibility of additional, alternative discharge mechanisms, particularly when *in vitro* conditions or nonphysiological secretagogues might alter or bypass the normal secretory mechanism.

The stage has been set for the consideration of the role of calcium in the secretory process by our review of the biological properties of calcium as well as the general features of the secretory process. Emphasis will now focus on the mechanism whereby calcium couples the membrane events that lead to the secretory process. However, when considering the actions of calcium on the various secretory processes, the complexities associated with calcium homeostasis, taken together with the heterogeneity of the pools of secretory product, dictate that caution should prevail when the temptation to make sweeping generalizations becomes overwhelming.

B. HISTORICAL DEVELOPMENT OF THE CONCEPT OF STIMULUS-SECRETION COUPLING

> Dwell on the past and you'll lose an eye. Forget the past and you'll lose both eyes
> (A. Solzhenitsyn)

Retracing the steps in the evolution of a concept can be justified from a number of perspectives. For example, this approach allows for the total picture to develop gradually, thus engendering a clearer understanding of the key elements of the system. Equally important, such an approach offers the research scientist an opportunity to view a given result as part of a continuum and not as an end in itself; this latter perspective is one that should always be kept in mind.

1. Cholinergic Nerves

The early investigator, constrained by limitations in theoretical knowledge and experimental tools, was left in large measure to his scientific imagination and intuition to draw correct conclusions from experiments that, by necessity, were limited in scope. This is readily observed by reading the reports of our scientific forefathers, who, by reason of these limitations, often produced papers that were mere duplications of previous studies. This is particularly true of the pioneer work that delineated a role for calcium in synaptic transmission.

It was the state of art during the latter half of the 19th century to study the effect of different constituents of blood on nerve–muscle preparations by incubating them in isotonic sodium chloride. Sidney Ringer (1883) accidently discovered that when sodium chloride was added to tap water, the solution was able to prevent a fall in the contraction of the ventricle. Ringer astutely perceived that a minute amount of calcium present in tap water, but lacking in distilled water, antagonized the injurious effects upon animal tissues of sodium chloride alone. The work of Ringer not only provided the first documented evidence for the critical role of calcium in biological systems, but also initiated the concept of ion antagonisms. The early physiologists became aware that cations exist in definite proportions in the tissue and introduced the concept of the physiological balanced salt solution (Loeb, 1900), although a debate raged as to why a physiological solution neutralized the "poisonous effects" of each constituent of the solution when added alone (Howell, 1905).

In another classical study, Locke (1894) found that within 20 min of immersing a nerve–muscle preparation in isotonic sodium chloride, single shocks to the nerve elicited no response and this effect could be counteracted by adding an appropriate amount of calcium. As a consequence, the role of calcium in synaptic transmission was revealed even before the concept of neurochemical transmission had been established. Other investigators at the time, including the budding scientific luminary Harvey Cushing (1901), repeated the experiments of Locke on the frog sciatic nerve muscle but left the preparation *in situ* and perfused it. Noting that during calcium deprivation the response to indirect (nerve) stimulation failed while that from direct (muscle) stimulation persisted, he concluded that the site of the defect must be at the nerve ending.

While accelerator nerves to the mammalian heart also were known to lose their actions in the absence of calcium, it was observed that the addition of very small amounts of calcium to the saline solution bathing the heart prevented the depression of cardiac contractility induced by vagal stimulation (Howell, 1905). Hagan and Ormond repeated these experiments in 1912, but the significance of their contribution lies in the fact that they postulated—before the concept of neurochemical transmission was proven—that the importance of calcium lies in its relation to "the production and conveyance of the vagus impulses to the heart muscle rather than to the inhibitory process itself which occurs within the heart muscle." These investigators arrived at this incisive analysis by the "principle of exclusion," for despite the lack of direct evidence, they had correctly observed that calcium in physiological amounts invariably causes an increase in contractility.

A similar accolade should be awarded to Mines, who in 1911 undertook to confirm and extend Locke's original experiment because he felt the most promising way to attack the problem of how calcium serves its role is to ascertain the biological effects of chemically related divalent cations. He confirmed not only that immersing the frog sartorius muscle in isotonic sodium chloride depressed the contractile response to nerve stimulation, but that the inhibitory action of vagal stimulation on the frog heart also required calcium. The cardioinhibitory action could be restored by strontium and barium, but the action of magnesium clearly stood apart from the other divalent cations in that it was ineffective in restoring the inhibitory response to vagal stimulation. Thus, Mines concluded that it was not their electric charge, but some special chemical property, that enabled these cations to interact with certain tissue constituents in a manner in which magnesium could not. So, here again, we have another example of a confirmatory study that added to our fund of knowledge by providing us with an important first step in our understanding of the nature of calcium's role in biological systems.

However, comprehension of the total picture required a clear understanding of the concept of neurochemical transmission, and there was some confusion at the time regarding the action of calcium upon the direct and indirect excitability of the nerve–muscle preparations. Ringer had discovered early on that frog muscle spontaneously twitched when bathed in a solution of pure sodium chloride, and that adding calcium abolished the twitching. Yet indirect excitability, as evoked by nerve stimulation, was increased by calcium. Hence, these early investigators were forced to reach what they thought were contradictory conclusions concerning the actions of calcium, when these actions were considered from the point of view of the response of muscle to nerve stimulation as opposed to direct muscle stimulation. So, with the rather limited scope of knowledge of these early investigators, the relation of calcium to excitability could not be expressed in general terms because the theory of neurochemical transmission had not yet been established and because the action of calcium on tissue excitability had also not been clearly defined. Thus, a clear interpretation of these early experiments had to await the discovery by Loewi (see Loewi, 1945) that the vagus nerve liberates a chemical substance (which he termed "Vagusstoff") before acetylcholine was identified as a chemical transmitter, and the concept of chemical transmission rested on firm grounds.

With the establishment of acetylcholine as the neurotransmitter at autonomic ganglia, parasympathetic effector sites, and the neuromuscular junction, by such stalwarts as Dale, Gaddum, Feldberg, and Vogt, it was only a matter of time before the mechanism by which calcium facilitates synaptic transmission was elucidated. In 1936 Feng suggested that the liberation of acetylcholine from a nerve–muscle preparation was increased by calcium. He reached this conclusion by showing that an increase in the calcium concentration restored the contractile responses evoked by single maximal shocks from partially curarized preparations. Again, these experiments carried out in the late 1930s were not unlike those conducted by Locke more than forty years before; however, the fact that the concept of neurochemical transmission was now a reality afforded Feng the opportunity to reach a more accurate conclusion.

A few years later Kuffler (1944) also studied the effects of calcium-deprivation at the neuromuscular junction, but this time the end-plate potential was monitored rather than mechanical contraction. Although a direct approach to this problem was still not possible at the time, his illuminating analysis clearly distinguished the actions of calcium on transmitter release from those on electrical excitability. He showed that during calcium deprivation the response of the end plate to applied acetylcholine was increased 100 to 1000 times, but that block of transmission eventually occurred, presumably due to the decreased release of transmitter. The effects of calcium on membrane properties that Kuffler observed resembled closely those seen in ganglia where a spontaneous rhythmic discharge of impulses was a concomitant of calcium deficiency.

However, the inability to measure acetylcholine release imparted a degree of uncertainty to the interpretation of these results, and it was the classical study of Harvey and MacIntosh (1940) that defined more precisely the role of calcium in synaptic transmission. Using the eserinized isolated cat cervical sympathetic ganglion and a bioassay for the measurement of acetylcholine, these investigators demonstrated that when calcium was removed from the perfusion medium the release of acetylcholine from the preganglionic nerve ending either during nerve stimulation or after the injection of potassium was either greatly diminished or completely abolished. The effects of calcium deprivation were ascribed to a direct effect on transmitter release and not to an effect on impulse conduction, since they noted that calcium lack produced hyperexcitability of the neuronal elements although synaptic transmission was blocked.

Katz and his associates subsequently employed the frog neuromuscular junction to study transmitter release despite the fact that acetylcholine was not measured directly. Using microelectrode techniques, they showed that the effects of calcium could be ascribed to the ability of this cation to increase the probability of the release of quanta of acetylcholine during depolarization by nerve impulses (Katz, 1958) Later, Katz and Miledi (1965) with the aid of a focal calcium electrode technique showed that the effect of calcium deprivation involved the release process directly and was not attributable to alterations in impulse conduction through the most distal nerve terminals. Moreover, they also demonstrated the ability of ionophoretically applied calcium to elicit transmitter release, but only when applied immediately following a depolarizing pulse (Katz and Miledi, 1967a) (Fig. 7). Quantitative analysis of the dependence of acetylcholine release on the external calcium concentration subsequently revealed that 3 or 4 calcium ions are required for the release of each quantal packet of transmitter by the nerve impulse (Rahamimoff, 1976), and, while strontium and barium are the only cations that can substitute for calcium, they are less effective. The power relation implies that a large fraction of receptor sites need to be occupied

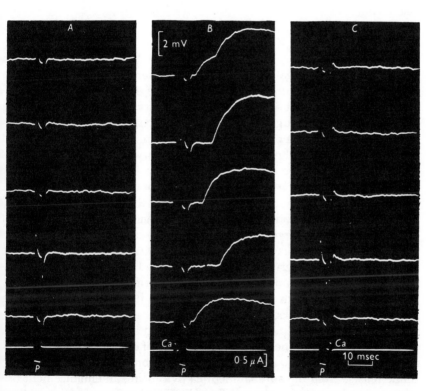

FIGURE 7. The timing of calcium action on the end-plate response of the frog sartorius. Depolarizing pulses (P) and calcium were applied from a twin-barrel micropipet to a region of the neuromuscular junction and intracellular recordings made from the end plate of the muscle fiber. Bottom traces show current pulses through the pipet. Column A, depolarizing pulse alone. B, calcium pulse precedes depolarizing pulse. C, depolarization precedes calcium pulse. Calcium pulses given before depolarization are necessary to evoke acetylcholine release as monitored by the end-plate response (Katz and Miledi, 1967a).

by calcium to elicit a given level of acetylcholine release. However, studies by Silinsky (1981) suggest that acetylcholine release does not obey the law of mass action with respect to occupancy of the hypothetical calcium receptor, but that in fact spare calcium receptors are present at motor nerve endings.

Clearly, this scenario as it relates to the action of calcium on transmitter release is similar to that observed in other fields, in that scientific investigators, though hampered by lack of fundamental knowledge and appropriate techniques, nurture an inexorable progression of knowledge and understanding, during which time basic concepts are constantly being reevaluated and modified. This scientific continuum will provide additional knowledge and novel concepts for those doing experiments years from now.

2. Adrenal Medulla

While the investigations on cholinergic nerves first clearly established the role of calcium in transmitter release, the possibility that this action of calcium might extend to other secretory systems was first considered by William Douglas beginning in the early 1960's. Douglas saw the role of calcium as not merely one of a mere factor in acetylcholine release but as a general mediator of the secretory process (see Douglas, 1968), and first selected the adrenal medulla as the system to test his theory. The medulla was an excellent choice for carrying out this initial study, since it was embryologically related to the neuron, concerning which previous work on the role of calcium had already been carried out; it also contains large stores of preformed catecholamines sequestered within cellular organelles that have a slow turnover rate; thus, any effects of calcium observed could be presumed to be entirely on the release *per se* and not an indirect effect on hormone synthesis. It was known at the time from the work of Katz and his associates (cf. Katz, 1958) that acetylcholine acted on the outer surface of the motor end plate to enhance membrane permeability to common species of extracellular cation, and with awareness of the importance of calcium for transmitter release, Douglas theorized that acetylcholine brought about catecholamine release from the adrenal medulla by causing an enhancement of calcium uptake into the chromaffin cell. The details surrounding the early experiments showing the calcium dependence of catecholamine release from the adrenal medulla have been recapitulated in some detail in a previous volume (Rubin, 1974a). Douglas (1968, 1975) has also presented several personal accounts of his endeavors. But for the purposes of continuity and completeness, a brief summary of these important findings is presented here.

The first and most basic experiment of course was to show that secretion in response to acetylcholine failed when calcium was omitted from the extracellular fluid and recovered promptly when calcium was restored. As the external calcium concentration was increased there was a corresponding rise in the secretory response to acetylcholine. By contrast, a decrease of external sodium or potassium did not inhibit acetylcholine-evoked catecholamine output but in fact enhanced it. Secretory responses to acetylcholine could even be elicited when the external medium consisted simply of isotonic sucrose and calcium, and this effect was lost when calcium was removed from the medium. The effects of cations on acetylcholine-evoked secretion could be mimicked by the elevation of external potassium which is known to act by depolarizing cell membranes by the resulting alteration in the ratio $[K]_i : [K]_o$.

While these experiments clearly delineated an important role for calcium in catecholamine release, it was still necessary to show that secretagogues such as acetylcholine or potassium were capable of promoting calcium conductance across the chromaffin cell membrane. The early findings by Douglas and Rubin (cf. Rubin, 1974a) that magnesium could inhibit acetylcholine-evoked catecholamine secretion provided indirect evidence for this thesis since magnesium was known to antagonize calcium actions and inhibit its transport in a variety of biological systems, but direct evidence was still needed. So, Douglas, together with Alan Poisner (Douglas and Poisner, 1962), demonstrated that acetylcholine's action was associated with an increase in ^{45}Ca uptake by the perfused adrenal gland. This finding was confirmed and extended in isolated adrenomedullary chromaffin cells by the demonstration of a causal relationship between calcium uptake and secretion (Kilpatrick et al., 1982).

Then, in an elegant series of electrophysiological studies, Douglas and his associates elucidated the mechanism by which acetylcholine sets in motion the train of events leading to catecholamine secretion. Not only did they demonstrate that acetylcholine and other medullary secretagogues depolarize the chromaffin cell membrane, they also found that the depolarizing current was carried by both calcium and sodium. But the critical point for our argument is that the calcium current is the essential factor for triggering secretion, for we have already noted that removing the external calcium abolishes the secretory response whereas a decrease in the sodium concentration accentuates it. The importance of the inward calcium current was confirmed by the findings that local anesthetics, which inhibited evoked secretion, blocked calcium uptake in the adrenal glands, but apparently did not impair the sodium current. One of the most critical pieces of evidence supporting the calcium hypothesis was that calcium itself was capable of eliciting a tremendous outpouring of catecholamines when introduced to chromaffin cells whose membranes were made leaky by exposing them to a calcium-free medium. This clearly demonstrated the ability of calcium to directly trigger the secretory response if permeability barriers are bypassed.

Thus on the balance of evidence Douglas viewed the sequence of the physiological events in the chromaffin cells as follows: acetylcholine producing an increase in membrane permeability → an influx of calcium → extrusion of catecholamine—a process he termed stimulus–secretion coupling (Douglas, 1968). The entry of calcium into the medullary chromaffin cell was perceived as promoting secretion in the same way as it promotes the release of acetylcholine from the motor nerve terminals (Katz, 1969). When the cationophore A23187 and the organic calcium antagonists became available in the 1970s, it was demonstrated that A23187 greatly enhances catecholamine output (Garcia et al., 1975) and D-600 blocks acetylcholine-evoked secretion (Pinto and Trifaro, 1976), and this evidence reinforced the view that calcium traverses the chromaffin cell membrane and activates the secretory process.

Douglas viewed stimulus–secretion coupling as a general concept and envisioned that "rather similar processes are at work in cells that, on first view, appear to have little in common—cells that differ widely in their secretory product, morphology, electrical excitability, and adequate stimulus" (Douglas, 1968). So he systematically undertook investigations of the submaxillary gland, neurohypophysis, and mast cell to demonstrate the ubiquity of calcium's action. Whether output of water, electrolytes, protein, peptide hormones or biogenic amines was being monitored, the scenario was always the same—the secretory response was depressed by calcium deprivation or high magnesium, but not by sodium or potassium lack; moreover, strontium and barium were generally capable of replacing calcium to activate the release process.

Thus based upon a wealth of evidence, Douglas concluded that in many secretory cells membrane activation is associated with an increase in membrane permeability causing calcium entry that somehow triggers the secretory response. But he also realized that there might be variations of this basic theme in that cells may also be able to utilize cellular calcium stores to sustain secretion. While the mast cell was known to respond to the antigen–antibody reaction with a calcium-dependent release of histamine, Douglas noted that prolonged calcium deprivation was required to depress the action of the histamine secretagogue 48/80, and he concluded that this polyamine induced secretion by mobilizing cellular calcium (Douglas, 1974a). He also provided direct evidence for the calcium hypothesis by eliciting histamine secretion from granules by the direct injection of calcium into the mast cell (Kanno *et al.*, 1973).

In acknowledging the alternate ways in which calcium can participate in promoting secretion, Douglas envisioned the now apparent parallels of stimulus–secretion coupling with excitation–contraction coupling in muscle, where the utilization of extracellular and cellular calcium by smooth, cardiac, and skeletal muscles varies widely. We will address this latter point in more detail in Chapter 3 but it should be stated here that Douglas perceived the fundamental similarity between the role of calcium in stimulus–secretion and excitation–contraction coupling, which Heilbrunn (1956) had speculated upon many years before. Through elegant and sophisticated experiments, present-day biologists are gathering convincing support for this view.

But Douglas was not satisfied with identifying the important role of calcium in stimulus–secretion coupling; in addition, he sought to define the underlying role of calcium in the secretory process, and in so doing he undertook morphological and biochemical approaches to develop the thesis that calcium action was linked to exocytosis. This view that calcium-activated exocytosis might be a general mechanism of secretion was presented at the First Gaddum Memorial Lecture delivered in Cambridge, England in 1967 (Douglas, 1968). And it is in large measure the result of his endeavors that the link between calcium and exocytosis has been unassailably made. Not only did he provide strong morphological evidence for the existence of exocytosis in the neurophypophysis and in the mast cell (Douglas, 1974b; 1974c), but he and his colleagues were the first to make a quantitative study of the simultaneous release of adenine nucleotides with catecholamines from the adrenal medulla, thereby establishing that the storage vesicles were the source of the secreted compounds (Douglas, 1968). Banks and Helle (1965) then demonstrated that the secretion from the adrenal medulla was accompanied by the release of soluble proteins from the chromaffin granules, which Kirshner and co-workers, and Blaschko, Winkler, and Smith identified as chromogranin and soluble dopamine-β-hydroxylase (Blaschko *et al.*, 1967; Viveros, 1975).

While it is clear from the evidence all too cursorily reviewed that acetylcholine triggers exocytotic catecholamine release from the adrenal medulla by increasing calcium entry into the cell, it is less clear how the alteration in membrane permeability is brought about. One theory suggests that acetylcholine causes a nonspecific increase in membrane permeability by displacing calcium from membranes (Douglas, 1975). On the other hand, acetylcholine is capable of eliciting tetrodotoxin-sensitive action potentials in cultured chromaffin cells (Biales *et al.*, 1976; Ritchie, 1979). Moreover, veratridine, which activates sodium channels and delays their inactivation, evokes a calcium-dependent release of catecholamines from perfused adrenal glands and isolated adrenomedullary chromaffin cells (Kirpekar and Pratt, 1979; Kilpatrick *et al*, 1982). Such experimental evidence may be used to support the argument that sodium action potentials are required to facilitate the activation of the voltage-dependent calcium channels. However, if tetrodotoxin-sensitive sodium channels play a role in the stimulation of catecholamine secretion, then the release response should be limited by tetrodotoxin. A survey of the existing literature reveals that tetrodotoxin, at best, produces only a partial inhibition ($< 50\%$) of the response to acetylcholine and little or no depression of the response to high K^+ (see, for example, Kirpekar and Prat, 1979; Kidokoro and Ritchie, 1980). Such findings, taken together with those demonstrating the potent blocking action of D-600, cobalt and manganese (Pinto and Trifaro, 1976; Ritchie, 1979), foster the conclusion that calcium influx is probably related primarily to the activation of the voltage-sensitive "slow calcium channels", although a proportion of these channels may be opened by a larger depolarization resulting from the activation of the voltage-dependent sodium channels. But it should be emphasized that while depolarization is an adequate stimulus for secretion, it is not, in and of itself, a sufficient stimulus and must be accompanied by a movement of calcium into the cell.

Another variable involves the existence of a receptor-operated channel in the chromaffin cell, since acetylcholine is capable of stimulating calcium-dependent secretion in the presence of a depolarizing concentration of potassium or when depolarization is prevented by the lack of extracellular sodium (Douglas, 1975). This discrepancy between catecholamine secretion and the level of depolarization based upon the operation of an acetylcholine receptor channel would also explain subtle differences noted in the mechanism of secretion induced by acetylcholine—which involves both depolarization and receptor activation—and that induced by potassium, which solely involves depolarization (Kidokoro and Ritchie, 1980). However, additional insight into the relative contributions of voltage-sensitive and receptor-operated channels to the secretory response of the adrenal medulla must await the biochemical characterization of these pathways, for little is known in biochemical terms about how the calcium signal is generated. One important obstacle along these lines is defining the chemical nature of the channels themselves. A greater understanding of the transduction steps involved in calcium gating may very well come from studies concerned with phospholipid hydrolysis, for there is an ever-growing body of evidence that phosphatidylinositol hydrolysis is linked in some way with the generation of the calcium signal.

So, kindled by the pioneering efforts of Douglas, a pattern was established regarding the role of calcium in secretion in the adrenal medulla, and, as we will see, was extended to the salivary gland, neurohypophysis, and mast cell; this pattern would be employed by numerous investigators probing other secretory systems. The profundity of Douglas's accomplishments and the impact that they have had on subsequent developments in this field will become abundantly clear as the features of each secretory system are analyzed.

3. Neurohypophysis

Neurosecretory cells differ from conventional neurons in that their secretory products, instead of acting on the synaptic junction with other neurons, are transported by the blood stream and act on target organs remote from their site of origin. In mammals the principal neurosecretory systems are found in the hypothalamus. The magnocellular system projects from parent cells localized to the supraoptic and paraventricular nuclei of the hypothalamus to the neural lobe of the pituitary gland where oxytocin and vasopressin are stored and secreted. Another system of neurons projecting to the median eminence of the hypothalamus secretes releasing and inhibiting hormones that influence the output of hormones of the anterior pituitary gland.

We will not attempt to recapitulate here the development of the concept of neurosecretion that was pioneered by the Scharrers except to state that they were the first to recognize that the neurohypophysis was the site of storage and release, rather than the site of production, of hormones (cf. Scharrer and Scharrer, 1945). They also perceived a dual role for neurosecretory cells as both neurons to conduct electrical activity and glandular elements to discharge product. But it was Bargmann, using the chrome–alum–hematoxylin method originally devised by Gomori, who actually demonstrated that hormone was produced in the perikaryon of the magnocellular neurons and was transported down the axons of these cells through the pituitary stalk to the posterior pituitary from which it could be released by appropriate stimuli such as dehydration or acute hemorrhage (Bargmann and Scharrer, 1951; Scharrer and Scharrer, 1954).

While exocytosis is considered the mechanism of pituitary hormone secretion, there is still some sentiment for the concept of release of extragranular hormone, which is based upon electron-microscopic evidence for electronlucent neurosecretory granules that have lost their core substance after stimulation, and the fact that hormone is found in granule-free supernatants of homogenates (Morris *et al.*, 1978). But compelling arguments favoring exocytosis are based not only upon morphological findings of such events occurring in the neurohypophysis but also upon the biochemical evidence that neurophysin, a cystine-rich protein present in neurosecretory granules that specifically binds oxytocin and vasopressin, is released in parallel with the hormone when the gland is stimulated *in vitro* or *in vivo* (for further information see symposium published in *Ann. N.Y. Acad. Sci.*, 1975, **248**).

Spanning this background of knowledge, the discovery by Douglas and Poisner (1964b) that the release of vasopressin from nerve endings of the hypothalamo-neurohypophyseal tract could be studied *in vitro* led to rapid progress in the understanding of the nature of the mechanisms involved in the release of neurohypophysial hormones. Using high potassium as the depolarizing stimulus for the axons and terminals of the isolated neural lobe, they clearly demonstrated that the ionic factors regulating secretion of posterior-pituitary hormones were the same as those regulating the secretion of chemical transmitters from conventional neurons and from the adrenal medulla. Douglas (1974b) has already reviewed the evidence in some detail so only a brief summary is required.

Douglas reasoned that the common developmental origin of the neurosecretory fibers of the neurohypophysis and the chromaffin cells of the adrenal medulla indicates the existence of a similar pattern in their secretory mechanisms; the finding that calcium was a sufficient as well as a necessary condition for potassium-evoked secretion of vasopressin, and that electrically stimulated release also required calcium supported a common mechanism. The observed increase in ^{45}Ca uptake elicited in potassium-depolarized glands was presumed to result from a transient decrease of the resting membrane potential of the nerve fiber terminals. Moreover, strontium and barium were able to substitute for calcium in triggering potassium-evoked release of pituitary hormone, thus delineating a calcium-receptive reaction similar to that found in cholinergic nerves and adrenal medulla (Ishida, 1968; Douglas, 1974b). The major difference emerging from all of these studies was that in medullary chromaffin cells a local depolarizing response triggered secretion, whereas, in nerves or neurosecretory cells, calcium entry occurred indirectly in response to a wave of all-or-none action potentials.

While calcium entry into the nerve terminal was obviously a critical event in triggering hormone release, the role of sodium in this process had to be addressed in light of its importance in the conduction of the action potential. Douglas and Poisner (1964b) found that the removal of sodium from the incubation medium not only did not block potassium-evoked pituitary hormone release, but, in fact, enhanced it, suggesting that one could dissociate action potential generating mechanisms from the release process. The additional finding by Dreifuss *et al.* (1971) that graded depolarizations by high potassium still induce a proportional enhancement of hormone output in the presence of tetrodotoxin provided further evidence that the activation of the release process could be uncoupled from the action potential generating mechanism. But Douglas and Sorimachi (see Douglas, 1974b), by showing that secretory response of isolated neural lobes to electrical stimulation could also be obtained in sodium-free medium containing tetrodotoxin, provided a clear indication that calcium entry is a sufficient condition to elicit neurohypophysial secretion. The alert reader will quickly relate this finding to those obtained at cholinergic terminals where tetrodotoxin is also able to abolish the action potential but does not affect the increased liberation of transmitter in response to local depolarization (Katz and Miledi, 1967b, 1969). The calcium entry associated with pituitary hormone release, being insensitive to tetrodotoxin and abolished by D-600, cobalt, or manganese, utilizes the tetrodotoxin-insensitive channel analogous to the "late calcium channel" described in squid axons (Dreifuss, 1975).

The conclusion appears valid despite such findings that veratridine evokes the release of oxytocin and vasopressin from isolated rat neurohypophysis and that this release is blocked by sodium deprivation, tetrodotoxin, or D-600 (Nordmann and Dyball, 1978). Do such data imply that coincident with stimulation, there is first a passive increase in sodium permeability that is responsible for the activation of the voltage-dependent calcium channel that is crucial for release? Or must we conclude that under physiological conditions the increase in calcium entry through the specific "slow" channels is not dependent on a previous increase in sodium permeability? This puzzle is seemingly only of heuristic interest since under physiological conditions there is in all probability an increase in sodium conductance concomitant with the electrical discharge of the nerve, which of course would lead to activation of the voltage-dependent calcium channel.

While the sequence involving stimulation, increased calcium uptake, and hormone release as formulated represents an attractive concept, one must also be cognizant of other variations of this theme that have been considered (Thorn *et al.*, 1978). While the findings of an increased calcium uptake after high potassium and an increase in oxytocin and vasopressin release after ionophore A23187 administration are all consistent with this simple concept, one must also acknowledge the difficulty in demonstrating an enhanced calcium uptake following electrical stimulation of isolated neural lobes; moreover, A23187 releases neurophypophysial hormone in the absence of external calcium, suggesting a mobilization of cellular pools. Thus, while calcium entry through the slow channels may be directly involved, it may also serve to trigger release of calcium from the binding sites on the plasma membrane or increase calcium release from the endoplasmic reticulum or mitochondria. Thus, the trigger calcium that enters the cell may be amplified in several ways by effects on intracellular calcium pools.

Sodium entry may also contribute to the rise in the intracellular calcium concentration by mobilizing calcium bound to intracellular organelles, such as the mitochondria. The ability of sodium accumulation by the cell to enhance calcium entry is amply documented by the finding that ouabain, an inhibitor of sodium-potassium ATPase, causes an increase in the output of oxytocin and vasopressin which is dependent on the presence of external calcium (Dicker, 1966). The ouabain-induced secretion which is widely observed in a variety of organs may be a consequence of the accumulation of intracellular calcium as a result of the activation of the sodium–calcium exchange mechanism caused by rising sodium levels in the cell. We will not be able to grapple with this problem directly until more sophisticated techniques become available for more accurately and precisely identifying various compartments of calcium within the cells of the neurohypophysis.

In conclusion, cholinergic nerves, adrenal chromaffin cells, and the neurophypophysis have all been considered from an historical perspective because not only the chronology of events dictates that we do so, but also the establishment of a pattern concerning the ionic events associated with the secretory response sets the stage for the consideration of other secretory systems. Ernst and Berta Scharrer addressed this point most perceptively in a review in 1945 when they noted that

> this process [secretion] is essentially similar in gland cells whose products serve functions as different as, for instance, the pancreas of a mouse, the ink gland of the squid, or the cutaneous poison glands of the newt. It follows that for the investigation of secretion as an intracellular process it is not necessary to establish first the functional significance of the secreted material.

Douglas, above all others, adopted this important concept and, as we have seen, proved its validity to a remarkable extent. As the data from other secretory systems are reviewed, one basic tenet will stand out: as with many other fundamental biological processes, there is a basic uniformity in the *modus operandi* by which nature brings about the extrusion from granules of secretory product, whatever the chemical makeup of the product may be; calcium occupies a pivotal role in this order of things.

C. OTHER NEURONAL SYSTEMS

1. Adrenergic Synapses

The observation that most of the catecholamine of the adrenal medulla is localized in chromaffin granules led to the discovery of similar particles storing norepinephrine in bovine splenic nerves in 1956 (von Euler and Hillarp, 1956), and morphological evidence for the presence of norepinephrine in the submicroscopic vesicular structures of adrenergic axons soon followed. Von Euler went on to win the Nobel Prize for his prior discovery of norepinephrine as the transmitter at adrenergic nerves, and Hillarp, although continuing to make important contributions to this field for the next ten years, died prematurely.

Advances in our knowledge of chromaffin granules have spawned parallel investigations of the nerve granules and the biochemical similarities between the two types of particles are very striking (Lagercrantz, 1976). On the other hand, nerve granules are considerably smaller than chromaffin granules, and while 80% of the total tissue catecholamine can be recovered in chromaffin granules, less than 50% is generally recovered in the vesicular fraction from sympathetic nerves, possibly resulting from the greater susceptibility of the smaller nerve vesicles to lysis during homogenization and centrifugation. Moreover, nerve granules are much more heterogeneous, showing marked variations in biochemical properties and in size and electron density (Winkler and Smith, 1975).

The discovery that medullary catecholamine release is brought about by a calcium-dependent exocytosis suggested that a common mechanism may exist in adrenergic nerves, in view of the common embryological origin of medulla and sympathetic nerves. Indeed, studies on sympathetically innervated organs have established the critical role of calcium in the release of norepinephrine from adrenergic nerves. It was Hukovic and Muscholl (1962) who first showed that the release of norepinephrine by stimulation of cardiac accelerator nerves was considerably depressed in low calcium solutions. In 1964 Burn and Gibbons, postulating a parallelism between the events in postganglionic sympathetic fibers and adrenal medulla, demonstrated that calcium plays a pivotal role in norepinephrine release by showing that the inhibition of pendular movements produced by stimulation of the nerve to the rabbit ileum depended on the external calcium concentration and could be prevented by magnesium. More detailed studies were carried out by Kirpekar and his associates studying norepinephrine release from cat spleen perfused *in situ* and by Boullin utilizing cat colon perfused *in vitro* (see Kirpekar,

1975). These two groups showed independently that removal of calcium markedly diminished evoked release, that norepinephrine output varied directly with the calcium concentration up to 7.5 mM, that magnesium antagonized release, and that barium and strontium were able to replace calcium in sustaining the release process.

In addition, adrenergic synapses were found to possess other basic characteristics similar to cholinergic synapses. We will recall that polyvalent cations including lanthanum, cobalt, and manganese depress or block transmission by decreasing transmitter release from presynaptic cholinergic nerve terminals. This antagonism is competitive in nature and is reversed by raising the extracellular calcium concentration. The findings by Kirpekar and his associates that the electrically-evoked release of norepinephrine from sympathetic nerves also requires calcium, and that magnesium, manganese, and lanthanum block release (Kirpekar, 1975) taken together suggest that a calcium channel also exists in sympathetic nerve terminals and is responsible for mediating norepinephrine release. In fact, when sodium and potassium currents are inactivated by tetrodotoxin and tetraethylammonium, respectively, depolarization of adrenergic nerve terminals results in a massive release of norepinephrine (Kirpekar and Prat, 1978), reinforcing the hypothesis that calcium entry through the voltage-sensitive calcium channel is a sufficient ionic event for activating the secretory apparatus.

In most of the previous systems that we have considered, the sympathetic effector organ is either perfused or incubated in a "Ringer-like" medium which is assayed for amine by either chemical or biological methods. However, just as in other synaptic junctions, presynaptic activity can be monitored at adrenergic synapses by analysis of the postsynaptic events. Junction potentials can be recorded by intracellular microelectrodes from various sympathetic neuro-effector junctions. The frequency and amplitude of the junction potentials can be decreased by sympathetic denervation as well as by the removal of calcium or the addition of excess magnesium (Kirpekar, 1975). Conversely, norepinephrine release, as monitored by the amplitude of the excitatory potential, varies directly with the external calcium concentration in the range of 0.8–2.5 mM (Bennett and Florin, 1975) (Fig. 8).

While conduction of nerve impulses into the terminal variocosities represents the primary stimulus for the release of norepinephrine from sympathetic nerves, evidence exists that transmitter release can also be modulated by drugs acting on presynaptic receptors (Starke, 1977; Westfall, 1977; Langer, 1980). Thus, it has been suggested that the decrease in norepinephrine release caused by activation of presynaptic α-adrenergic, muscarinic, and prostaglandin receptors is mediated by dim-

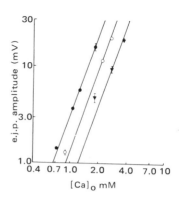

FIGURE 8.   The effect of calcium on neuromuscular transmission in the mouse vas deferens. The dependence of norepinephrine release on the external calcium concentration as monitored by the electrically-induced excitatory junction potential (e.j.p.) is shown at three different magnesium concentrations: ●, 1.2 mM; ○, 6.2 mM; ▼, 11.3 mM (Bennett and Florin, 1975).

inshing the availability of calcium ions for stimulus–secretion coupling (Göthert, 1977; McAfee *et al.*, 1981). α-Adrenergic and muscarinic agonists decrease norepinephrine release in response to various types of calcium-dependent stimulation—electrical stimulation, high potassium, and nicotinic agonists—while sparing calcium-independent tyramine-induced release. The fact that the inhibitory effect on calcium-dependent release is antagonized by increasing the calcium concentration in the perfusion fluid implies that the inhibitory effect of these presynaptic blocking agents is exerted by impeding the calcium influx attending the depolarization of the terminal varicosity. However, it must be also kept in mind that since calcium entry is the final common pathway in the sequence of events involving transmitter release, any factor that interferes with any event in this sequence may be affected by changes in the calcium concentration.

In contrast to the requirement for calcium in the mechanism of norepinephrine release from sympathetic nerves, the expendability of sodium was established by the finding that the removal of extracellular sodium did not block potassium-induced norepinephrine release. In fact, sodium deprivation produced an enhancement of spontaneous norepinephrine release that was not dependent on the presence of external calcium. Exocytotic catecholamine release can also be provoked by sodium deprivation in the adrenal medulla (Lastowecka and Trifaro, 1974) by a mechanism involving an increase in the availability of intracellular calcium, although sodium substitution elicits secretion in certain species only when extracellular calcium is available (Nishimura *et al.*, 1981). The latter results support the view that sodium deprivation may also promote calcium entry either by activation of calcium influx linked with sodium efflux or by competition between the two cations for uptake sites in the cell membrane.

The apparent demonstration that norepinephrine release from adrenergic nerve endings is calcium-dependent and occurs in the form of multimolecular packets, of course, provides testimony for the concept that an exocytotic process is at work here as well, and biochemical evidence supports this hypothesis. Soon after it was demonstrated that dopamine-β-hydroxylase and chromogranin are released in parallel with catecholamines from the adrenal medulla, studies were undertaken to determine whether a similar sequence of events occurs in sympathetic effector organs. Not unexpectedly, stimulation of sympathetic nerves by electrical current, high potassium, or veratridine elicited a concomitant discharge of norepinephrine and dopamine-β-hydroxylase by a calcium-dependent mechanism (Johnson *et al.*, 1975; Thoa *et al.*, 1975) (Fig. 9). The finding that tyramine and amphetamine induce release by a calcium-independent nonexocytotic process is a very important one (Thoenen *et al.*, 1969; Chubb *et al.*, 1972) because it provides additional justification for linking calcium and exocytosis.

Discrepant evidence does, however, exist in this system in the sense that release of catecholamine and soluble protein are not always tightly coupled (Weinshilboum, 1979). But this is not surprising in light of the convincing evidence that the noradrenergic vesicles, like those found in cholinergic nerve endings and perhaps endocrine organs as well, are

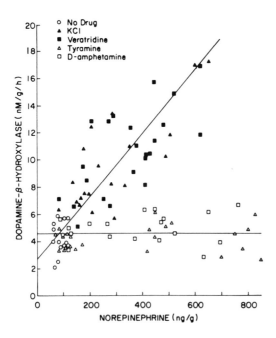

FIGURE 9. Release of norepinephrine and dopamine-β-hydroxylase from guinea pig vas deferens induced by depolarizing agents and sympathomimetic amines. Exocytotic secretion is elicited by high potassium and veratridine as evidenced by the high correlation between norepinephrine and dopamine-β-hydroxylase release. By contrast, norepinephrine release evoked by tyramine and amphetamine is not accompanied by parallel increases in dopamine-β-hydroxylase output (Thoa *et al.*, 1975).

heterogenous with regard to size and biochemical composition (Smith, 1972). Although the small dense-cored vesicles, like the large dense-cored vesicles, contain soluble protein (Bisby *et al.* 1973; DePotter and Chubb, 1977), the subpopulations of vesicles may sequester differential amounts of protein. Consequently, norepinephrine release not accompanied by a parallel discharge of soluble protein might only reflect the loss of transmitter predominantly from the small vesicles that are relatively deficient in soluble protein (Dahlström, 1973). A similar explanation can be offered to counter the arguments of those who would want us to believe that acetylcholine released from nerve endings is primarily derived from a free cytoplasmic pool.

The physiological significance of an extravesicular store of norepinephrine is still open to question. This store can be increased after inhibiting norepinephrine degradation by blocking monamine oxidase, making it analogous to surplus acetylcholine that accumulates in cholinergic neurons after inhibition of acetylcholine esterase, but is not available for release by nerve stimulation (Birks and MacIntosh, 1961). The fact that with appropriate pharmacological manipulations, the extravesicular pool of norepinephrine can be discharged by a non-calcium-dependent mechanism indicates that it does not directly participate in the release process under normal physiological conditions. This pool could, however, act as a storage form.

2. Other Peripheral Synapses (Excitatory and Inhibitory)

The emerging pattern of calcium's action on neurosecretory organs provides cause for entertaining the hypothesis that the action of calcium on neurotransmitter release reflects a general control mechanism, rather than mere examples of interactions of calcium with isolated events in certain specific neuronal systems. When one delves more deeply into the existing evidence, this hypothesis rests on very firm grounds, for chemically-mediated synaptic transmission—whether it is excitatory or inhibitory—universally depends on the presence of calcium in the extracellular fluid.

There is substantial evidence that in invertebrates, amino acids can function as either excitatory or inhibitory neurotransmitters (Usherwood, 1978). L-Glutamate serves as the excitatory transmitter at crayfish, lobster, and locust neuromuscular junctions, while γ-aminobutyric acid serves as the transmitter substance in peripheral inhibitory neurons. The demonstration of terminal synaptic vesicles and excitatory junctional potentials in insect and crustacean muscle fibers supports the thesis that neuromuscular transmission is quantal in nature in these lower organisms (Osborne, 1975; Atwood, 1976). It has also been established that calcium is a necessary element for excitatory neuromuscular transmission in these tissues (Bracho and Orkand, 1970).

Moreover, Miledi and Slater (1966) demonstrated that in low calcium a depolarizing pulse to the nerve ending of the squid stellate ganglion was unable to evoke a postsynaptic potential, indicating a lack of transmitter release, but transmission could be restored by ionophoretic application of calcium to a localized area of the presynaptic nerve ending (Fig. 10). These studies confirmed the belief that calcium's action was directly on the release mechanism and not indirectly mediated through effects on the electrical properties of the axon, since it was clearly demonstrated that during calcium deprivation the nerve impulse still reached the most distal terminals. A similar conclusion was reached by Berlind and Cooke (1971), who demonstrated that in the isolated pericardial organ of the crab the electrically-induced release of a cardioexcitatory peptide was blunted when the external medium was free of calcium, although the propagated electrical activity was not noticeably altered.

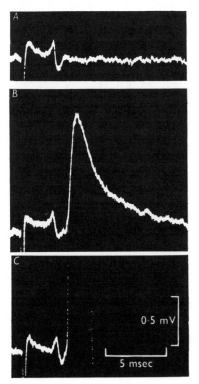

FIGURE 10. Effect of calcium on synaptic transmission in the squid giant synapse. (A) Intracellular recording from the postsynaptic axon of the ganglion. In calcium-free solution, stimulation of the preganglionic nerve evokes only a presynaptic spike. (B) Eight minutes after starting perfusion with a physiological solution containing 9 mM calcium, the presynaptic spike is followed by a postsynaptic potential (PSP). (C) Forty-two minutes later the postsynaptic potential has fully recovered and generates an action potential that is off the screen. The changes in calcium concentration which have a profound effect on the PSP do not greatly modify the presynaptic spike recorded by an electrode in the postsynaptic axon (Miledi and Slater, 1966).

Raising the magnesium concentration also reduced release of the peptide. The calcium "receptor" was similar to those found in nerves of
higher organisms in that barium substitution increased spontaneous
transmitter release and maintained electrically-induced transmitter release.

Sensory synapses display similar properties. Synaptic transmission
between photoreceptors (rods and cones) and horizontal cells (second
order neurons) in turtle and skate retina is also chemical in nature.
Although the anatomical appearance of the neuronal contacts does not
strictly conform to that of conventional chemical synapses, the presence
of vesicles in receptor terminals, spatial separation between junctional
membranes, and the latency between pre- and postsynaptic responses
provide evidence that synaptic transmission is chemically mediated
(Cervetto and Piccolino, 1974). It has been amply documented that
changes in the ionic composition of the extracellular medium such as a
decrease in calcium, an increase in magnesium, or the addition of cobalt,
block the release of an unknown transmitter as measured by the electrical
responses of the horizontal cells to light (Dorsling and Ripps, 1973). The
additional demonstration that strontium ions can substitute for calcium
in the release process confirms that the synapses between photoreceptors and horizontal cells share properties that are characteristic of more
conventional chemical synapses (Piccolino, 1976). In vertebrate retina,
calcium is important not only for synaptic transmission from photore-
ceptor cells to horizontal cells but also as a mediator for sensory transduction by which the photochemical apparatus in response to light transmits its signal to the presynaptic membrane. This unusual mechanism
involves the mobilization of cell calcium, which causes hyperpolarization
of the rods by impeding sodium influx and results in a diminution of
the release of an unidentified transmitter (Yoshikami and Hagins, 1978).

Carotid body chemoreceptors—a complex formed by glomus cells
and afferent nerve endings—have the morphological appearance of a
sensory synapse and are activated by low oxygen and high carbon diox-
ide tensions in blood. The mechanism of chemoreceptor signal transduction is quite complex, but ultrastructurally the glomus cells of the
carotid body share morphological and physiological features with neurosecretory cells and adrenal medullary chromaffin cells, for they contain
numerous cytoplasmic vesicles that store biogenic amines—particularly
dopamine—which are released by calcium-dependent exocytosis (Eyzaguirre and Fidone, 1980. Of physiological relevance is the finding that
the chemoreceptor response to changes in pH and electrical stimulation
is influenced by the extracellular calcium concentration. It thus appears
likely that calcium plays a crucial role in the transduction mechanisms

in the carotid body, although the picture here is clouded because the exact role of the glomus cells in the transmission of the signal to the afferent nerve has not been elucidated.

The effects of calcium on synaptic transmission are not confined to excitatory synapses but extend to inhibitory synapses (Usherwood, 1978; Nicoll and Alger, 1979). The pioneering work in this area was conducted by Kravitz and his associates (cf. Otsuka *et al.*, 1966), who demonstrated that, in low calcium, nerve stimulation failed to produce inhibitory junctional potentials in lobster muscle fibers, and while action potentials were still recorded from the nerve, no release of γ-aminobutyric acid was detectable. Thus the fundamental effect of calcium on synaptic transmission has been demonstrated in peripheral neurons up and down the phylogenetic scale, encompassing excitatory as well as inhibitory synapses. The importance of calcium in mediating synaptic transmission is now so well established that the use of calcium deprivation and magnesium to block transmission is an accepted standard for defining the nature of synaptic transmission as chemical rather than electrical.

3. Central Synapses

There is also now a wealth of neurophysiological and neurochemical evidence suggesting that synaptic transmission in the central nervous system involves a fundamentally similar process to that in the peripheral nervous system. Even so, a number of amino acids have been proposed as neurotransmitters in the mammalian central nervous system in addition to the known biogenic amines that serve as transmitters at peripheral synapses. An important criterion for neurotransmitter candidates is that they be released upon depolarization by a calcium-dependent process; early studies carried out on slices of cerebral cortex and spinal cord showed that the potassium-induced release of γ-aminobutyric acid, glycine, and glutamine are greatly reduced in the absence of extracellular calcium (cf. Mulder and Snyder, 1974). More recent studies using high potassium concentrations as a depolarizing stimulus and ionophore A23187 reveal a similar calcium-dependent mechanism operative in central neurons with regard to putative neurotransmitters including acetylcholine, 5-hydroxytryptamine, glutamine, γ-aminobutyric and aspartic acids (Clark and Collins, 1976; Collins, 1977; Waller and Richter, 1980).

In addition to measurement of the release of putative transmitter by chemical or radiotracer analysis, analysis of electrical activity in olfactory cortex and cerebellum has also been used to assess neurotransmitter activity. In these systems calcium and magnesium have antagonistic effects (Richards and Sercombe, 1970; Hackett, 1976). In the cortex, the amplitude of the postsynaptic potential is proportional to the external calcium concentration, and an analysis of postsynaptic events revealed that a cooperative action of 3 calcium ions is involved in the release of each quantal packet of transmitter in prepiriform cortex. This compares favorably with a cooperative action of 3–4 calcium ions for release of each quantum of acetylcholine at the neuromuscular junction. In the cerebellum, the order of potency of divalent cations in blocking synaptic transmission (Mn > Co > Mg) was similar to that observed for the neuromuscular junction, implying that a similar process is at work in these two widely-separated systems. A lesser impairment of the secretory response to electrical stimulation or veratridine by calcium deprivation is not particularly disquieting (Szerb, 1979; Cunningham and Neal, 1981), because in these situations it is likely that either residual extracellular calcium is available or that intracellular calcium is released by sodium influx. Thus, continuous flow of a superfused system (Waller and Richter, 1980) offers an important advantage over static incubations by curtailing the opportunity for calcium ions leached from cells and connective tissue to accumulate in the extracellular medium.

Calcium-dependent neurotransmitter release is also demonstrable at the level of the spinal cord (Erulkar et al., 1974). Before a neurotransmitter role was assigned to γ-aminobutyric acid in the amphibian spinal cord, a calcium-dependent release by depolarizing stimuli had to be demonstrated. Accordingly, potassium-evoked release of γ-aminobutyric acid was inhibited by low calcium, as well as by magnesium, manganese, and cobalt; these agents were also able to block synaptic transmission (Collins, 1974; Adair and Davidoff, 1977). The fact that the properties of depolarization-evoked release of γ-aminobutyric acid, as well as glycine, correlate well with the properties of the synaptic transmission not only lends support for a transmitter role for these amino acids in frog spinal cord but also provides us with important information regarding the control mechanisms governing their release.

The lack of suitable preparations for physiological investigation hampered advances in our knowledge regarding transmitter release processes at synapses of the central nervous system until the advent of the synaptosome. The development of synaptosomal preparations (pinched-off nerve endings) (DeBelleroche et al., 1972; Blaustein et al., 1972) for the measurement of neurotransmitter metabolism represents a significant breakthrough for analyzing presynaptic molecular events taking place in the central nervous system. It is possible to measure the transmembrane potentials in synaptosomes by indirect chemical methods, and there is a potassium diffusion potential across their surface membranes, as evidenced by the fact that graded release of transmitter can be observed with graded increases in potassium concentration (Blaustein and Goldring, 1975). At a given calcium concentration, a higher potassium concentration produces a greater depolarization and thus elicits a greater release of neurotransmitters such as acetylcholine (Blaustein, 1975). The high potassium concentration also promotes an increase in ^{45}Ca uptake into the synaptosomes. Thus, in the central nervous system, as in peripheral nerve, calcium influx appears to be controlled by the membrane potential.

Further similarities among the release mechanism of the central nervous system, the release of acetylcholine at the neuromuscular junction, and of catecholamines from the adrenal medulla are striking and argue that similar mechanisms control the release of the central neurotransmitters. In addition to the calcium requirement for potassium-induced transmitter release from isolated brain synaptosomes, there seems to be an involvement of the late calcium channel since manganese, cobalt and magnesium, but not tetrodotoxin, inhibit potassium-facilitated release (Blaustein, 1975). The voltage-dependent calcium channel in synaptosomes is not particularly sensitive to channel blockers such as D-600 (Nachsen and Blaustein, 1979). However, a similar calcium-sensitive cellular receptor seems to exist in the central neurons as in peripheral neurons since barium and strontium substitute for calcium in stimulating the release of putative transmitters such as γ-aminobutyric acid from brain synaptosomes (Levy et al., 1974). Taken together with the finding that potassium-induced depolarization of synaptosomes is accompanied by the calcium-dependent release of ATP (White, 1978), the evidence is compatible with a calcium-dependent exocytotic mechanism operating in the central nervous system.

Clearly, the development of brain synaptosomal preparations has provided us with profound insight into the physiological and biochemical mechanisms of synaptic transmission in the central nervous system that is essential for understanding brain function. In addition, the synaptosomal preparation has fostered a greater understanding of the mechanism of action of a variety of pharmacologic agents that affect the central nervous system. One important locus of action of pharmacologic agents in the central nervous system would appear to be the synapse, and indeed there is abundant evidence that drugs interfere with the release of neurotransmitters in the brain by inhibiting calcium-dependent neurotransmitter release. Barbiturates, ethanol, and tranquilizers, in particular, inhibit potassium-induced calcium influx and transmitter release by synaptosomes (Haycock et al., 1977; Olsen et al., 1977; Harris and Hood, 1980). The lack of information on transmitter release induced by electric currents notwithstanding, the potential significance of such experiments lies in the fact that, at least with regard to the barbiturates, concentrations that block potassium-evoked calcium uptake and release approximate those that are employed *in vivo* (Blaustein and Ector, 1975). The site of barbiturate action may be the voltage-dependent calcium channel, since this group of agents does not impair calcium influx or transmitter release promoted by ionophore A23187 (Haycock *et al.*, 1977).

But even greater potential significance must be attributed to the endeavors of E.L. Way and his colleagues regarding morphine's interactions with calcium metabolism in synaptosomal preparations. The association between the actions of morphine and calcium metabolism has been known for some time, since it was demonstrated that the analgesic effects of morphine could be reversed by the intracerebral injections of calcium (see Chapman and Way, 1980) and that morphine blocks neurotransmitter release in peripheral nerves (Paton *et al.*, 1971) by competitively antagonizing the action of calcium (Opmeer and van Ree, 1979). Judiciously employing the synaptosomal preparation, Way and his associates found that acute morphine administration produces a decrease in calcium influx that appears to be a specific opiate agonist action as revealed by the fact that naloxone, a morphine antagonist, prevents these effects of morphine (Chapman and Way, 1980). The concept that calcium mediates pain perception is supported by findings that agents that impair calcium entry into cells, such as lanthanum and EDTA, are capable of producing analgesia. Perhaps even more intriguing is the apparent association between elevated intracellular calcium levels and the development of tolerance. This probing analysis has afforded us new insights into the mechanism of action of opiates, and with the advent of the synaptosomal preparation, as well as cultured brain cells, powerful tools are available for accruing additional important knowledge about the nature of physiological and pharmacological mechanisms that alter or regulate brain function.

Isolated systems represent a particularly important probe for elucidating brain function because of the dramatic explosion in the number of putative neurotransmitters that have recently been identified, including gut-brain peptides, substance P, and vasoactive intestinal polypeptide, as well as the enkephalins and endorphins. Even hormones such as insulin, corticotropin, and angiotensin have been found in the brain and may play modulatory roles as neurotransmitters (Snyder, 1980). The physiological roles of many of these newly discovered brain peptides are being debated. Still, the enkephalins and endorphins possess opiate-like activity, and like morphine, β-endorphin decreases neurotransmitter release and decreases synaptosomal uptake of ^{45}Ca prior to the development of tolerance and increases ^{45}Ca uptake when tolerance has developed (Guerrero-Munoz et al., 1979).

Such findings imply that these opioid peptides play some physiological role in pain perception and drug tolerance. The behavior of these peptides is characteristic of other substances thought to be released by neurotransmitters in the central nervous system since calcium is necessary for their evoked release (see Section I.3., this chapter). Enkephalins or endorphins may also serve as neurotransmitters or neuromodulators in the peripheral nervous system as evidenced by the findings that endogenous opiate ligands in longitudinal muscle strip of guinea pig ileum can be released by electrical stimulation by a calcium-dependent mechanism (Oka and Sawa, 1979). The localization of Met- and Leu-enkephalin to the chromaffin granules of the adrenal medulla, taken together with their calcium-dependent discharge into the systemic circulation along with other soluble granular constituents, prompts additional speculation that enkephalins may also serve as "modulatory hormones" modifying the response of target organs or effector cells to humoral agents or neurotransmitters, respectively (Viveros et al., 1980).

This cursory overview should certainly strengthen the conviction that regardless of the position of the animal in the phylogenetic hierarchy, the peripheral or central location of the nerve terminals, the identity of the transmitter, or the function of the synapse involved (sensory or motor, excitatory or inhibitory), a basically similar calcium-dependent process is at work. We will subsequently see that this unitary mechanism for neurotransmitter release extends to a multiplicity of other systems.

D. HISTAMINE RELEASE FROM MAST CELLS

Many of the allergic manifestations associated with the response of a sensitized animal when challenged with an appropriate antigen (allergen) are due to the release from cells of one or more mediators. Histamine, the first such mediator to be identified, is present in high concentrations in the mast cell of connective tissue, as well as in the basophil leukocyte of blood, and is released during allergic and anaphylactic reactions. To be active, the antigen must bridge two sites linking adjacent-antibody immunoglobulin E molecules bound to specific receptors on the surface of the cell (Ishizaka et al., 1979; Foreman, 1981). Release can also be triggered by non-anaphylactic responses induced by endogenous substances, including some complement factors (C3a, C5a) and such a large number of diverse substances as to make any attempt at classifying them rather futile. These agents include polymeric amines (compound 48/80), polysaccharides (dextran), certain antibiotics (poly-myxin B), A23187, concanavalin A, and even ATP (Henson *et al.*, 1978; Kazimierczak and Diamant, 1978). In contrast to antigen, all of these agents, including activated complement, induce histamine release at their first contact with tissues.

The mast cell offers a particularly useful model for studying stim-ulus–secretion coupling mechanisms since the large size of its secretory granules allows light microscopic observation of granule discharge, which occurs by exocytosis. The large secretory granules impressed Paul Ehrlich, and to describe these cells he coined the term "Mast Zelle," which translates from the German into "a cell fattened or engorged with food." Another advantage afforded by the mast cell is that in contrast to excitable cells such as the medullary chromaffin cell this model cir-cumvents the added complications of a potential-sensitive apparatus associated with a depolarizing stimulus. Thus, the mast cell provides a model system for demonstrating that calcium alone is a sufficient stim-ulus for secretion and that concomitant alterations in membrane activity and/or ion movements are unnecessary.

Chronologically, evidence implicating calcium in histamine release was offered at a relatively early date. In 1958, Mongar and Schild ob-served that histamine release from mast cells in chopped lung stimulated by an antigen–antibody reaction required the presence of extracellular calcium. This observation was subsequently confirmed in isolated rat mast cells and in basophils in blood (Lichtenstein and Osler, 1964; Fo-reman and Mongar, 1972). The antigen–antibody reaction associated with histamine release leads to an increase in membrane permeability; calcium then enters the cell to evoke the release of the contents of the histamine-containing granules without a concomitant loss of other cy-toplasmic constituents such as lactate dehydrogenase and potassium.

In addition to the antigen–antibody reaction, dextran, ATP, and concanavalin A stimulate histamine release from mast cells by a process that requires extracellular calcium (Sugiyama, 1971; Foreman *et al.*, 1976; Ennis *et al.*, 1980). Optimal calcium concentrations for most of these systems center around 0.5–1.0 mM (Foreman *et al.*, 1976), but increases in the extracellular calcium concentration above these levels attenuate release. Extracellular calcium is not essential for the secretory action of the polyamine compound 48/80 and this fact casts doubt on whether the mast cell indeed represented a valid model for stimulus–secretion cou-pling (Uvnäs and Thon, 1961). However, subsequent findings that in-cubations containing calcium chelating agents resulted in the abolition of the response to compound 48/80, and that secretory activity could be restored by the addition of calcium, added to the rapidly growing wealth of information that certain secretagogues mobilize intracellular calcium pools rather than enhance calcium entry into the cell (Cochrane and Douglas, 1974; Payne and Garland, 1978).

Douglas, who had already contributed so much to the development of the calcium hypothesis with his work on the adrenal medulla, neurohypophysis, and salivary glands, productively utilized the rat peritoneal mast cell to further substantiate this theory. His endeavors focused on establishing that calcium was a sufficient stimulus for eliciting exocytosis, and he showed that the simple introduction of calcium into the cell by (a) adding the calcium ionophore A23187 (Cochrane and Douglas, 1974; actually first demonstrated by Foreman et al., 1973), (b) the injection of calcium into mast cells through a micropipette (Kanno et al., 1973), (c) the reintroduction of calcium after a period of calcium deprivation (Douglas and Kagayama, 1977), and (d) the delivery of calcium to the cell interior by using phospholipid vesicles (liposomes) as a carrier (Theoharides and Douglas, 1978) all triggered histamine release. While each of these experiments may be subject to individual criticism, all of these data taken together provide strong support for the view that calcium ions either from the extracellular compartment or from somewhere in the mast cell are a sufficient stimulus for initiating the exocytotic response.

John Foreman and his colleagues also provided us with insight into the molecular mechanism of histamine release by an incisive correlation of the secretory response with other biochemical and physiological parameters associated with secretion. One feature that secretory cells have in common is that the resting membrane permeability to calcium is low, so that they are generally resistant to variations in the extracellular calcium concentration. Foreman and his associates demonstrated that the antigen–antibody reaction initiated histamine secretion by raising the mast cell membrane permeability to calcium (Foreman, 1981). Moreover, strontium substitutes for calcium in this role as it does in other secretory systems. But strontium behaves in an unusual manner in mast cells, by activating histamine release in the absence of any stimulus (Foreman, 1977). This effect of strontium on spontaneous histamine release was a particularly important finding, for the degree of spontaneous release correlated temporally and quantitatively with the amount of strontium accumulated in the mast cell, thereby fortifying the link between the uptake of divalent cations into the cell and activation of secretion (Foreman et al., 1977b) (Fig. 11).

Other correlations between calcium uptake into the cell and the enhancement of histamine release from mast cells were obtained. Goth and his colleagues (1971) first showed that phosphatidylserine selectively

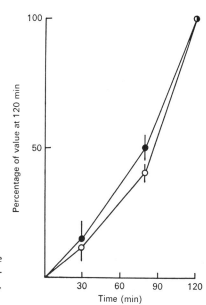

FIGURE 11. Relationship between the time course of ^{45}Sr uptake (●—●) and histamine release (○—○) from rat mast cells (Foreman *et al.*, 1977).

potentiated antigen-induced histamine release from mast cells, and Foreman and his co-workers demonstrated that this potentiation by phosphatidylserine was associated with an increase in ^{45}Ca uptake into rat peritoneal mast cells (Foreman *et al.*, 1976). This potentiating action of phosphatidylserine extends to other mast cell secretagogues including concanavalin A, a lectin from the jack bean, which elicits a calcium-dependent release of histamine by an action on the cell membrane. The biochemical basis of this action may relate to the formation of lyso-phosphatidylserine (Martin and Lagunoff, 1979) and a resulting impairment of calcium channel inactivation. Although the mast cell does not possess voltage-dependent calcium channels similar to those observed in excitable cells, it does possess receptor-operated channels, and a calcium inactivation mechanism exists to limit the accumulation of calcium ion within the cell (Foreman and Garland, 1974), just as this mechanism operates in secretory cells with voltage-dependent calcium channels (Baker and Rink, 1975; Nordmann, 1976). The inactivation mechanism must play a rather crucial role in modulating the amount of free calcium in the cytosol, since mitochondria and endoplasmic reticulum are not prominent in mast cells, although a plasma membrane calcium pump may function to reduce levels of cytosolic calcium (Cooper and Stanworth, 1976).

Cyclic AMP is an inhibitor of histamine release from mast cells;

Foreman and his associates showed that cyclic AMP or its analogues inhibit ^{45}Ca uptake into mast cell induced by antigen or other secretagogues, further strengthening the link between calcium entry and histamine release (Foreman et al., 1976). There is also available evidence for a correlation between the inhibition by local anesthetics of A23187-induced ^{45}Ca flux into mast cells and the inhibition of histamine release (Johnston and Miller, 1979). Lanthanum also inhibits antigen-induced histamine release and this inhibition is reversed by increasing the extracellular calcium concentration (Foreman and Mongar, 1973). Additionally, in concentrations greater than 10 μM lanthanum increases the spontaneous release of histamine, an effect demonstrating the dual action of lanthanum on the mast cell. The ability of lanthanum to activate the secretory apparatus in the mast cell as well as in certain other secretory cells (see Chapter 1, Section G.2) implies that the role of calcium in activating histamine release is basically similar. Foreman and Mongar (1972) documented this well-founded conclusion by demonstrating that the calcium receptors in mast cells and neuromuscular junction are pharmacologically similar. We make this comparison here in order to emphasize once again the basic similarity in the mechanisms of release of a diverse group of secretory cells.

This basic pattern extends to the role of cellular metabolism in histamine release. The direct participation of available endogenous energy (in the form of ATP) in the mechanism of histamine release has been amply demonstrated by the ability of glycolytic or respiratory inhibitors to impair the action of a variety of histamine liberating agents (Peterson, 1974; Johansen, 1979). The presence of glucose will completely counteract the inhibitory effects of oxidative phosphorylation, so anaerobic glycolysis is sufficient to furnish the requisite ATP. Since the involvement of energy is at a step in the release process beyond the uptake of calcium into the cell (Foreman et al., 1976, 1977a), ATP is presumed to act in concert with calcium to elicit secretion.

While the anaphylactic release of histamine from mast cells has a requirement for metabolic energy, the ability of ATP to directly stimulate exocytotic release of histamine is not a consequence of this general metabolic action, but appears to be a direct action of the nucleotide on the plasma membrane to increase cell membrane permeability (Dahlquist, 1974; Bennett et al., 1981). While a calcium-magnesium-activated ATPase has been demonstrated on the outer surface of the plasma membrane of rat peritoneal mast cells, no change in ATPase activity during or after histamine release induced by compound 48/80, dextran, or ATP has been demonstrable (Chakravarty and Echetebu, 1978), and this evidence favors an action of ATP on a specific receptor (purinergic?) on the mast cells surface rather than by way of an ecto-ATPase (Cockcroft and Gomperts, 1979). Nevertheless, we cannot ignore models proposing that hydrolysis of plasmalemmal ATP promotes the fusion of the granule membrane with the cell membrane (Woodin and Wieneke, 1970; Poste and Allison, 1973). We will consider these models in more detail in the next chapter.

While attention has been focused on the role of calcium in histamine release, some attention should be paid to the effects of sodium, not only because it is the major extracellular cation, but also because alterations in sodium metabolism exert effects on calcium disposition. Cochrane and Douglas (1976) found that sodium deprivation stimulated the spontaneous release of histamine from mast cells, even when the extracellular calcium concentration was reduced, although pretreatment with calcium-chelating agents prevented this release. After this pretreatment with calcium-chelating agents, calcium—but not magnesium—restored the secretory response to sodium deprivation, implying that sodium deprivation brings about a mobilization of cellular calcium. The mobilization of cell calcium may result from the activation of sodium–calcium countertransport mechanism of the kind found in certain other secretory organs, where a lowering of the extracellular sodium concentration effects a decrease in calcium efflux, the net effect being an increase in the concentration of free intracellular calcium.

The inability of sodium deprivation to compromise antigen-evoked histamine release is in general agreement with findings obtained in other secretory organs. In fact, lowering the extracellular sodium concentration in certain instances enhances stimulus-evoked release, although prolonged sodium deprivation, in general, impairs evoked release (cf. Banks et al., 1969; Hellman et al., 1974a). It is difficult to ascribe these inhibitory effects of prolonged cation deprivation to direct actions of sodium on the release process *per se,* although it has been proposed that a final stage of the process of histamine release involves an ion-exchange mechanism with an equivalent amount of sodium taken up in exchange for the histamine lost (Uvnäs, 1973).

In certain instances, the effects of sodium deprivation seem to relate to interference with the replenishment of vesicular stores of neurotransmitter by direct actions on neuronal uptake in adrenergic nerves or on acetylcholine synthesis in cholinergic nerves. This is supported by evidence that sodium deprivation is much more effective in attenuating sustained than acute secretion (Birks, 1965; Parsons, 1970), where the replenishment of stores of secretory product becomes more consequential. We also, of course, cannot ignore the adverse effects of prolonged deprivation of sodium—the major extracellular cation—or of the osmotic substitutes (sucrose, choline, and lithium), on the general metabolic properties of the cell. Such general effects on cell metabolism would of course lead to adverse effects on the specialized functions of the cell, including secretion. With regard to the mast cell in particular, the fact that sodium enters the mast cell during stimulation of the release reaction is ample justification for assuming that this cation would in some indirect way influence the secretory response. Alternatively, it would not be difficult to envision certain adverse effects produced by the various "foreign" substances that replace sodium when they accumulate in the cell as a result of the increase in membrane permeability associated with evoked histamine.

We cannot leave the subject of histamine release without also considering another commonly used model, the basophil leukocyte. Granules of basophils closely resemble those of mast cells in that they contain histamine and heparin. Since basophils respond to allergic and stress phenomena and account for a large proportion of histamine present in the peripheral circulation, they are considered "circulating analogues" of mast cells that are derived from and localized to connective tissue. Although there are some differences found between these two models, a basically similar underlying mechanism is involved. Thus, basophils are rendered reactive to antigenic stimulation by having IgE antibodies, complement factors (C5A), or formylmethionyl-peptides bind to their surface (Lichtenstein, 1971; Siraganian and Hook, 1977; Henson *et al.*, 1978). The release of histamine attending exposure to a suboptimal dose of secretagogue can be inhibited by removing extracellular calcium (Lichtenstein, 1975). Moreover, exogenous cyclic AMP or agents which elevate endogenous cyclic AMP levels depress the response to antigen in basophils just as they do in mast cells (Lichtenstein, 1971).

However, more careful analysis has demonstrated that while cyclic AMP decreases the maximal obtainable discharge of histamine, it enhances the initial rate of release (Foreman *et al.*, 1980). This apparent paradox has been interpreted to mean that cyclic AMP is a positive modulator of the release response in basophils just as it is in so many other systems, but also exerts an inhibitory action by increasing the rate of channel inactivation. This notion is supported by the finding that A23187-provoked histamine release—which does not manifest desensitization—is not blocked by cyclic AMP (Foreman *et al.*, 1977a). So, at least in cells derived from connective tissue or myeloid elements, cyclic AMP exerts a positive modulatory action on calcium movements, perhaps not unlike the one it is purported to exert on calcium channels in excitable cells (see Chapter 4, Section A.4).

To round out this aspect of our discussion, we can conclude that there exists a certain uniform pattern associated with the mechanism of histamine release from several diverse cell types, including the rabbit platelet, since the secretory process of this system exhibits a calcium requirement similar to that of the mast cell and basophil (Tidball and Scherer, 1972).

E. LYSOSOMAL ENZYME RELEASE FROM NEUTROPHILS

While the primary function of the neutrophil in host defense reactions is intimately related to its ability to engulf and destroy foreign materials, i.e., phagocytosis, it must also be considered a secretory organ, for the process of intracellular degranulation into the phagocytic vacuole is one of secretion, although in this particular case it is internal secretion. Granule constituents may also be discharged to the cell exterior by exocytosis to provide a source of proteolytic enzymes for the inflammatory process (Weissmann et al., 1976). Another way to induce extracellular granule discharge is with the use of the fungal metabolite, cytochalasin B. This agent inhibits phagocytosis but potentiates the release reaction in some ill-defined manner—perhaps by disrupting microfilaments—thereby dissociating these events concerned with phagocytosis from those concerned with the secretory process (Weissmann et al., 1976).

Viewing the neutrophil as a secretory cell requires knowledge of the nature of the organelles that secrete their contents during stimulation, and two major types of enzyme-containing granules have been identified in the cytoplasm of neutrophils: (1) primary or azurophil granules—containing lysosomes and (2) secondary or specific granules containing lysozyme and alkaline phosphatase. Azurophil granules are formed earlier and are larger and denser than the specific granules, while the specific granules are more accessible for mobilization and exocytosis (Bainton and Farquhar, 1968). Contents of both types of granules are released by complement components (C3, C5a) and various immunoglobulin subclasses (IgG) by acting on their respective receptors, but only when the cells are pretreated with cytochalasin B (Chenoweth and Hugli, 1978; Henson et al., 1978). N-formyl-methionyl and related synthetic oligopeptides also induce granule enzyme secretion and aggregation of neutrophils by interacting with surface receptors (Becker, 1979). Whereas contents of both primary and secondary granules are released by immunological stimuli and by chemotactic peptides in the presence of cytochalasin B, concanavalin A, phorbol esters and low concentrations of A23187 provoke the release of only the contents of specific granules

(Wright *et al*, 1977). Hence, the neutrophil granules can be differentiated by functional, as well as by morphological and biochemical, criteria.

Another form of nonimmunological release is exerted by leukocidin, a product of staphylococci, which interacts with rabbit neutrophils from peritoneal exudates to release the granule constituents. Although the eventual effect of leukocidin is cytotoxic, the lysosomal enzyme release represents a secretory response preceding the cytotoxic effect. Hence, the use of this secretagogue disclosed valuable information concerning the biochemical events associated with secretion from neutrophils. Although this early work of Woodin and his collaborators (cf. Woodin, 1968) has been superseded somewhat by more recent studies involving the use of more physiological stimuli, the basic contributions of these investigators still remain.

Since Woodin's investigations took place in the 1960s, when the importance of calcium in the release of adrenomedullary catecholamines was being established, both endeavors proceeded along parallel lines. When neutrophils were treated with leukocidin in the presence of calcium, proteins present in the secretory granules such as β-glucuronidase were released into the medium, whereas soluble (cytoplasmic) proteins were retained within the cell. The omission of calcium, while inhibiting the stimulated exocytotic discharge of granule proteins, increased the permeability of the cell to the soluble constituents, thereby clearly dissociating the action of calcium on the secretory apparatus from that on membrane permeability. The events associated with cell activation were clearly distinguished from those associated with calcium-dependent protein secretion by pretreating neutrophils with leukocidin in the absence of calcium and then with antibody to terminate the action of leukocidin; the subsequent addition of calcium promoted protein secretion (Woodin and Wieneke, 1970). The ability of calcium to restore the secretory response decreased as the time of incubating the leukocidin-treated cells in calcium-free medium was increased, but the efficiency of the release response could be restored by adding nucleoside triphosphates or diphosphates to the medium, and this result established that both calcium and ATP are required to give optimal secretion (Fig. 12). On the basis of these data, a model for secretion was formulated that proposed a calcium-dependent adherence of granules to the surface membrane following stimulation and a subsequent removal of calcium brought about by high concentrations of orthophosphate generated by the hydrolysis of ATP (cf. Woodin and Wieneke, 1970). While we now focus our attention on subsequent studies that employed perhaps more sophisticated methodology as well as more physiological stimuli, the work of Woodin and his associates should not be disregarded, because much of

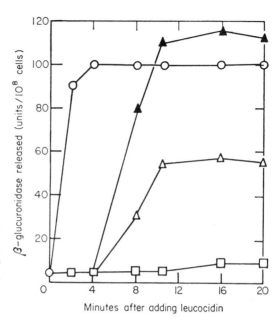

FIGURE 12. The stimulatory effect of ATP on calcium-induced lysosomal enzyme release from rabbit neutrophils. Neutrophils in calcium free medium containing EDTA (100 μM) were pretreated with leukocidin. ○, Calcium (1 mM) and ATP (1 mM) were present at zero time; △, calcium (1 mM) added at 4 min; ▲, calcium (1 mM) and ATP (1 mM) added at 4 min; □ no calcium or ATP added (Woodin, 1968).

what will be considered hereafter will, in many ways, be an elaboration of these pioneering investigations.

While numerous substances, both particulate and soluble, are capable of releasing granule-associated enzymes from neutrophils, the synthetic oligopeptides such as F-Met-Leu-Phe are potent soluble degranulating agents and have proven very useful in elucidating the physiological mechanisms associated with degranulation of neutrophils. The discovery of the similarities in structure between F-Met-Leu-Phe and an enzyme-releasing chemotactic factor from bacteria affirms the relevance of employing F-Met-Leu-Phe for the study of regulation of granule enzyme release. Just as lysosomal enzyme release from rabbit neutrophils induced by the bacterial chemotactic factor is enhanced by external calcium, but not magnesium, F-Met-Leu-Phe-induced secretion has similar properties (Smolen and Weissmann, 1981; Sha'afi and Naccache, 1981). The removal of extracellular calcium in many cases shifts the dose–response curve of agonist but does not completely abolish secretory responsiveness, and this effect suggests a more complex sequence of ionic events than merely a facilitation of calcium entry (Smith *et al.*, 1980; Hoffstein *et al.*, 1981).

Changes in the membrane potential of cells is one of the earliest events in cell activation, and stimulation of the neutrophil by various secretagogues induces a rapid depolarization followed by a hyperpolarization of the cell (cf. Jones *et al.*, 1981). The sodium and calcium dependency of depolarization, taken together with the fact that ionophore A23187 also induces depolarization, suggests that changes in ionic permeability are a consequence of neutrophil activation. The hyperpolarization response, which precedes enzyme release, requires sodium (Korchak and Weissmann, 1980). While changes in the membrane potential as measured indirectly using fluorescent probes provide evidence that ion fluxes occur during activation, a group headed by Elmer Becker and R. I. Sha'afi at the University of Connecticut demonstrated a close correlation between stimulated cation fluxes, particularly of calcium, and the secretory activity of rabbit peritoneal neutrophil. Thus, while cytochalasin B enhances the secretory activity of F-Met-Leu-Phe and ionophore A23187, it produces a rapid and parallel enhancement of calcium and sodium uptake induced by these secretagogues (Naccache *et al.*, 1977a,b). Although a correlation was observed between the percent change in cellular calcium and sodium produced by F-Met-Leu-Phe and the percent change in enzyme release, these ionic fluxes may not be the cause but rather a consequence of the degranulation process (Sha'afi and Naccache, 1981).

Low concentrations of F-Met-Leu-Phe can also mobilize cellular stores of calcium, as evidenced by the fact that it induces a transient decrease in the steady state levels of exchangeable calcium (Petroski *et al.*, 1979) and decreases fluorescence of neutrophil-associated chlorotetracycline (Naccache *et al.*, 1979). The plasma membrane is considered the likely membranous component involved since the cytoplasm of neutrophils is characteristically devoid of mitochondria and endoplasmic reticulum, two potential sites of calcium storage. The additional finding that lanthanum and cationic local anesthetics are capable of blocking ^{45}Ca uptake and neutrophil functions including enzyme release supports an important role for calcium in signal transmission from cell surface receptors (Boucek and Snyderman, 1976; Goldstein *et al.*, 1977; Showell *et al.*, 1977).

These findings provide the basis for the postulate that neutrophil activation by cell surface stimuli such as chemotactic factors involves a graded displacement of bound calcium (cf. Smolen and Weissmann, 1981), which Woodin and Wieneke (1970) had proposed some time ago as an essential step in the exocytotic process. The resulting increase in the calcium concentration in the cytosol brought about by the increase in membrane permeability participates in the activation of the contractile apparatus leading to cell movement (locomotion and phagocytosis) as well as degranulation. Extrusion of calcium by an ATP-driven calcium pump mechanism is likely to follow, in light of the absence of other intracellular calcium-binding organelles and the existence of ATPase activity associated with neutrophil membrane fractions that are stimulated specifically by micromolar calcium concentrations (Jackowski et al., 1979).

It is apparent that the multiple actions of calcium in the neutrophil make it difficult to pinpoint the specific pool or compartment of calcium involved in the release of granule enzymes. One approach to disentangling the various calcium pools involved is, of course, to dissociate granule enzyme secretion from other neutrophil functions. This appears to be a feasible approach in the equine neutrophil as opposed to rabbit or human neutrophil, since chemotactic peptides stimulate granule enzyme release in this particular species without a concomitant effect on chemotaxis (Snyderman and Pike, 1980). Additionally, one may focus on enzyme secretion by employing a pharmacological agent such as cytochalasin B, which inhibits phagocytosis but potentiates enzyme secretion, or ionophore A23187, which stimulates enzyme secretion without enhancing phagocytosis. The use of this latter agent, however, is somewhat limited because of its widespread effects on both extracellular and intracellular calcium pools.

Phorbol myristate, a tumor promoter, also enhances neutrophil degranulation except that this agent stimulates predominantly lysozyme secretion distributed in the specific granules and does not significantly enhance lysosomal enzyme (β-glucuronidase) release from primary granules (Goldstein et al., 1975). Thus, phorbol myristate (and other phorbol esters) represent a potentially interesting tool for studying the release mechanism of the neutrophil since they exert a differential effect on granule secretion and act in a different manner from that of chemotactic peptides in that pretreatment with cytochalasin B does not significantly enhance releasing activity. Moreover, phorbol myristate is only weakly chemotactic and its action does not depend upon extracellular calcium. It apparently promotes the redistribution of calcium within intracellular pools, as evidenced by the fact that the intracellular calcium antagonist TMB-8* blocks lysozyme release induced by this agent (Smith and Iden, 1979). But because the potential calcium-binding sites in the neutrophil are so limited by the virtual absence of mitochondria and endoplasmic reticulum, the pool of calcium utilized by phorbol myristate, like the pool of bound calcium utilized by chemotactic factors, is probably localized to the plasma membrane, but involves a more tightly-bound calcium fraction. So, although the relative contributions of extra-

---

* 8-(N,N-diethylamino) octyl 3,4,5-(trimethoxybenzoate)

cellular calcium in lysosomal enzyme secretion depend on the concentration of calcium in the extracellular fluid, the nature of the stimulus, and the level of stimulation, the final common denominator is an increase in the cytoplasmic calcium concentration, regardless of the nature of the stimulus and the source of the calcium utilized.

A possible role of extracellular sodium in the mechanism of enzyme release from neutrophils has been explored in some depth, and extracellular sodium appears required for optimal release (Showell et al., 1977). However, as with other secretory systems it is unclear whether sodium plays an active or a supportive (ancillary) role. In the absence of extracellular sodium, release induced by F-Met-Leu-Phe or ionophore is reduced but not prevented when calcium is present. A secondary role for sodium in the neutrophil is indicated by the fact than monensin, an ionophore selective for sodium, and ouabain, which causes an intracellular accumulation of sodium by inhibiting the sodium–potassium pump, are not effective secretagogues (Korchak and Weissmann, 1980; Smolen and Weissmann, 1981). On the other hand, sodium permeability is enhanced during the action of chemotactic peptides and ionophore A23187. The calcium dependency of this sodium accumulation connotes the existence of a calcium–sodium exchange mechanism in the neutrophil. Supporting the possibility of such a mechanism is the knowledge that removing extracellular sodium results in enhanced spontaneous enzyme release, which is greater in the presence of extracellular calcium (Showell et al., 1977).

The fact that F-Met-Leu-Phe induces a dramatic increase in potassium efflux, which is closely related to the extent of degranulation, is consistent with a calcium-sensitive potassium efflux mechanism also functioning in the neutrophil (Naccache et al., 1977b), as does the fact that cells incubated in potassium-free media are much less sensitive to chemotactic factors (Showell et al., 1977). But whatever the role of potassium, it appears to be intimately linked to the primary actions of calcium. For associated with the actions of membrane stimulating agents, there appears to be a graded release of calcium from membrane stores, as well as an increase in the permeability of the membrane to calcium, sodium—and potassium; these ionic events link membrane activation to the discharge of lysosomal enzyme and the subsequent cessation of activity.

The effects of calcium on neutrophil function must also be considered in concert with possible actions of other cellular mediators including cyclic nucleotides. For example, a close temporal relationship has been reported to exist between calcium influx, cyclic GMP accumulation and β-glucuronidase secretion, suggesting a cause-and-effect relationship involving these three processes (Ignarro, 1978). Knowledge that calcium stimulates guanylate cyclase activity in various tissues and increases cyclic GMP accumulation (see Chapter 4) might suggest that the sequence of events involves an increase in calcium permeability leading to an increase in calcium influx and an accumulation of cyclic GMP that triggers the secretory process. However, while such a mechanism may in fact exist, it is also possible that alterations in calcium flux and cyclic GMP levels are merely occurring in parallel.

Work has also centered around the possible role of cyclic AMP in the release reaction from neutrophils, and, as with the blood platelet, it appears that an increase in cyclic AMP levels fosters an inhibition of enzyme release (Jackowski and Sha'afi, 1979). Such findings may tempt one to conclude that a functional "dualism" exists in the neutrophil with regard to the proposed reciprocal actions of cyclic AMP and cyclic GMP. However, this theory, while having attracted proponents during the middle 1970s, has now fallen into disrepute and can only be rehabilitated by persuasive experimental evidence to support it (see Chapter 4).

Early work by Weissmann and his associates also demonstrated that certain of the classical prostaglandins (PGE and PGF) are inhibitors of the release process in neutrophils, a conclusion that prompted consideration of the possibility that these agents also acted as mediators of the release response (Weissmann et al., 1976). These inhibitory effects of prostaglandins on enzyme release from neutrophils, like those of cyclic AMP, parallel their inhibitory effects on the platelet release reaction. But as the field of prostaglandins developed and new arachidonic acid metabolites were discovered, it became clear that a theory imparting a simple inhibitory role for prostaglandins in lysosomal enzyme release was simplistic.

Neutrophils metabolize arachidonic acid during the degranulation response by two distinct pathways: (1) the cyclooxygenase pathway converting arachidonic acid to the more classical prostaglandins, PGE and $PGF_{2\alpha}$; and (2) the lipoxygenase pathway, which is the primary one in the neutrophil, and involves the formation of HPETE and subsequent conversion to HETE and the leukotrienes (Samuelsson et al., 1980). The ability of rabbit and human neutrophils to release lysosomal enzyme is clearly dependent on the integrity of the lipoxygenase pathway (Naccache et al., 1981). Arachidonic acid and lipoxygenase metabolites enhance ^{45}Ca influx and enzyme release, and the inhibition of arachidonic acid metabolism produces a parallel inhibition of enzyme release. We may thus infer that a close correlation exists between stimulated calcium flux, arachidonic acid metabolism, and the secretory activity of the neutrophil (Fig. 13).

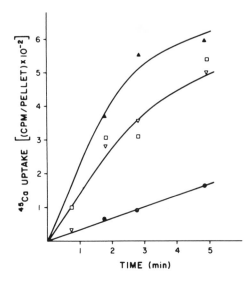

FIGURE 13. Effect of arachidonic acid, ∇, the arachidonic acid metabolite leukotriene B$_4$, $\square$, and the oligopeptide F-Met-Leu-Phe, $\blacktriangle$, on the time course of ^{45}Ca uptake into rabbit neutrophils. $\bullet$, Control (Naccache *et al.*, 1981).

This relationship must not only be viewed in terms of the involvement of arachidonic acid metabolism in the mechanisms regulating calcium permeability, but also in terms of the importance of calcium in regulating phospholipase A$_2$ activity, since A23187 is capable of stimulating phospholipase A$_2$ and arachidonic acid metabolism by bringing about the cellular redistribution of calcium (Samuelsson *et al.*, 1980). These findings aver that multiple pools of calcium are involved in the events associated with activation of neutrophils (see page 159).

However, it is still an open question whether arachidonic acid metabolites are essential to the secretory process, are only modulatory in nature, or exert indirect effects on secretion by way of primary actions on other neutrophil functions such as aggregation. In a more complex system such as the neutrophil where several different related calcium-dependent events are triggered by stimulating agents, it is much more difficult to disentangle the role of calcium in the secretory process from its other closely related actions. However, future studies should be able to define a mechanistic role for this sequence of complex biochemical reactions in the process associated with calcium-dependent enzyme release from the neutrophil.

While the interrelationships among calcium, cyclic nucleotides, and arachidonic acid metabolites are at the center of present concepts concerning nonmuscle cell activation, secretion of lysosomal enzymes from neutrophils, as well as chemotaxis and phagocytosis, may depend on one or another of the so-called contractile mechanisms of the neutrophil, particularly since these functions are characterized by active amoeboid motion. Findings in accord with this concept are the isolation of acto-myosinlike proteins from neutrophils (Crawford *et al.*, 1980) and the calcium-dependent contraction of glycerinated rabbit neutrophils by ATP (Hsu and Becker, 1975). The simplest and possibly most appealing hypothesis is that an increase in the intracellular calcium concentration brings about an increase in neutrophil locomotion involving the micro-filamentous and/or microtubular systems.

In view of the complex actions of such agents as colchicine, vin-blastine, and cytochalasin B on neutrophils (Smolen and Weissman, 1981), stronger evidence than their ability to impair neutrophil function is needed for implicating cellular contractile mechanisms. Additionally, the fact that cytochalasin B inhibits locomotion and chemotaxis and interferes with microfilament function, yet potentiates stimulus-evoked secretion, justifies some temperance in assigning to the microfilament and microtubular systems a direct role in calcium's action to promote secretion, although it would certainly not rule it out. But whatever the role of the contractile elements in enzyme release, insight into the action of calcium will probably be found when a better understanding of the secondary roles of cyclic nucleotides and arachidonic acid metabolites is achieved. For further discussion of this problem, the reader is directed to Chapters 3 and 4.

F. BLOOD PLATELET RELEASE REACTION

Blood platelets, like neutrophils, are derived from myeloid elements of bone and perform an important physiological role in hemostasis. In the presence of the proteolytic enzyme thrombin, platelets adhere to one another, triggering the so-called release reaction. Platelet interaction with subendothelial connective tissue components, most notably colla-gen, also induces the release reaction. As in many other secretory sys-tems, the platelet membrane is of prime importance in initiating a release reaction that exhibits a temperature dependence and utilizes ATP as an energy source (Holmsen, 1975; Feinstein, 1978a). Moreover, strontium and barium are able to substitute for calcium in this system, but they are considerably less effective (Sneddon, 1972).

The products discharged during the release reaction are contained within several types of storage granules (Holmsen, 1975; Kaplan, 1981). Specific granules, more commonly termed α-granules, contain many proteins, some of which are specific for the platelet and some of which are not. Electron-dense granules are also found that sequester mainly

5-hydroxytryptamine, as well as a nonmetabolic pool of adenine nucleotides and calcium. Organelles storing lysosomal enzymes also can be identified. The electron-dense granules are generally the first to be released, so unlike the neutrophil where lysosomal enzyme release is monitored, outputs of 5-hydroxytryptamine, adenine nucleotides, as well as calcium, are generally measured to determine secretory activity of the platelet.

The biochemical basis of platelet physiology has been extensively investigated; although there are many variables to consider, there is one final common pathway in all cases—the redistribution of calcium. Holmsen (Holmsen *et al.*, 1969; Holmsen, 1975) first formulated the theory that all platelet responses to stimulation are regulated by intracellular free calcium levels. Others, including Maurice Feinstein and J. G. White, followed this path and have worked and written extensively on the subject of the role of calcium in platelet function (Feinstein, 1978a, b; Feinstein *et al.*, 1981; Gerrard *et al.*, 1981). Calcium acts as the primary intracellular messenger for transmitting information from the surface of the platelet—where most stimulating agents exert their effects—to the internal machinery. Depending on the species and the nature of the stimulus, secretion may require extracellular or intracellular calcium. But the critical importance of calcium in the platelet release reaction is reinforced by the ability of ionophore A23187 to mimic almost all of the effects of thrombin on platelet function, including the release reaction (Feinman and Detwiler, 1974).

Establishing whether extracellular or intracellular calcium plays the principal role in the platelet release reaction has been a prodigious task in light of the fact that the requirement of extracellular calcium varies not only with the nature of the stimulus, but with the species under scrutiny, as well as even the temperature of the incubation medium. Whereas thrombin-induced secretion from human and calf platelets proceeds unabated in the absence of extracellular calcium (Holmsen and Day, 1970; Robblee and Shepro, 1976), the action of thrombin is greatly reduced under similar calcium-deprived conditions when rabbit and rat platelets are the test objects (Mustard and Packham, 1970; Bennett *et al.*, 1979). On the other hand, the release of 5-hydroxytryptamine from rat platelets induced by collagen and concanavalin A is also extremely sensitive to the extracellular calcium concentration (Bennett *et al.*, 1979). This differential sensitivity to extracellular calcium deprivation is not difficult to fathom if one assumes that certain stimuli exert a primary action on the cell membrane to increase calcium entry while others act through intracellular messengers to mobilize intracellular calcium. Yet, the finding that the thrombin-induced platelet release reaction is more dependent on extracellular calcium at 15° than at 37° (Grette, 1962) is more puzzling and defies a simple explanation; however, we know that platelet function is complex, with certain platelet responses occurring in parallel with one another, i.e., aggregation and secretion, and it is apparent that factors affecting aggregation will, in turn, influence the degree of secretory activity.

Another approach that is commonly utilized to gain knowledge of the calcium movements that occur during the release reaction is to monitor ^{45}Ca uptake during platelet activation. The complexity of the events occurring during platelet activation tends to obscure the specific calcium movements associated with the release reaction since several types of calcium movement appear to occur during platelet activation including: (1) an increase in calcium permeability produced by alterations in the plasma membrane, (2) calcium release from the plasma membrane into the cytoplasm, and (3) the mobilization of calcium from cellular organelles (Massini *et al.*, 1978). Despite variations in the calcium requirement for various stimuli, ^{45}Ca uptake is associated with the release reaction irrespective of the nature of the release-inducing agent.

The increase in calcium exchange that occurs during the release reaction is the combined result of a discharge of a large pool of calcium stored within the secretory granules and the mobilization of calcium from other intracellular binding sites. Since both calcium influx and calcium efflux may be enhanced during platelet activation, it then becomes difficult to specifically identify a component of calcium uptake that might be associated with the release reaction (Massini and Lüscher, 1976). Moreover, calcium entry decreases whenever the release response is prevented, suggesting that calcium uptake may, at least in part, be a consequence rather than the cause of platelet activation (Massini and Lüscher, 1976). Thus, an initial phase of calcium uptake is not required for the first manifestations of the release reaction, although other events associated with platelet function such as clot retraction seem to depend on the transmembrane movement of calcium from the suspending medium (Massini, 1977).

Still, the increase in the permeability of the plasma membrane to calcium may modulate the secretory response in some other way, as for example, by permitting a long-term response otherwise curtailed by the limited availability of the intracellular cation. Alternatively, the entry of calcium may induce a regenerative release of calcium from intracellular stores, or mobilize a cellular messenger such as prostaglandin endoperoxides and/or thromboxane A_2 that could release calcium from intracellular sites (see Chapter 4). This latter possibility should be given careful consideration since platelet secretagogues act through several different pathways to bring about a mobilization of calcium (Feinstein, 1978a, Feinstein et al., 1981; Gerrard et al., 1981).

Evidence supporting the concept that cellular pools of calcium are directly involved in the primary events associated with the release reaction first emerged in 1962 when Grette isolated from platelets a factor that inhibited superprecipitation of platelet actomyosin—an action which could be reversed by calcium. This "relaxing factor" was later shown to consist of membranous vesicles which accumulated calcium in the presence of ATP. These findings imply that an internal calcium store critical to platelet function is found in an intracellular membrane system that may be a structural correlate of the internal sarcoplasmic reticulum membrane system in muscle.

Morphological evidence later revealed that blood platelets contain two distinct channel systems: the open canalicular system, which consists of invaginations of the platelet surface, and a dense tubular system, which presumably arises from the endoplasmic reticulum (White, 1972). The open canalicular system is thought to be a conduit for the extrusion of secretory material (exocytosis) as well as for the ingestion of particulate matter (endocytosis). Activator calcium may be conveyed internally along the dense tubular system by way of the open-canalicular system (Gerrard et al., 1978). Implicit in this concept is the notion that the internal calcium store critical for platelet activation is found in the platelet dense-tubular system. Pharmacological and biochemical evidence supports this postulate. Certain secretagogues including thrombin and A23187 mobilize cellular stores of calcium in platelets, and local anesthetics prevent the secretagogue-induced mobilization of these stores (Feinstein et al., 1976). The additional finding that a platelet microsomal fraction is capable of actively accumulating calcium (Käser-Glanzmann et al., 1978) supports the platelet–muscle analogy and also provides support for the hypothesis that contractile physiology dominates the functional activity of platelets.

The platelet relaxing factor originally described by Grette was probably a membrane-associated ATPase resembling the sarcotubular relaxing factor of muscle. Platelets also contain contractile proteins with physicochemical properties similar to those of muscle. Platelet actomyosin, a contractile protein with Ca-Mg-ATPase activity, has been isolated from platelets and separated into actin and myosin-like compounds (Lüscher et al., 1972; Hanson et al., 1973). Since the regulation of muscle contraction involves calcium modifying the interaction of actin and myosin, the mechanism by which calcium controls contractility in platelets is probably analogous to that of smooth muscle cells. In fact, of all the secretory systems that have been surveyed, the blood platelet presently represents the best model for drawing parallels between the action of calcium on secretion and muscle contraction. We can extend this analogy by capitalizing on the fact that platelets not only possess calcium-activated ATPase activity, but also contain both calcium-dependent and cyclic AMP-dependent protein phosphorylation reactivities that are stimulated by activators of the release reaction (Haslam *et al.*, 1979). Hence, phosphorylation mechanisms may be important in regulating platelet function (Feinstein *et al.*, 1981) and platelet and muscle contraction (Adelstein, 1980). But regardless of the nature of the as yet unknown biochemical mechanisms involved, the activation of platelet function appears to be an event that is dependent upon a contractile process involving calcium-activated actomyosin. Apart from this action of calcium on the contractile machinery, a direct effect of this cation on the exocytotic process to facilitate membrane fusion cannot be excluded.

Thus, although calcium plays a key role in the platelet release reaction as it does in other secretory systems, the specific sites from which calcium is mobilized are still unknown and the mechanisms responsible for its release by stimuli remain to be elucidated. Despite the obvious complexities associated with activation of platelet function, a multidisciplinary approach to this problem should help to disentangle the various events associated not only with the platelet release reaction, but with other aspects of platelet function as well. Toward this end, the multiple pools of calcium in the various platelet functions must be dissected out to obtain a clearer understanding of the specific role of calcium in the platelet release reaction. Attention must also focus on cyclic AMP, as well as the prostaglandins and other arachidonic acid metabolites, in light of the rather persuasive evidence favoring their participation in this system. Cyclic AMP and certain of the prostaglandins play negative modulatory roles, while other prostaglandins, most notably thromboxane A_2, function as positive modulators (Gorman, 1979). An intriguing— and possibly complicating—factor in interpreting the roles of these putative intracellular messengers relates to the fact that the inhibitory actions of prostaglandins appear to be expressed through the adenylate cyclase-cyclic AMP system (Feinstein *et al.*, 1981). More detailed consideration of the interactions of calcium with cyclic AMP and the prostaglandins as they pertain to the blood platelet will be found in Chapter 4, and a scheme depicting the possible mechanisms for the mobilization of calcium during platelet activation is presented in Figure 14.

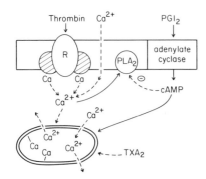

FIGURE 14. The mobilization of calcium during platelet activation. The interaction of thrombin with membrane receptors (R) increases cellular calcium by enhancing calcium influx or by releasing calcium bound to the membrane. A rise in the cytoplasmic calcium concentration may elicit additional calcium release from intracellular organelles such as the dense tubular system. The mobilization of calcium also activates membranous phospholipase A_2 (PLA$_2$) to release arachidonic acid which serves as a substrate for synthesizing prostaglandins. Thromboxane A_2 (TXA$_2$), a metabolite of the cyclooxygenase pathway, may release calcium from intracellular storage sites. Prostacyclin (PGI$_2$) produced by vascular endothelial cells may act as a negative modulator of these events by promoting the formation of cyclic AMP which either enhances calcium sequestration or directly inhibits phospholipase A_2 activity.

G. EXOCRINE SECRETION

1. Exocrine Pancreas

While a great variety of cells synthesize, package, and discharge exportable products, they do so by somewhat different routes; in endocrine cells, for example, the product enters the circulation by way of the extracellular fluid, whereas in the paracrine cells the product enters the extracellular space but acts on adjacent tissues. In exocrine glands the functional unit is the acinar cells grouped around a narrow lumen to form an acinus; the secretory product is discharged into the lumen and transported through a duct system onto an epithelial surface. The major function of pancreatic acinar cells is to secrete digestive enzymes, while cells of the ductular system secrete a bicarbonate-rich fluid important in neutralizing intestinal gastric acid. Over 20 digestive enzymes are secreted by the pancreas, including those involved in the breakdown of carbohydrate (amylase), protein, (trypsin and chymotrypsin), fat (lipase) and nucleic acids (RNAase, DNAase) (Williams, 1980a).

For many years, the primary modulator of pancreatic secretion was believed to be the autonomic nervous system, mediated through the vagus nerve; however, this concept had to be modified with the discovery of gastrointestinal hormones. The physiological stimulus to secretion is the presence of food in the upper gastrointestinal tract, which by initiating nervous reflexes and by stimulating the release of cholecystokinin-pancreozymin (CCK-PZ) and secretin from the duodenal mucosa, acts on acinar cells to accelerate enzyme discharge and evoke secretion of electrolytes and water (Case, 1978). Besides CCK-PZ, a number of peptides including bombesin, caerulein, and physalaemin, which are isolated from amphibian skin, and eledoisin, which is isolated from the octopod salivary gland, also accelerate secretion. Cholinergic (muscarinic) agonists, CCK-PZ, and bombesin-like peptides express their effects through specific receptors on the acinar cell membrane, although all three types of agonists appear to exert their effects on protein secretion through a common mechanism to be discussed below. These receptors are presumably confined to the basolateral portion of the acinar cell, while the product is discharged from the luminal portion of the cell, implying that a messenger system exists for transmitting the signal from one portion of the cell to another. Cells within individual acini are linked together by gap junctions, permitting the dissemination of various electrical and chemical messages through the acinus and neighboring acini (Petersen and Iwatsuki, 1978).

While Hokin first biochemically isolated the secretory granules of the exocrine pancreas and demonstrated that they contain amylase (Hokin, 1955), the essential features of the synthesis, intracellular transport, and packaging of pancreatic enzymes in secretory granules were defined by the classical studies of Palade and his associates (Jamieson and Palade, 1971), who also provided morphological evidence that acinar cell secretion of digestive enzyme occurs by exocytosis (Palade, 1959). Since water and ion release occurs by a nonexocytotic mechanism, we will stress in this discussion the control mechanisms involved in the exocytotic release of protein from the acinar cells. It must however be realized that protein and water secretion are physiologically linked in the sense that water and electrolytes must be secreted in sufficient amounts to purge the secreted protein from the ductal system.

Indications that the basic model of stimulus–secretion might apply to the exocrine pancreas stemmed from early studies revealing that under certain conditions evoked amylase release was depressed in calcium-free medium. Initially Hokin (1966) studied pigeon pancreas and found that presoaking slices in calcium free, EDTA-containing media completely abolished the secretory response to acetylcholine, and he logically concluded that the pancreas utilized extracellular calcium for secretion. Additional support for this concept was provided by the findings that cholinergic agents and CCK-PZ depolarize acinar cells (Matthews *et al.*, 1973). However, upon closer scrutiny it was established that the exocrine pancreas differs somewhat from the basic model of stimulus–secretion coupling in that high potassium which depolarizes the acinar cell is unable to directly promote enzyme release. However, in situations where neurotransmitter release is enhanced by high potassium, then acinar cell secretion is likewise stimulated (Benz et al., 1972; Pearson et al., 1981a).

Additionally, as additional experiments were conducted under different experimental conditions and with a variety of different preparations, including intact perfused pancreas, pancreatic pieces or fragments, isolated acini or isolated cells, it became clear that in contrast to certain other secretory tissues such as the adrenal medulla and neurohypophysis—which show an almost absolute dependence on extracellular calcium—a requirement for extracellular calcium was difficult to demonstrate for amylase secretion during the initial phase of secretion (Chandler, 1978). By contrast, a requirement for extracellular calcium was demonstrable during sustained stimulation (Petersen and Ueda, 1976). The inability to elicit a clear-cut effect of extracellular calcium on stimulated enzyme release from the exocrine pancreas prompted consideration that cellular rather than extracellular calcium was a prime mediator, so investigators began resorting to the use of pharmacological tools in their attempt to elucidate the relative importance of intra- and extracellular calcium in pancreatic amylase secretion.

Heisler and Grondin (1973), who were among the earliest proponents of the concept that intracellular calcium supported protein secretion from stimulated exocrine pancreas, reported that carbachol increased the rate of ^{45}Ca efflux from pancreatic fragments. Chandler and Williams (1974) demonstrated that lanthanum reduced the total calcium content of pancreatic acinar cells with a concentration dependence similar to that seen for the inhibitory effect of lanthanum on enzyme release. Since lanthanum displaces tightly bound calcium from membranes, the conclusion was tentatively reached that a membrane-bound "trigger calcium" might be implicated in pancreatic enzyme release. Subsequent papers by this group (for example, Williams and Chandler, 1975; Dormer et al., 1981) presented additional evidence that pancreatic stimulants might release intracellular stored calcium. The inability of calcium channel blockers such as D-600 and verapamil to exert inhibitory effects on ^{45}Ca uptake and evoked release of enzyme militated against the existence of voltage-sensitive calcium channels in pancreatic acinar cells that might provide a route for calcium entry (Case, 1978; Dormer et al., 1981), although such data do not, of course, rule out the possible existence of receptor-operated channels in the plasma membrane that might serve a similar function (Putney, 1978).

Other significant variations from the basic stimulus–secretion model were also revealed by the inability of a tenfold elevation in the magnesium concentration to inhibit evoked enzyme release; while strontium was found to partially substitute for calcium in sustaining evoked secretion, barium was ineffective in promoting spontaneous enzyme release or in sustaining stimulated release induced by cholinergic and peptide agonists (Case, 1978; Scheele and Haymovits, 1979). The inability of barium to activate the hypothetical calcium receptor responsible for triggering secretion is possibly a very significant deviation from the central dogma, because it implies that the nature of the underlying mechanism governing protein secretion may differ qualitatively from that existing in the adrenal medulla, the prototype model for stimulus–secretion coupling, for example. On the other hand, the inability of barium to trigger protein secretion may merely be a reflection of its inability to gain access to the critical activator site. Nevertheless, this important discrepancy remains unresolved.

While the exocrine pancreas does not adhere to the prototypical model of stimulus–secretion coupling, deviation from this model does not inevitably connote a secretory system operating via a non-calcium-dependent mechanism. This was clearly demonstrated by the ability of ionophore A23187 to promote an increase in ^{45}Ca influx and efflux and enzyme secretion from a variety of pancreatic preparations (Williams, 1980a), a result that provided evidence for the idea that translocation of cell calcium is an effective stimulant of protein secretion. On the other hand, the limitations of the use of this pharmacological tool rest with the fact that it may not be mimicking the actions of the more physiological stimuli with regard to the pools of calcium that are utilized. For example, while acetylcholine exerts a primary action on the cell membrane, the ionophore is able to penetrate intracellularly and release calcium from intracellular stores as well as promote calcium entry (Ponnappa and Williams, 1980). So while A23187 possesses certain actions which are similar to those of the physiological secretagogues in that it enhances pancreatic enzyme release and promotes the efflux of ^{45}Ca, conclusions drawn from experiments utilizing the ionophore as a secretagogue cannot be arbitrarily applied to the elucidation of the mechanisms by which cholinergic agents and peptides bring about pancreatic enzyme release. A direct measurement of intracellular free calcium concentration has not been made in pancreatic acinar cells during the secretory response, and it is therefore not possible to state unequivocally that acetylcholine and/ or CCK-PZ elevates the free calcium concentration of the acinar cell.

While a role for calcium in pancreatic amylase release seems likely, on the basis of evidence cited up to now, one cannot convincingly argue that the mobilization of either extra -or intracellular calcium is the pivotal triggering event. However, electrophysiological studies, particularly those conducted by O.H. Petersen and his colleagues, have rendered valuable information regarding the mode of action of pancreatic secretagogues, as well as the possible origin of the calcium that initiates enzyme release (cf. Petersen, 1980). While early studies demonstrated that activation of acetylcholine, CCK-PZ, or bombesin receptor sites causes membrane depolarization in acinar cells (Matthews *et al.*, 1973), the relationship of the electrophysiological changes to the calcium-dependent activation of the secretory process appears problematic in light of the fact that membrane depolarization produced by calcium deprivation or high potassium was ineffective in promoting enzyme release. Moreover, the electrical responses to secretagogues were blunted by the absence of extracellular sodium but not calcium, and this result demonstrated that sodium was the principal current carrier.

However, some relationship between the electrophysiological events and the secretory responses was revealed by the findings that membrane depolarization, like ^{45}Ca efflux, preceded amylase release, and the dose–response curves of acetylcholine-evoked membrane depolarization were similar to those for ^{45}Ca efflux and the secretory response (Petersen, 1980). The ability of acetylcholine to increase sodium permeability in the pancreatic acinar cell is not a novel finding, but it does involve a more complex mechanism than that related to its action at certain other cholinergic sites, as for example the adrenomedullary chromaffin cell, where acetylcholine increases membrane permeability to commonly occurring cations (see Putney, 1978).

Although the calcium component of the inward current during acetylcholine action on the pancreatic acinar cell is not detectable by the usual electrophysiological techniques, Petersen and his colleagues reasoned that if an increase in cellular calcium mediates the actions of acetylcholine, it should be possible to mimic the effects of this secretagogue whose actions are localized to the cell surface by introducing calcium directly into the cell. Indeed, they found that by inserting a $CaCl_2$-filled micropipette into an acinus it was possible to demonstrate that short pulses of calcium injection mimicked the effect of short pulses of acetylcholine applied to the outside of the cell in that there was a decrease in membrane resistance, depolarization, and electrical uncoupling of neighboring cells (Petersen and Iwatsuki, 1978, 1979). These elegant studies thus demonstrated that the electrical changes induced by acetylcholine, like its secretory effects, are mediated by a rise in the free intracellular calcium concentration. The acetylcholine-evoked changes in membrane potential are explained by the action of elevated intracellular calcium on membrane channels, which is to increase sodium and chloride conductances (Putney, 1978).

Important clues toward furthering our understanding of calcium's role in amylase release were obtained by the demonstration that while short pulses of secretagogues were capable of eliciting short bursts of enzyme release in the absence of extracellular calcium, sustained stimulus-evoked secretion required extracellular calcium. These results, taken together with the ability of short pulses of magnesium or manganese to evoke transient bursts of enzyme secretion delivered in the absence of extracellular calcium, gave rise to the notion that a releasable store of membrane bound calcium—not readily exchangeable with the extracellular fluid—is mobilized during the early stage of secretagogue action. In addition, an increase in calcium entry from the extracellular fluid is indicated, by the evolving dependence of acetylcholine-evoked depolarization and secretion on extracellular calcium (Laugier and Petersen, 1980).

While all of the aforementioned results are consistent with the hypothesis that secretagogues of the exocrine pancreas initiate an increase in membrane permeability to calcium, the validity of this hypothesis would be considerably strengthened by the demonstration of an increase in calcium uptake into the pancreatic acinar cell during activation of the secretory process. While early studies using pancreatic fragments and superfused tissue were unable to detect an increase in ^{45}Ca uptake into pancreatic acinar cells during secretory activity, with one exception (Heisler and Grondin, 1973), subsequent studies using isolated pancreatic acinar cells have indeed demonstrated an increase in ^{45}Ca uptake associated with the membrane activation of pancreatic acinar cells by various secretagogues. The advantages of utilizing single cells for such studies not only relate to the use of a homogeneous cell population but also to the rapid and precise manipulation of the constituents of the extracellular medium.

Irene Schulz and her associates (Schulz, 1980) demonstrated that the addition of acetylcholine or pancreozymin to ^{45}Ca preloaded cells results in a quick release of ^{45}Ca followed by an increase in ^{45}Ca uptake and concluded that calcium fluxes during secretagogue-stimulated enzyme release are biphasic due to two processes, one involving efflux and the other influx. These events are triggered by the mobilization of a pool of cellular calcium and subsequent extrusion of the cation by a calcium pump. There is also an associated increase in the permeability of the plasma membrane to calcium and other cations.

Designating the source of the "trigger" calcium as the plasma membrane is most appealing because the release of bound calcium from the plasma membrane during the actions of secretagogues could likely lead to an increase in membrane permeability that would bring about the subsequent increase in calcium influx or mobilize cellular calcium. On the other hand, fluorometric analysis of dissociated pancreatic acini pre-labeled with chlorotetracycline and subcellular fractionation studies offers evidence favoring the mitochondria (Williams, 1980a) or the endoplasmic reticulum (Ponnappa *et al.*, 1981) as potential intracellular sites from which calcium is released during stimulation. But, despite the uncertainty regarding its source, the pool of "trigger" calcium is in all probability utilized for short-term secretion, while the extracellular calcium pool is required for sustained secretion either to directly activate the secretory process or in turn to replenish the depleted reservoir pools within the cell.

We have avoided consideration of the possible role of sodium in the secretory response of the exocrine pancreas up to this point in order to focus on the role of calcium. Although early studies demonstrated the inhibitory effects of sodium-deprivation on enzyme secretion from pancreatic fragments or perfused pancreas (Case and Clausen, 1973; Kanno *et al.*, 1977), these effects were later explained by a necessity for a background secretion of electrolytes and water required to transport enzymes through the duct system (Petersen and Ueda, 1976; Williams, 1980b). The expendability of sodium in the mechanism of pancreatic enzyme release is underscored by the augmentation of basal amylase release from isolated acini during sodium deprivation. These findings, among others, weaken the credibility of hypotheses that implicate sodium entry as a prerequisite for the direct activation of enzyme release. While it is clear that secretagogues evoke an increase in sodium conductance, the entry of sodium into the cell is a calcium-mediated process—the purpose of which is to activate the mechanisms involved with fluid, rather than enzyme, secretion.

While a large body of evidence now supports the concept of a "trigger" pool of calcium initiating the sequence of events associated with enhanced pancreatic enzyme release, the preceding and subsequent steps in the sequence remain obscure. Michell's hypothesis that phosphatidylinositol turnover is responsible for calcium mobilization is a most attractive one, particularly as it pertains to exocrine glands, and must be seriously considered. With regard to the steps that are distal to those involved with calcium movements, these may be expected to be more complex than in some other secretory systems in light of the polar arrangement of the acinar cell. Another intriguing question that must be addressed is how the increase in the free intracellular calcium concentration near the base of the cell provides the signal for amylase secretion at the opposite pole near the lumen. The depth of our ignorance on this subject allows the elaboration of only the most rudimentary hypotheses that encompass other putative cellular mediators.

While cyclic AMP has been implicated in the mechanism of action of secretion to control bicarbonate secretion by the tubular epithelium, this cyclic nucleotide was originally thought to play a negligible role in pancreatic enzyme secretion (Case, 1978; Schulz and Stolze, 1980), as compared to its role in the more closely related parotid gland, or in certain endocrine glands, including the adrenal cortex. Cyclic AMP levels do not consistently increase in response to enzyme secretagogues in rat and mouse pancreas; conversely, cholera toxin, which markedly increases cyclic AMP levels in the pancreas, does not invariably elicit an increase in enzyme release and also does not affect the ability of acetylcholine or CCK-PZ to evoke enzyme secretion (Case, 1978). But in isolated guinea pig acinar cells the activation of certain peptide receptor sites increases cyclic AMP levels and evokes enzyme secretion (Robberecht et al., 1976; Pearson et al., 1981b). Therefore, as in the parotid gland (see below), there are two mechanisms operating in the exocrine pancreas, at least in certain species—a calcium-requiring non-cyclic AMP-dependent system and a cyclic AMP-dependent system that does not require extracellular calcium (see Pearson et al., 1981a) (Fig. 15).

In contrast to the variable effects of pancreatic secretagogues on cyclic AMP levels, cholinergic agonists or CCK-PZ and its analogues more consistently augment cyclic GMP levels (Gardner, 1979). Moreover, dose–response relationships between an increase in cyclic GMP levels,

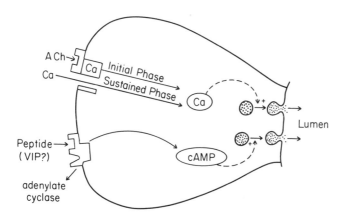

FIGURE 15. Schematic representation of a pancreatic acinar cell depicting mechanisms involved in receptor-mediated amylase secretion. During the initial phase of acetylcholine (ACh) (or cholecystokinin) action calcium is released from superficial membrane sites, and during sustained stimulation, membrane permeability is enhanced, bringing about the influx of extracellular calcium. In certain species, stimulation of peptidergic receptors—perhaps by vasoactive intestinal peptide (VIP)—mediates amylase secretion by the activation of the adenylate cyclase-cyclic AMP system (for further details see text).

stimulation of calcium efflux, and enhanced enzyme release are found to be similar and appear to be initiated by agonist interaction at the same receptor sites; but the association between the increase in calcium release and cyclic GMP is not known. The rise in cyclic GMP may be a consequence of the increase in the free calcium concentration in the cell (Singh, 1980); alternatively, and probably less likely, the increase in cyclic GMP might promote the release of cellular calcium.

But regardless of the exact sequence involved, the close association between the increase in intracellular calcium and stimulation of tissue cyclic GMP levels is underscored by the ability of the ionophore A23187 to elevate cyclic GMP levels in pancreatic acinar cells (Christophe *et al.*, 1976; Gardner *et al.*, 1980). Thus one might be tempted to postulate that the increase in cellular calcium brings about an increase in cyclic GMP levels, which, in turn, through some unknown biochemical pathway, triggers enzyme release.

However, the role of cyclic GMP as a potential mediator of pancreatic enzyme secretion is still quite unclear, for in the guinea pig pancreas following calcium withdrawal, the cyclic GMP response to cholinergic agonists or pancreozymin remains intact (Scheele and Haymovits, 1979), although the stimulation of enzyme release is nearly abolished. Such experiments suggest that the action of calcium is distal to the generation of cyclic GMP formation. The very rapid increase in cyclic GMP levels in the pancreatic acinar cells in response to CCK, cholinergic agonists or electrical stimulation of noncholinergic, nonadrenergic secretomotor nerves, occurring as rapidly as within 10–15 sec, certainly renders support for a primary action of cyclic GMP in the pancreatic secretory process (Christophe *et al.*, 1976; Pearson *et al.*, 1981b). So it is imperative that the role of cyclic GMP in this process be clearly defined. But the dauntless researchers who undertake this task must be forewarned that it will not be an easy one; for while the consensus of evidence suggests that the translocation of calcium is required for promoting cyclic GMP formation, cyclic GMP is probably not an obligatory modulator of enzyme secretion, in light of the dissociability of elevated cyclic GMP levels and enhanced amylase release (Heisler and Lambert, 1978).

With regard to the possibility that prostaglandins or other arachidonic acid metabolites may play a role in the control mechanisms involved in pancreatic exocrine secretion, there is only rudimentary evidence to sustain such a proposal (Marshall *et al.*, 1981). So despite the author's bias in its favor, we must await further evidence for the participation of these substances in the control mechanism governing secretion from the exocrine pancreas, or for that matter, from any other secretory organ. We will deal more comprehensively with this problem in Chapter 4.

2.   Salivary Glands

The salivary glands, like the exocrine pancreas and the gastric glands, are storage sites for digestive enzymes. The presence of zymogen granules in the acinar cells of exocrine glands has been known since the classical studies of Heidenhain and Langley. In fact, the basis of our knowledge of the secretory process was first elaborated in the latter part of the nineteenth century by Heidenhain, who was the first to recognize that secretory cells exhibit morphological and functional changes indicative of a "secretory cycle." Heidenhain and Langley, working independently, found that the number of secretory granules decreased after stimulation and that the histological picture correlated with the concentration of enzymes in the acinar cells of salivary and gastric glands (cf. Babkin, 1944). In addition to Heidenhain and Langley, Claude Bernard is another famous name from the past who made fundamental contributions to the field of salivary gland physiology, and the reader is referred to an interesting and entertaining review of the early history of developments in the field of salivary secretion, authored by Garrett (1975).

Three paired glands, the parotid, submandibular (submaxillary), and the sublingual, constitute the salivary glands in mammals. While the parotid acinar cell, like the pancreatic acinar cell, synthesizes, stores, and secretes mainly amylase and therefore is designated a serous cell, the rat submandibular and sublingual glands contain mucous cells that store and secrete appreciable amounts of mucins; these are high-molecular-weight glycoproteins rich in sialic acid, which serve to coat and lubricate the food bolus in its sojourn from mouth to stomach. Heterogeneity also exists with regard to the neural influences of the sympathetic and parasympathetic systems. Thus, while secretomotor parasympathetic fibers seem to innervate all mammalian salivary glands and are capable of producing wide changes in the composition and rate of secretion, salivary glands of some species including dog and cat exhibit little or no secretory response to sympathetic nerve stimulation (Schneyer et al., 1972; Jirakulsomchok and Schneyer, 1979). The rat parotid gland, which has dual innervation, has been most commonly utilized as the preparation for studying the molecular mechanisms governing salivary secretion; while valuable information has accumulated from the formidable array of evidence obtained from utilizing this preparation, it must be kept in mind that the rat parotid is not representative of salivary glands in general, and these exceptions will be appropriately noted.

Our previous volume (Rubin, 1974a) alluded to the truism that "scientific progress depends on the prudent selection of a basic problem which is susceptible to being solved at the current stage of scientific advancement and the selection of a proper system to study this problem," a tenet that Douglas followed when he elucidated the mechanism of catecholamine release from the adrenal medulla. The same dictum applies to the work of Schramm and his associates, who, by employing the isolated rat parotid slice preparation, laid the foundation for the understanding of the molecular mechanisms of protein secretion from salivary glands. They perceived the rat parotid gland as a system which would yield valuable data because it stored within its zymogen granules relatively large amounts of the cell protein in the form of a single enzyme, i.e. amylase (Schramm and Danon, 1961). While others had demonstrated that cholinergic agonists were adequate salivary secretagogues, the parotid slice preparation manifested a very low basal rate of secretion or "leakage" and responded to catecholamines with a brisk and extended discharge of amylase (Batzri and Selinger, 1973).

Employing both pharmacological and biochemical methods, Schramm and Selinger demonstrated that catecholamines act on the β-adrenergic receptor to cause the specific discharge of exportable protein, while α-adrenergic agonists, including the physiological neurotransmitter norepinephrine, act on α-adrenoreceptors to increase potassium and water release (Schramm and Selinger, 1974). Although these two effects are mediated by diverse receptor mechanisms, they are linked physiologically, since whenever protein secretion occurs it will always be accompanied by both potassium release and water secretion to convey the concentrated exportable protein away from the acinar cell into the duct system. Activation of cholinergic receptors, on the other hand, enables the gland to produce large volumes of saliva with lesser amounts of protein—a finding which Langley had made many years earlier. A group of undecapeptides (including Substance P, physalaemin and eledoisin) also exert effects primarily on water and electrolyte secretion. Each of the above described agonists interacts with its own unique receptor to bring about salivary secretion. Thus the parotid acinar cell with at least four different receptors (α-adrenergic, β-adrenergic, muscarinic cholinergic, and peptidergic) presents an opportunity to evaluate the various signals transmitted by these diverse receptors. We must, however, focus our attention on the β-adrenergic component since this appears to be the primary route by which enzyme release is brought about.

By 1970, calcium had been implicated in the mechanism of secretion in a variety of diverse systems, and Douglas and Poisner (1963) had already demonstrated the calcium dependency of acetylcholine-evoked enzyme release from the perfused cat submandibular gland. By measuring both water and protein release they found that protein output was more severely impaired by calcium deprivation. They also found that when the calcium concentration was increased fourfold the secretory response to acetylcholine was augmented, although this only involved an increase in protein and not water output. Magnesium produced a modest depression of the secretory response to acetylcholine, although again the inhibitory effects were more marked on protein than on water secretion. While they also showed that fluid secretion induced by nor-epinephrine was a calcium-dependent process, they did not present evidence that protein secretion is blunted by the omission of calcium. On these grounds, Douglas and Poisner proposed that calcium might act as a coupling agent in the salivary gland, although they were im-pressed by the fact that the effects of varying the calcium concentration and magnesium concentration were less dramatic in the salivary gland than, for example, in the adrenal medulla. While they did not necessarily exclude the possibility of calcium entry being the factor necessary to elicit the secretory response, they did consider alternative explanations of the role of calcium in salivary secretion, one of which was that calcium might influence the membrane events associated with extrusion of elec-trolytes and proteins. The prescient comment in this early paper that "calcium influences sodium transport by affecting sodium permeability" is also worthy of note.

Selinger and Naim published a short report in 1970 that has often been cited as evidence for a role for calcium in the mechanism of enzyme secretion from the rat parotid in response to β-adrenergic receptor stim-ulation. But it was necessary to incubate the tissue slices for prolonged periods of time in calcium-deprived media containing a calcium-chelat-ing agent to disclose a partial depression of amylase release in response to epinephrine. These investigators did not show that this inhibitory effect was reversed by the reintroduction of calcium, a fact that posed the question whether the inhibitory effects of calcium deprivation were the result of nonspecific effects on tissue viability or responsiveness, rather than a direct action on the release process *per se*.

Schramm and Selinger were also looking into the possible role of cyclic AMP as a mediator of enzyme release, since Sutherland and his colleagues had provided evidence, first in the liver and then in other tissues, that the actions of epinephrine and other catecholamines, par-ticularly related to β-adrenergic receptor activation, were mediated through cyclic AMP (cf. Robison *et al.*, 1971). Indeed, the role of cyclic AMP as a putative mediator of catecholamine-induced parotid amylase secretion was favored by the findings that cyclic AMP accumulated in parotid glands in response to epinephrine, that inhibitors of phospho-diesterase augmented the secretory response to epinephrine, and that derivatives of cyclic AMP mimicked the secretory activity of β-adrenoceptor agonists (Batzri *et al.*, 1973). Moreover, the findings that the secretory activity of α-adrenergic and cholinergic agonists were much more susceptible to calcium deprivation than β-adrenergic agonists seemed to cast a cloud over the potential role of calcium in the β-adrenergic receptor-mediated events associated with parotid amylase secretion. While Schramm and Selinger agreed that cholinergic stimulation, which was clearly a calcium-dependent event, did elicit enzyme secretion, they perceived the cholinergic pathway in the rat parotid gland as a minor pathway, in contrast to the exocrine pancreas, since the rate of amylase secretion brought about by muscarinic agonists was only about 10% of the rate produced by stimulation of β-adrenergic receptors. Not only was it difficult to show a reduction in catecholamine or cyclic AMP-induced enzyme release during calcium deprivation, but the ionophore A23187 induced a powerful stimulation of potassium release with only a very modest amount of protein secretion from the rat parotid gland (Selinger *et al.*, 1974).

Schramm and Selinger clearly expressed their skepticism about the possible role of calcium in enzyme secretion from the parotid gland in a review article published in 1975. They proposed that cyclic AMP, not calcium, served as the primary messenger of β-adrenergic receptor-mediated salivary amylase release, although they did concede that calcium might be required in some subsequent reaction which leads to the final step in the secretory response. A hypothesis proposing a dual action of the physiological neurotransmitter on the two adrenergic receptors in the parotid cell for the optimal response was promulgated, wherein norepinephrine acts at both α- and β-adrenergic receptors to elicit two distinct though related responses, resulting in the secretion of both protein and water and electrolyte. α-Adrenergic and cholinergic agonists were considered to elicit mainly water and electrolyte secretion, by utilizing a common calcium-dependent pathway, while β-adrenergic receptors were purported to play the major role in the regulation of amylase release through a cyclic AMP, non-calcium-dependent pathway (Fig. 16).

Once a fundamental concept has taken hold, it is sometimes very difficult to acknowledge that exceptions may exist; this tendency stems from the natural desire of scientists to apply unifying principles to all situations. The generality of calcium as an essential component in the secretory response is now so inculcated into our scientific thought that all existing experimental avenues must be explored before the possibility can be seriously entertained that calcium is not involved as a primary messenger in a given secretory response mediated by physiological stim-

FIGURE 16. Schematic model depicting the mechanisms regulating the secretory response of the salivary (parotid) gland. Catecholamines such as epinephrine (EPI) stimulate β-adrenergic receptors which in turn enhance adenylate cyclase (AC) activity. The cyclic AMP formed releases sequestered calcium (CaX) to elicit amylase secretion. Calcium entry into the cell is promoted by muscarinic, peptidergic, or α-adrenergic receptor activation by acetylcholine (ACh).

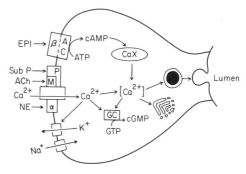

Substance P, or norepinephrine (NE), respectively Enzyme release is activated by calcium through a direct effect of the secretory apparatus or by enhancing cyclic GMP formation. Secretion is terminated by the sequestration of calcium. The increase in the calcium concentration of the acinar cell also leads to the release of water and electrolyte.

uli, particularly since we have already seen that certain secretory systems may very well be able to utilize intracellular calcium to mediate the response. So consideration must be given to the possibility that certain physiological secretagogues of the parotid gland may also prefer to utilize a cellular pool of calcium for a second messenger. Let us then consider the evidence favoring such a concept (see also Butcher and Putney, 1980).

The first of the usual criteria that one applies in an investigation concerned with the possible role of calcium as a mediator of parotid amylase secretion is, of course, that calcium deprivation blunts the response to the primary stimulus. We have seen that this criterion is not convincingly satisfied for β-adrenergic stimuli, since omitting calcium from the incubation medium and even adding EGTA may only modestly reduce the elevation in amylase secretion attending the stimulation of β-adrenergic receptors, while the secretory response to α-adrenergic and cholinergic agonists as well as the ionophore A23187 are markedly depressed or abolished. However, such experiments do not rule out the possibility that an intracellular pool of calcium, not readily exchangeable with the extracellular fluid, transmits the signal from the β-receptor. This possibility is strengthened by the finding that during prolonged calcium deprivation, with magnesium added to the medium to nullify the deleterious effects on the cell membrane of prolonged calcium derivation, the acceleration of amylase secretion in response to β-adrenergic stimulation is depressed (Putney *et al.*, 1977b).

A second criterion to establish the second messenger role for calcium in parotid amylase release is to demonstrate that calcium is a sufficient stimulus to elicit secretion. This problem, of course, may be approached in several ways. The readmission of calcium to the perfusion fluid after a period of calcium deprivation has proven to be a sufficient stimulus to evoke secretion in certain secretory systems, including the adrenal medulla and endocrine pancreas. While the reintroduction of calcium in cat submandibular and cockroach salivary glands elicits fluid secretion and electrophysiological responses indicative of enhanced potassium permeability, stimulation of protein secretion has not been demonstrable (Ginsborg et al., 1980). The pharmacological approach to increasing intracellular levels of calcium employs the ionophore A23187; while this agent increases enzyme release (Rossignol et al., 1974), it is generally less efficacious than β-adrenergic agonists such as isoproterenol. The fact that ionophore A23187 is unable to increase cyclic AMP supports the possibility that cyclic AMP and calcium act in concert to produce optimal secretory activity, an idea that is in harmony with the finding that calcium deprivation, which reduces the increase in cyclic AMP levels promoted by β-adrenergic receptor stimulation (Harfield and Tenenhouse, 1973), also reduces amylase release.

A third approach that has been used to investigate the putative role of calcium in amylase secretion relates to the effects of various agonists on calcium metabolism, for if calcium is involved in salivary amylase secretion, calcium fluxes should be modified by stimulants of amylase release. Indeed, it has been demonstrated in several different laboratories that isoproterenol and cyclic AMP are capable of stimulating the release of ^{45}Ca from perfused cat submandibular glands and isolated parotid acinar cells and slices, presumably by mobilizing bound or sequestered cellular calcium (Butcher and Putney, 1980). On the other hand, whereas α-adrenergic and cholinergic agonists also increase calcium uptake, β-adrenergic agonists generally have no effect (Koelz et al., 1977). The increase in calcium uptake into isolated parotid acinar cells that has been demonstrated in response to β-adrenergic agonists (Miller and Nelson, 1977; Putney et al., 1978a) represents either pinocytotic uptake or a replenishment of cellular calcium stores since it lags behind the secretory response. The finding that prolonged β-adrenergic receptor stimulation is associated with a decrease in the calcium content of the parotid acinar cell supports the need for such a restorative mechanism.

If indeed β-adrenergic receptor activation is associated with the mobilization of intracellular stores, then it should be possible to label subcellular fractions and determine the intracellular source of calcium needed for the regulation of enzyme release. Microsomes and mitochondria prepared from parotid glands accumulate calcium by an ATP-dependent process and epinephrine increases the rate of ^{45}Ca uptake into both mitochondria and microsomes prepared from rat parotid glands (Dormer and Ashcroft, 1974); in preloaded parotid acinar cells, catecholamines reduce the amount of ^{45}Ca found in both mitochondria and microsomes (Kanagasuntheram and Randle, 1976). These results support the contention that intracellular organelles are an important site of calcium sequestration and that β-adrenergic agonists mobilize calcium from these sites. The measurement of an increase in the free intracellular calcium concentration of the parotid acinar cell during activation of amylase release by β-adrenergic stimulation has not been reported, so the crucial experiment in support of this postulate is still lacking.

Since the events triggered by β-adrenergic receptor stimulation probably involve an activation of the adenylate cyclase-cyclic AMP system with no associated changes in membrane permeability, but a resulting increase in the intracellular calcium concentration by the mobilization of intracellular stores, the electrophysiological alterations observed with β-adrenergic agonists cannot bear much relation to the events associated with secretory response. Isoproterenol elicits a small depolarization of the acinar cell membrane of unknown significance (Petersen, 1976; Roberts and Petersen, 1978). A hyperpolarization is usually elicited by α-adrenergic or cholinergic agonists, although a depolarization may be observed in certain cells that possess a resting membrane potential that is lower than the acetylcholine-equilibrium potential (Roberts and Petersen, 1978; Wakui and Nishiyama, 1980). These electrical responses are manifestations of an increase in the passive permeability to cations since they are blunted by the removal of monovalent cations (Roberts *et al.*, 1978) or by metabolic inhibitors such as dinitrophenol (Petersen, 1971). The additional important finding that the amplitude of the secretory potential to various secretagogues is not markedly affected by calcium deprivation (cf. Ginsborg and House, 1980) not only annuls any direct association between the secretory potential and amylase secretion, but also rules out the possibility that the effects of calcium deprivation on secretion are expressed by alterations in receptor responsivity. Despite the limitations imposed upon extrapolating electrical events to the secretory response, the electrophysiological changes accompanying cholinergic and α-adrenergic receptor activation are clearly consonant with findings utilizing other approaches which indicate that these agonists, in contrast to the β-adrenergic receptor agonists, increase membrane permeability to cations.

Perhaps a better perspective of the relevance of electrical events to secretory activity in salivary glands can be gleaned by briefly considering tissues from submammalian species such as the molluscan salivary gland, where stimulation is associated with regenerative electrical activity rather than local potential changes (Kater *et al.*, 1978). In this tissue, calcium—which mediates the secretory response elicited by nerve stimulation—functions conjointly with sodium to produce an all-or-none action potential. Accordingly, we may speculate that at the lower echelon of the phylogenetic scale, the importance of electrical events in disseminating information is greater than in higher animals, in which they have been superseded by more complicated biochemical mechanisms for transmitting information.

Accepting the premise that both calcium and cyclic AMP are required for optimal amylase release provoked by catecholamines, one may then propose that cyclic AMP acts to direct calcium to the critical cellular site necessary for promoting secretion. Extending this line of reasoning, agents such as α-adrenergic and cholinergic agonists are less efficient inducers of amylase secretion because, while they increase the free intracellular calcium concentration of the cell by enhancing membrane permeability, the calcium that enters is not localized to the critical site for triggering enzyme secretion, at least not in the rat parotid gland. However, in the mouse parotid gland, acetylcholine releases amylase in amounts comparable to those released by catecholamines; so in this species, the amount of amylase released by cholinergic stimulation is sufficient to be considered physiologically significant (Watson *et al.*, 1979). The underlying significance of the findings in the mouse parotid for our present argument is that irrespective of the specific pool activated, calcium is able to trigger exocytotic protein discharge from the salivary gland, just as it appears to do in the exocrine pancreas.

The reader may logically conclude from this line of argument that β-adrenergic receptor stimulation brings about enzyme release through a cyclic AMP-dependent mobilization of cellular calcium, while α-adrenergic and cholinergic receptor stimulation mediates salivary amylase release by promoting calcium entry into the cell. Does this then imply that the entry of extracellular calcium into the cell evokes secretion without the interposition of another intermediate? This postulate may perhaps be simplistic, for while β-adrenergic receptor stimulation is associated with increased formation of cyclic AMP, stimulation of muscarinic receptors is associated with an increased formation of cyclic GMP. The additional findings that the stimulatory effect of cholinergic agonists requires the presence of extracellular calcium (Spearman and Pritchard, 1979), that ionophore A23187 mimics the effect of muscarinic agents on cyclic GMP levels and enzyme release (Butcher, 1975), and that cyclic GMP stimulates enzyme release in certain salivary gland preparations (Albano *et al.*, 1976; Butcher and Putney, 1980) arouses speculation that cyclic GMP serves as a critical intermediate in the process associated with salivary secretion. However, there are several results that temper enthusiasm for implicating cyclic GMP as an obligatory mediator of amylase secretion from the salivary gland (cf. Butcher and Putney, 1980). For example, substance P does not increase cyclic GMP levels yet stimulates enzyme release; conversely, sodium azide increases parotid cyclic GMP levels without increasing enzyme release. Thus, although cyclic GMP formation might be considered a calcium-mediated event in the salivary gland, we have much more to learn about its possible role in the regulation of secretion by the salivary gland.

Finally, for the sake of completeness, a brief consideration of the role of calcium in the mechanism of mucin secretion from salivary glands should also be included. The acinar cells of the rat submandibular gland—which are innervated by both the sympathetic and parasympathetic systems—also synthesize and store in separate cells high molecular weight glycoproteins (mucins) (cf. Quissell, 1980). Secretion occurs by an exocytotic mechanism and the intracellular reactions leading to the release of mucin are probably similar to those for secreting amylase. β-Adrenergic agonists effectively increase mucous secretion, whereas α-adrenergic or cholinergic agonists enhance the rate of release of saliva that has a low organic content. The omission of calcium and the addition of EGTA does not completely attenuate catecholamine-induced mucin secretion from isolated submandibular cells and a high concentration of extracellular magnesium does not impair the secretory response of these cells (Quissell and Barzen, 1980). On the basis of these experiments the conclusion was reached that some intracellular store of calcium, which is not very rapidly depleted, is essential for mucin secretion.

Results from mouse sublingual glands present a different picture. Adrenergic nerves are absent, and catecholamines induce only a slow calcium-dependent secretion from the sublingual gland *in vitro*. Moreover, no direct relationship between cyclic AMP and mucin secretion appears to exist, despite the relatively high activity of both adenylate cyclase and phosphodiesterase in glandular preparations (Nieuw-Amerongen *et al.*, 1980). Likewise, cholinergic agonists, which stimulate mucin secretion more effectively than norepinephrine, require extracellular calcium and are unable to elevate tissue cyclic AMP levels (Vreugdenhil and Roukema, 1975). This evidence reinforces the concept that calcium is a prime mediator of enzyme release from the salivary gland, whether cyclic AMP is implicated or not.

While the arguments presented here would support the concept that cyclic AMP requires calcium for optimal discharge of enzyme, there is still the haunting possibility—which only future experimental evidence can refute—that an effective cyclic-AMP-mediated mechanism can obviate the requirement for calcium. But this contingency is too perplexing to entertain for an extended period of time, so buoyed—perhaps deceptively so—by the supposition that arguments favoring a pivotal role for calcium in salivary enzyme release have been supported by more fact than fancy, we will proceed to a discussion of the lacrimal gland.

3. Lacrimal Glands

The relatively few studies devoted to characterizing the properties of the lacrimal gland reveal that it behaves similarly in many respects to other exocrine glands. Tears resulting from exposure of the eye to cold, wind, and foreign bodies, and in response to emotional stress, are mediated through muscarinic receptor activation transmitted by nerve impulses traveling along parasympathetic secretomotor fibers (Botelho, 1964). Just as in the exocrine pancreas and salivary gland, the lacrimal gland stores protein in secretory granules and discharges the product by a process that resembles exocytosis (Herzog et al., 1976). This process requires energy and is inhibited by low temperature. Although a prealbumin is the major constituent of the secretory granule, peroxidase or lysozyme is also present. With the use of cytochemical and electron microscopic techniques, Herzog and his associates (1976) demonstrated that the sojourn of this enzyme through the lacrimal acinar cell is basically similar to that of exportable proteins in other exocrine cells, for they found that the peroxidase reaction was confined to the lumen of rough endoplasmic reticulum, elements of the Golgi apparatus, and the secretory granules.

Protein release induced by cholinergic agonists from the exocrine pancreas and parotid gland is severely impaired when calcium is omitted from the medium, and the lacrimal gland behaves no differently. Not only is peroxidase release in response to cholinergic agonists depressed by calcium deprivation (Putney et al., 1977a), but these stimulants also induce an increase in ^{45}Ca uptake and release from rat lacrimal glands in vitro and from isolated cells (Keryer and Rossignol, 1976; Putney et al., 1978b; Parod et al., 1980a). The electrophysiological studies that have been carried out on the lacrimal gland reveal marked similarities to events observed in salivary glands and support the thesis that activation of muscarinic receptors in lacrimal glands increases membrane permeability to sodium and potassium (Iwatsuki and Petersen, 1978a; Parod and Putney, 1980; Parod et al., 1980b). Calcium entry into the cell, while possibly contributing to a minor degree to the depolarization, is more critically involved in triggering enzyme secretion by an action on the machinery controlling the exocytotic process and in eliciting electrolyte and water secretion by an action on the channels controlling sodium and potassium permeability (Parod and Putney, 1980).

The physiological significance of sympathetic innervation in the regulation of lacrimal gland secretion is still in question. Pharmacologic studies have demonstrated the existence of both α- and β-adrenergic receptors in the lacrimal gland of several species; however, sympathetic nerve stimulation does not directly increase secretion from cat and rabbit glands (Bromberg, 1981). In these species, adrenergic fibers do not impinge on secretory cells, indicating that their function is confined to the control of tone in vascular smooth muscle of the gland (Ruskell, 1975). Consequently, alterations in the secretory rate of the lacrimal gland of cat and rabbit induced by sympathetic nerve activity may best be explained by changes in the hemodynamics of the gland. In the rat, α-adrenergic receptor stimulation mimics the effect of muscarinic receptor stimulation in that there is an increased permeability to cations and an increase in protein secretion (Putney et al., 1978b; Parod et al., 1980a). Still, if calcium acts as a second messenger, it must be released from some cellular store, in light of an early transient phase of agonist-induced potassium release that is refractory to calcium deprivation (Parod and Putney, 1978). Moreover, the increase in ^{45}Ca efflux induced from slices of rat lacrimal glands by either carbachol or epinephrine is greatly reduced when the other agonist is subsequently added, suggesting a common pool of calcium utilized by both agonists. This phenomenon is identical to the situation for the salivary gland and can best be explained by the release of a pool of cellular calcium with limited reserves (Parod and Putney, 1979).

In summary, in mammals the flow of tears is comprised of water, electrolytes and protein and is regulated primarily through the parasympathetic nervous system, although at least in the rat the sympathetic nervous system also may play some direct role. The activation of muscarinic and α-adrenergic receptors of the rat lacrimal gland provokes protein discharge by mobilizing a cellular pool of calcium and by increasing membrane permeability to calcium ions, and to sodium and potassium as well. Thus, the rat lacrimal gland strongly resembles other exocrine secretory processes that apparently utilize a similar calcium-dependent pathway to trigger the secretory response (Iwatsuki and Petersen, 1978b). The same calcium influx responsible for activating exocytosis is also responsible for the release of electrolytes and water, and this fact leads us to the conclusion that the lacrimal acinar cell is capable of executing both enzyme and potassium release by utilizing the same receptor and second messenger (calcium) to mediate these two seemingly diverse functions—exocytosis and ion and water transport.

After summarizing the salient information that establishes the pivotal role of calcium in the mechanisms controlling the secretion of saliva and tears, we would be negligent to overlook the secretion of sweat, where secretory activity is completely dependent upon extracellular calcium (Prompt and Quinton, 1978). While we reaffirm the important distinction between the effects of calcium on macromolecular (protein) secretion and water and electrolyte secretion, it is clear that, despite the apparent underlying differences in their respective modes of discharge, nature has seen fit to endow calcium with the ability to be a prime mediator in both processes. The teleologists among us can grapple with this problem while we move on to consider the role of calcium in gastric secretion, which also represents a composite of water, electrolytes, and protein.

H. GASTRIC SECRETION

While calcium plays a pivotal role in the mechanism whereby macromolecules packaged in secretory organelles are discharged from the cell by exocytosis, this cation also plays a principal role in regulating membrane permeability and transport of ions across epithelial cell membranes. Calcium augments passive cation flow by a direct action on ion channels in the cell membranes of a variety of secretory cells (Putney, 1978). This cation also influences the rate and duration of ion transport in epithelial cells of rabbit ileum and colon, as demonstrated by the ability of the cholinergic agonists to increase chloride secretion by a calcium-mediated process (Hubel and Callanan, 1980) and the calcium ionophore A23187 to cause secretion of sodium and chloride (Field, 1979).

The regulation of gastric acid secretion, which ranks as one of the more complex secretory processes, also involves the active secretion of hydrogen and chloride ions by parietal (or oxyntic) cells and the proteolytic enzyme pepsin by the peptic (or zymogen) cells. While we could easily ignore this aspect of cellular secretion because the secretion of electrolyte and water is seemingly not compatible with the stimulus–secretion coupling model, who among us would summarily disregard this aspect of calcium action, if our ultimate aim is to gain a complete understanding of calcium's action on biological systems relat-

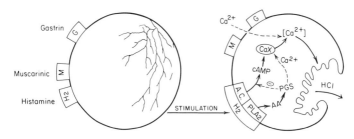

FIGURE 17. Neurocrine, paracrine, and endocrine pathways of hydrochloric acid secretion from the gastric parietal cell. Acetylcholine released from the vagus nerve interacts with muscarinic receptors to permit the entry of calcium from the extracellular fluid (neurocrine). Histamine locally released from the enterochromaffin cells in the gastric mucosa activates H_2 receptors to stimulate adenylate cyclase (A.C.) and elevate cyclic AMP levels which promote acid secretion by mobilizing intracellular calcium (paracrine). Prostaglandins—formed by the availability of free arachidonic acid (AA) resulting from the activation of phospholipase A_2 (PLA_2)—depress the effect of histamine by an inhibitory action on the adenylate cyclase-cyclic AMP system, or on calcium action or availability. Gastrin released from the antrum of the stomach into the circulation stimulates acid secretion by a still unknown mechanism (endocrine). The morphological changes associated with acid secretion involve the branched collapsed tubular system expanding into canaliculi, and the subsequent discharge of acid from the cell.

ing to secretory phenomena, whether they involve macromolecules or electrolytes and fluid?

The stimulus for gastric secretion is of course the presence of food in the stomach and upper small intestine, triggered by acetylcholine released by the vagus nerve, histamine released from enterochromaffin-like and/or mast cells, and gastrin released from the antrum of the stomach. This complex control mechanism of gastric secretion through neural (vagal), endocrine (gastrin) and paracrine* (histamine) pathways converges at specific receptors on the parietal cell (Soll and Walsh, 1979) (Fig. 17). The actions of the three gastric secretagogues are mutually interdependent, despite the fact that they utilize different second messenger pathways to trigger hydrogen ion secretion. Thus, while the effects of histamine to elicit a copious secretion of gastric juice are mediated through a direct action on the parietal cells by a cyclic AMP-dependent mechanism, the presence of an intact vagus nerve permits a higher rate of secretion. A similar potentiating effect of the histamine response can be demonstrated with gastrin. The interdependence of gastric secretagogues has fascinated physicians, as well as researchers,

---

* Paracrine secretion refers to diffusion of the chemical messenger from its cell of origin across intercellular spaces to its nearby target.

because these interactions are of fundamental significance for optimizing therapeutic efficiency in patients with peptic ulcers.

Morphologic evidence has compounded the complexities of this system. Whereas the peptic cell contains secretory granules which discharge pepsin by exocytosis, the process whereby hydrochloric acid is discharged from the parietal cell is completely distinct. Electron microscopic analysis depicts a complex intracellular array of membranes (tubulovesicles) in the parietal cell, and coincident with stimulation of gastric acid secretion, the cytoplasmic tubulovesicles decrease in number; simultaneously, intracellular vacuoles develop and expand, followed by the formation and expansion of secretory canaliculi (Dibona et al., 1979). The site of acid secretion, then, is the secretory canaliculus, and the mechanism involves a complex membrane fusion process which includes the pinching off of vesicles (Berglindh et al., 1980a) (Fig. 17).

This rather unique mode of secretion engenders interest regarding the factors involved in its regulation. Although it is clear that the secretory rate is controlled by the availability of ATP (Hersey 1974; Berglindh et al., 1980b), details concerning the mechanism by which gastric acid is produced are fragmentary. Potassium is required for hydrogen ion secretion, and due to this fact, combined with the finding of a gastric Na^+-K^+-ATPase localized to microvilli of the secretory canaliculus, it has been proposed that the stimulation of acid secretion is brought about by an increase in potassium chloride flux into the tubular lumen, followed by an exchange of potassium for hydrogen ion (Berglindh et al., 1980a).

The uniqueness of this secretory process does not exempt calcium from participation in a key modulatory role, as evidenced by the fact that changes in the calcium concentration of the medium affect gastric secretion in experimental animals, as well as in man. Early studies carried out on the frog gastric mucosa revealed that exposure to calcium-free solutions bathing the nutrient and secretory sites of the mucosa leads to a decrease in the transmembrane potential, in membrane resistance, and in the rate of acid secretion (Forte and Nauss, 1963). Although these effects were for the most part ascribed to the depletion of intercellular calcium affecting the binding between cells, Jacobson and his colleagues (1965) suggested that during an early phase of calcium deprivation, the rate of acid secretion was reduced by a direct effect on the secretory mechanism due to the loss of cellular calcium.

The results of more detailed studies were later described by Kasbekar (1974) during stimulation of gastric secretion when the frog mucosa was depleted by repeated washings with a calcium-free, EDTA-containing solution until acid secretion had ceased. The important contribution of Kasbekar was finding that the readdition of 1 mM calcium to the serosal bathing medium produced almost complete reversal of inhibition of the secretory response to histamine, pentagastrin, or acetylcholine, and this finding enabled him to postulate the existence of intracellular calcium compartments related to the process of acid secretion in addition to those in the junctional complexes. The use of isolated rabbit gastric glands and rat gastric mucosa have afforded additional insight into the role of calcium in acid secretion by disclosing that the actions of acetylcholine as an acid secretagogue depends at least in part on extracellular calcium, but that the actions of histamine and dibutyryl cyclic AMP are independent of external calcium (Main and Pearce, 1978; Berglindh *et al.*, 1980c; Chew *et al.*, 1980). According to this view, cholinergic stimulation modifies the permeability of the plasma membrane to calcium while histamine acts by a parallel pathway involving cyclic AMP, and presumably intracellular calcium (Fig. 17).

Experiments employing calcium-deprived conditions are subject to variable interpretations in light of the diverse effects of calcium on permeability properties of membranes. Therefore, more convincing experiments implicating calcium as a mediator of gastric acid secretion would involve conditions under which an increase in the intracellular calcium might be effected, as for example by increasing the extracellular calcium concentration. Indeed, the finding that calcium infusions in man elicit an increase in gastric acid secretion points to a possible role of calcium in the mechanisms governing gastric acid secretion (Barreras, 1973). However, while the induction of hypercalcemia in man by the intravenous administration of calcium salts is usually associated with an increase in volume, acidity, and pepsin content of gastric acid secretion, an increase in the circulating gastrin accompanies these changes and may be the chief mediator of the increase in the secretory response. The possibility that the primary effect of the hypercalcemia may be mediated through the calcium-dependent release of gastrin is strengthened by the knowledge that the release of gastrin from Zollinger–Ellison tumor tissue is very sensitive to alterations in serum calcium. Moreover, the increased prevalence of peptic ulcer disease in patients with hyperparathyroidism (Hotz *et al.*, 1975) appears to be accounted for by the coexistence of gastrinomas residing in the pancreatic islets of Langerhans, and from this fact a marked sensitivity of gastrin-secreting cells to calcium may be inferred. To add to the complexities involved in interpreting such studies, the stimulating effect of hypercalcemia on human gastric secretion is blocked by muscarinic or ganglionic blocking agents, suggesting that calcium is acting to increase acetylcholine release from the vagus nerve.

Attempts to gain a better understanding of the effects of calcium on gastric secretion have been hindered by the unfortunate fact that hypercalcemia inhibits rather than stimulates gastric acid secretion in the dog and rat (Barreras, 1973), although this inhibition may only be a reflection of a greater stimulation of bicarbonate transport (Flemström and Garner, 1980). Fortunately, however, the infusion of calcium into cats causes a dose-dependent increase in gastric secretion, and as in man the hypercalcemia parallels the increase in serum gastrin levels (Becker *et al.*, 1973; Christiansen *et al.*, 1979). These stimulatory effects are also blocked by the anticholinergic substance atropine, so it appears that in cat and man a cholinergic reflex is important for calcium-induced elevations of gastrin, although we cannot exclude the possibility that gastrin release and acid secretion are parallel, independent responses to calcium. The same considerations apply when comparing the differential actions of calcium on pepsin and acid secretion, as clearly illustrated by the finding that the dielectric breakdown of the plasma membrane induced by high voltage discharge across the gastric gland results in a massive discharge of peptic granules in the presence of calcium, but not in its absence, suggesting that calcium is the second messenger for exocytosis by the peptic cell. By contrast, the same level of extracellular calcium inhibits acid secretion, implying that the role of calcium in these two events is different (Berglindh *et al.*, 1980b).

Consequently, because of the obstacles associated with such studies, including a complex mixture of cell types that exists in the gastric mucosa, the nature of calcium's role in gastric acid secretion remains uncertain. Calcium deprivation experiments on isolated gastric mucosa and gastric glands suggest that calcium is a necessary component in the process of gastric secretion (Black, 1979; Berglindh *et al.*, 1980c). Supportive evidence for this postulate may be gleaned from the additional findings that cyclic AMP and theophylline (Kasbekar and Chugani, 1974) stimulate calcium efflux from frog gastric mucosa, indicating that cyclic AMP plays some intermediate role in transmitting the signal from the membrane to evoke gastric secretion, perhaps by mobilizing calcium from cellular stores or by enhancing the cellular actions of calcium in some unknown manner.

Evidence along these lines is persuasive with regard to the action of histamine. In isolated parietal cells, histamine, but not cholinergic agonists or gastrin, stimulates adenylate cyclase and increases cyclic AMP accumulation; phosphodiesterase inhibitors potentiate histamine-induced cyclic AMP formation and hydrochloric acid secretion (Soll and Walsh, 1979; Soll *et al.*, 1981; Thompson *et al.*, 1981). Prostaglandins, on the other hand, specifically block histamine-stimulated hydrochloric acid secretion *in vitro* and inhibit acid secretion *in vivo*. The utilization of isolated parietal cells has shed some light on the nature of this inhibitory mechanism by demonstrating a depressant action of prostaglandins on the adenylate cyclase–cyclic AMP system (Soll and Whittle, 1981). Additional evidence from other secretory systems verifies that prostaglandins are capable of inhibiting adenylate cyclase (Gorman, 1979; Carchman *et al.*, 1980). So, prostaglandins may either inhibit or activate adenylate cyclase, depending upon the species of prostaglandin employed and the type of secretory organ under scrutiny.

Thus, while the actions of cholinergic agonists and gastrin are mediated by intracellular messengers that are different from those mediating the action of histamine, it is likely that the final common denominator in the actions of all of these secretagogues is calcium, whether mobilized by perturbations in calcium flux across the plasma membrane, as with acetylcholine, or by translocation from cellular pools, as with histamine (Fig. 17). Nevertheless, much more still has to be learned about the mechanism of gastric acid secretion and the possible role of calcium in this unique secretory mechanism. Advances made in the development of techniques for isolating parietal cells should provide the means to probe this system more effectively, but whether such information will yield further insight into the action of calcium in the mechanism of exocytotic secretion is open to question.

I. ENDOCRINE SECRETION

As this narrative has progressed, it has become clear that neurosecretory cells and conventional neurons possess voltage-dependent calcium channels that are activated by electrical or chemical events; calcium enters the cell through these channels to initiate the steps involved in activation of the secretory process. We have also seen that other secretory cells that are derived from blood-forming or mesenchymal elements, and epithelial cells and parenchymal cells originating from entoderm have more complex mechanisms for mobilizing calcium when the appropriate signal is provided. The endocrine system represents another diverse group of secretory organs, and in response to humoral and neural factors it discharges hormones into the circulation which are destined for distant target organs. The role of calcium in endocrine secretion will again be seen as a divergent one, for in certain systems, such as the pancreatic β cell, the prototypical stimulus–secretion coupling model will be represented; while in the thyroid and parathyroid glands, a divergence from this general theme will be portrayed.

1. Insulin, Glucagon, and Somatostatin

The islets of Langerhans constitute a very small proportion of total tissue in the mammalian pancreas, yet are of enormous importance in endocrine function because of the primary role that they play in the regulation of carbohydrate metabolism. The pancreatic islets are composed of at least three major cell types: α, β, and δ cells, which synthesize, store and secrete glucagon, insulin, and somatostatin, respectively (Wellman and Volk, 1977). The β cells, which constitute a large proportion of the cell population in islets, are localized in the central portion of the islet; the α cells are generally situated peripherally, with the δ cells more diffusely distributed or localized to the periphery of the islets, occasionally interposed between the peripheral α cells and central β cells. This structural arrangement of the endocrine cells of the islet suggests some functional link among them. Somatostatin is thought to play a role in regulating the secretion of both insulin and glucagon and hence has been proposed as a model of paracrine secretion.

A multiplicity of hormonal mechanisms exists for elevating the blood glucose concentration, but just one hormone, insulin, brings about a decrease in the blood glucose concentration, which is accomplished by a stimulation of peripheral glucose uptake and an inhibition of glucose liberation from the liver. D-glucose seems to be the predominant factor controlling β-cell function and insulin release, although many additional factors are known to affect secretory activity. Sympathetic and parasympathetic activity also modulates insulin release in that activation of α-adrenoreceptors decreases insulin release, while activation of β-adrenoreceptors or muscarinic cholinergic receptors increases insulin release (Woods and Porte, 1974; Gerich and Lorenzi, 1978). Fluorescence microscopy and light microscopic autoradiographic studies have also identified 5-hydroxytryptamine and dopamine in pancreatic islets, and these substances generally exert a depressant effect on insulin release; however, the presence of these monoamines appears to be species-dependent, occurring most commonly in rodents (Lambert, 1976). Insulin release can also be evoked by other carbohydrates and related metabolizable substances (mannose, sorbitol, dihydroxyacetone), certain amino acids and long chain fatty acids, various pharmacological substances, most notably the oral hypoglycemic agents (sulfonylureas), and even ATP (Gerich et al., 1976; Hedeskov, 1980).

However, the key and central role of glucose in insulin release is underscored by the fact that the effects of most insulin secretagogues are transitory or even absent unless a threshold concentration of glucose is also present. The differential effects of various fuels to promote insulin release in various species and even in different stages of ontological development emphasize the close correlation between energy utilization and insulin release. The ability of nutrients to stimulate insulin release appears to be expressed through a stimulus recognition site(s) rather than an indirect effect associated with the metabolic requirements of the release process *per se.*

Advances in our knowledge of the mechanism of insulin release progressed greatly during the 1970s mainly due to the development of radioimmunoassay techniques and methods for separation and isolation of pancreatic islets. The following preparations all have been used to study insulin secretion: pieces of adult or fetal pancreas, isolated perfused pancreas, and islets isolated by proteolytic digestion and/or microdissection. The isolated islet preparation has become particularly popular since it may employ static incubations or perifusions. While it is difficult to prepare a completely pure β-granule fraction from the small amount of islet tissue available from mammals, it has been established that the secretory granules present in the β cell possess insulin and its precursor proinsulin, as well as a pool of adenine nucleotides; morphological and biochemical studies have clearly established that insulin secretion from the β cell occurs by exocytosis (Leitner *et al.*, 1975; Orci and Perrelet, 1977; Hedeskov, 1980) (see Fig. 6, p. 49).

Glucose-induced insulin release occurs in two phases, a short transient phase lasting approximately 5 min, and a second more prolonged phase. Since the second phase can be partially inhibited by cycloheximide or puromycin it was originally thought to represent *de novo* insulin synthesis (Curry *et al.*, 1968a). However, later studies utilizing static incubations and perifused islets revealed that there is little newly synthesized insulin secreted before 2 hr, suggesting that newly synthesized hormone is not responsible for insulin release observed during either of these two phases (Gerich *et al.*, 1976).

Grodsky (1972) described a model for the dual action of glucose in which insulin is envisioned to be stored in "packets" that respond in an all-or-none fashion when their threshold to glucose is reached. The hypothesis that insulin is released in unit packets, as is acetylcholine for example, is an attractive one and has been generally accepted. Nevertheless, insulin release provoked by glucose is a process of considerable complexity, as evidenced by nonhomogeneous storage of insulin in labile and stable pools and by the biphasic pattern of insulin release. Glucose also exerts a priming effect on subsequent exposures to glucose in eliciting insulin release, suggesting that it not only acts as a secretagogue but also produces long term effects on the β cell that facilitate subsequent responses of the cell.

Despite these inherent complexities in the mechanism governing insulin secretion, a veritable wealth of information appeared during the 1970s which has afforded us insight into the nature of the molecular events associated with insulin secretion. Our understanding evolved mainly because of the multidisciplinary approach used in the investigations. The data accumulated by these numerous and divergent studies have been congruent in that they have led to the inescapable conclusion that the primary action of glucose, as well as other insulin secretagogues, is principally exerted by the increased accumulation of calcium ions in the β cell. We will enumerate the major pieces of evidence to substantiate this conclusion.

Early studies by Hales and Milner (1968a) using pieces of rabbit pancreas and by Grodsky and his associates using the intact rat pancreas (Curry et al., 1968b) clearly demonstrated that the stimulatory action of glucose depends on the presence of extracellular calcium. The dynamics of the secretory response and the role of calcium were more clearly delineated in the perfused or perifused system where it was disclosed that a "square-wave pulse" of glucose triggered a biphasic insulin release and that both the early and late phases were diminished in the absence of extracellular calcium (Grodsky, 1972); the requirement for extracellular calcium appeared in these early studies to be greater for the early phase of secretion, although, curiously, this is still a subject of some debate (cf. Siegel et al., 1980). A direct regulatory role for the cation in the secretory process was established by the findings that the ionophore A23187 augments insulin secretion (Karl et al., 1975; Charles et al., 1975) and that calcium is a "sufficient" stimulus for insulin release in the absence of stimulating concentrations of glucose (Hellman et al., 1979) (Fig. 18).

The key role of calcium in insulin secretion suggested that the coupling mechanism might involve selective changes in the permeability of the β-cell membrane upon recognition of the glucose signal. Electrophysiologists (Matthews, 1977; Meissner et al., 1979) provided convincing evidence in support of this postulate by demonstrating that glucose elicits depolarization and spike activity in pancreatic β cells. They not only showed that the electrical potentials of the β-cell plasma membrane were reliable indicators of glucose sensitivity, but they also obtained essential information about the membrane conductance changes associated with the secretory response by demonstrating a close quantitative and functional relationship between glucose-induced electrical activity and insulin release. An increase in the glucose concentration causes an increase in the frequency and duration of spike activity coincident with an increase in insulin secretion. A further association between the effects

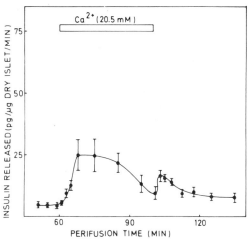

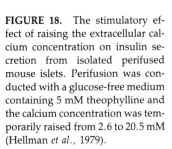

FIGURE 18. The stimulatory effect of raising the extracellular calcium concentration on insulin secretion from isolated perifused mouse islets. Perifusion was conducted with a glucose-free medium containing 5 mM theophylline and the calcium concentration was temporarily raised from 2.6 to 20.5 mM (Hellman *et al.*, 1979).

of glucose on electrical activity and insulin release was afforded by the finding that mannoheptulose, an inhibitor of glucose metabolism and insulin release, also blocks electrical activity induced by glucose. The initiation of electrical activity is a direct consequence of a voltage and time-dependent calcium influx, as evidenced by the dependency of the membrane potential of the pancreatic islet cell upon the external potassium concentration and by the ability of D-600 or manganese, but not tetrodotoxin or low sodium, to abolish glucose-induced electrical activity.

The use of radiotracer analysis to substantiate these seemingly clear-cut findings has led to a bewildering array of data that in certain respects defy clear interpretation. Nevertheless, such studies, combined with morphological findings using the pyroantimonate technique (Herman *et al.*, 1973), have demonstrated that glucose increases the accumulation of calcium in the β cell. Some of the earliest studies on calcium handling by islets showed an enhanced ^{45}Ca uptake after stimulation with glucose (Hellman *et al.*, 1971; Malaisse-Lagae and Malaisse, 1971). Verapamil, D-600, and magnesium inhibit glucose induced ^{45}Ca uptake and insulin release, showing a correlation between these two phenomena (Malaisse *et al.*, 1975, 1978). The glucose-induced uptake of ^{45}Ca by islet cells for the most part parallels other parameters of islet cell function, including glucose oxidation and cyclic AMP production. However, the precise temporal relationship of these phenomena and the nature of their interactions have not been defined.

Further attempts to delineate the various calcium pools affected by glucose stimulation have yielded inconclusive findings, although employment of the lanthanum wash technique made it possible to demonstrate that glucose acts on two pools of calcium which respond rapidly enough to conceivably provide the initial trigger for insulin release (Hellman et al., 1978). The specific localization of these two calcium pools has not been ascertained although the lanthanum nondisplaceable pool, which is increased by various insulin secretagogues including glucose, leucine, and sulfonylureas is thought to represent intracellular calcium stores; whereas, the lanthanum displaceable pool is more superficially localized, perhaps to the plasma membrane. Glucose also decreases the intensity of a calcium-dependent fluorescence signal monitored in or near the β-cell plasma membrane (Sehlin and Taljedal, 1979), so it is likely that the action of glucose is associated with the turnover of calcium within the plasma membrane. Glucose also initiates complex alterations in ^{45}Ca efflux from perifused islets, causing first a fall and 3 to 4 min later a secondary rise in ^{45}Ca efflux; the effluent radioactivity appears to originate entirely from the lanthanum nondisplaceable pool (Hellman, 1978; Herchuelz and Malaisse, 1980). The temporal and quantitative coincidence between the secondary rise in ^{45}Ca efflux and insulin secretion implies that these two phenomena are tightly coupled. While the glucose-induced decrease in calcium efflux probably reflects the trapping of the cation by intracellular organelles (Hellman et al., 1979), the ^{45}Ca efflux triggered by glucose reflects an increase in calcium exchange (Herchuelz et al., 1980; Abrahamsson et al., 1981). Both effects represent sustained phenomena operative throughout the period of exposure to glucose so that multiple processes are apparently being activated, thus making it difficult to unravel each of the individual components.

Some insight into the complex regulation of calcium by the β cell may be gleaned by focusing on the role of cyclic AMP in the regulation of insulin secretion. Early studies by Malaisse and associates (Brisson et al., 1972; Malaisse et al., 1975) demonstrated that cyclic AMP and theophyline increase calcium efflux from the perfused pancreas in the absence of extracellular calcium—and even in the absence of glucose—suggesting that cyclic AMP might act to induce translocation of calcium within the β cell. Ultrastructural cytochemical studies have revealed adenylate cyclase in the plasma membrane of the β cells, which is stimulated by hormones such as glucagon, corticotropin, and pancreozymin, and inhibited by calcium (cf. Hedeskov, 1980). The glucose-induced increases in cyclic AMP formation in islets, which require glucose metabolism and the presence of extracellular calcium, occur in parallel with insulin secretion (Grill and Cerasi, 1974; Gerich et al., 1976).

However, the increase in cyclic AMP is not a sufficient condition for insulin release. Ionophore A23187 increases insulin secretion without any enhancement of cyclic AMP formation; conversely, cyclic AMP levels markedly increase in the presence of phosphodiesterase inhibitors without any corresponding increase in insulin release (Hellman et al., 1974b; Charles et al., 1975). One may therefore consider cyclic AMP as a positive modulator of secretion, rather than as a primary trigger (Hedeskov, 1980). A clue to a better understanding of the role of cyclic AMP in insulin release may be derived from studies on hamster insulinoma cells, which possess a cyclic-AMP-potentiated insulin secretory system analogous to that found in normal islets. In this system, which does not respond to glucose, high potassium or ouabain increases calcium uptake and enhances insulin release, whereas theophylline or cyclic AMP are unable to increase calcium uptake, and these findings indicate that insulin release mediated through cyclic AMP involves the redistribution of an intracellular pool of calcium (Shapiro et al., 1977).

The experimental findings presented here are reminiscent of those encountered in the studies of amylase release from the parotid gland, wherein both cyclic AMP and calcium appear necessary for an optimal response. Cyclic AMP action may be described as a sensitizer of the stimulatory action of calcium entering the β cell (Hellman et al., 1979); the effects of cyclic AMP may be expressed by an increased responsiveness of the secretory machinery to the calcium signal or by an increase in the effective free calcium concentration in the cell. The latter alternative would be a more likely possibility. Mitochondria- and microsome-enriched subcellular fractions of islet homogenates accumulate ^{45}Ca by an ATP-dependent process, and calcium uptake by a mitochondria-enriched fraction is inhibited by cyclic AMP (Howell et al., 1975). It thus seems likely that the mitochondria, endoplasmic reticulum, and even secretory organelles play an important role in the regulation of cytoplasmic calcium levels in the β cell, although the mitochondria appear to be the most important of the particulate calcium pools in this regard (Kohnert et al., 1979; Hahn et al., 1980).

On the other hand, a potential link between glucose metabolism, calcium uptake, and the rate of insulin release was forged by evidence that ATP-dependent calcium uptake by islet mitochondria is profoundly inhibited by the glycolytic intermediate, phosphoenolpyruvate (Sugden and Ashcroft, 1978). The evidence for cyclic AMP-induced translocation of intracellular calcium, particularly from mitochondria, must be somehow integrated with the other biochemical reactions activated by receptor recognition of the glucose stimulus, including modification of the NAD:NADH concentrations and cyclic GMP formation (Malaisse et al., 1978; Hahn et al., 1980). We particularly cannot ignore the possibility that cyclic GMP may also participate in the events associated with insulin secretion, since glucose is able to increase cyclic GMP formation in isolated islets, and nitroprusside and ascorbic acid, which are also potent activators of insulin release, also elevate cyclic GMP levels in the absence of increased cyclic AMP formation (Laychock, 1981). However, while the experiments utilizing nitroprusside and ascorbic acid to increase insulin release substantiate the concept of a nonobligatory role for cyclic AMP in the process of insulin secretion, they by no means denote a critical functional role for cyclic GMP.

While glucose-mediated events taking place at the level of the cell membrane involve the activation of voltage-dependent calcium channels, the possibility that other cations are also intimately involved in this facet of glucose action was initially raised by the flux measurements of Sehlin and Taljedal (1974) and confirmed by the electrophysiological studies that revealed that potassium permeability was reduced in the presence of glucose (Atwater et al., 1978, 1979). Quinine and its optical isomer quinidine, which block calcium-sensitive potassium efflux in other systems, powerfully potentiate the insulinotropic effect of glucose, and this fact strongly supports the postulate that the action of glucose is associated with a decrease in potassium efflux. The initial fast depolarization occurring during glucose stimulation results from an increase in resistance produced by a decrease in potassium permeability; while as the plateau response commences, resistance decreases resulting from an increase in permeability to potassium by a calcium-sensitive potassium efflux pathway (Ribalet and Beigelman, 1979, 1980). The calcium-activated potassium permeability thus provides a sensitive control mechanism for the periodic oscillations of the membrane potential and insulin secretion. The picture that emerges is that the handling of calcium by the β cell—and thus its action on the secetory process—is closely linked to potassium turnover. The process of glucose-induced insulin release thus involves: a decrease in potassium permeability → membrane depolarization → gating of voltage-dependent calcium channels → insulin release. The calcium-stimulated increase in potassium permeability, which is also voltage dependent, is responsible for repolarization of the β-cell membrane.

Hales and Milner (1968b) first suggested that a rise in the sodium concentration of the β cell may also be a fundamental event in the stimulation of insulin release. Not only is the release of insulin evoked by glucose and other secretagogues decreased at low extracellular sodium concentrations, but conditions known to increase intracellular sodium, such as the addition of ouabain or veratridine, all increase insulin release—although the increase in intracellular sodium also increases calcium influx by activating the voltage-dependent calcium channels. The fact that prolonged sodium deprivation inhibits glucose oxidation, even more than does the omission of calcium, suggests that at least some of the effects of sodium deficiency on insulin release are due to the inhibition of glucose oxidation, rather than a direct effect on the secretory machinery *per se* (Hellman *et al.*, 1974a). This possibility is strengthened by the knowledge that sudden deprivation of extracellular sodium has no marked effect on glucose-induced insulin release; in fact, insulin release may be initially increased by sodium withdrawal because of the ability of extracellular sodium to inhibit calcium uptake or accelerate its extrusion via sodium–calcium exchange. The same interpretation may apply to experiments that employed potassium deprivation to produce a marked inhibition of glucose metabolism (Henquin and Lambert, 1974; Sener *et al.*, 1980), making it difficult to dissociate the effects of potassium deprivation on glucose metabolism from those on the secretory mechanism. However, the failure of tetrodotoxin to affect electrical activity and insulin release evoked by glucose argues against the idea that sodium accumulation within the β cell is an obligatory component of the physiological mechanism regulating insulin secretion (Matthews, 1977; Pace, 1979).

On the balance of evidence, then, we must focus our attention on the key role of calcium in the mechanism of insulin secretion and, in fact, consider it a prototypical system in the stimulus–secretion coupling model. The hypothetical calcium "receptor" can be considered analogous to that of the adrenal medulla, wherein strontium and barium will replace calcium in activating the secretory machinery, and the stimulant actions of barium are depressed by calcium (Hedeskov, 1980). Moreover, magnesium, manganese, and lanthanum inhibit glucose-induced insulin release (Malaisse *et al.*, 1975, 1978; Flatt *et al.*, 1980). While magnesium is completely devoid of any stimulant actions, manganese is capable of promoting insulin release, presumably by its ability to displace calcium from membrane-binding sites (Hermansen and Iversen, 1978). The inescapable conclusion obtains that the accumulation of calcium in the β cell represents an obligatory step in the release process just as it does in the adrenal medulla.

While up to this point we have mainly confined our discussion to the stimulant actions of glucose because of the relative complexities of the response of the β cell to this stimulus, it is necessary to also consider other secretagogues, mainly because they may afford further insight into the nature of the mechanisms controlling insulin secretion. Still, the effects of glucose cannot be completely divorced from the actions of these other secretagogues because, with the exception of the amino acid leucine, they require the presence of nonstimulating concentrations of glucose for expressing their actions.

The sulfonylureas have been used to treat diabetic patients since the 1950s, when the role of the pancreas in the hypoglycemic action of sulfonylureas was established by the pioneer work of Loubatieres (1977). Tolbutamide is the most comprehensively studied of this group of compounds, and it has been clearly demonstrated that these agents do not penetrate β cells but reversibly bind to the plasma membrane and, like glucose, produce a reversible depolarization of the β cells (Matthews, 1977). The actions of tolbutamide require extracellular calcium, are blocked by D-600, and are associated with an increase in ^{45}Ca uptake into the cell (Henquin, 1980).

While the effects of tolbutamide are, in many respects, similar to those of glucose, in contrast to glucose-induced insulin release where there is a biphasic and prolonged response to glucose, tolbutamide produces a short, monophasic release response (Grodsky et al., 1977). The initial activation of calcium entry followed by a progressive inactivation of the channels may account for the monophasic calcium entry and the transient secretory response. Some correlation actually exists between the hypoglycemic and bioelectric effects of the sulfonylureas (Matthews, 1977). For example, glibenclamide possesses the most potent and persistent bioelectrical effects and is an extremely potent hypoglycemic agent, facts that prompt the conclusion that sulfonylureas induce a permeability change in the membrane, ionic fluxes, action potential discharges, and insulin release, just as does glucose. Since sulfonylureas depolarize β cells, their secretory actions are probably dependent on the gating of voltage-dependent calcium channels (Henquin, 1980; Hellman, 1981). However, studies on artificial membrane models suggest that these oral hypoglycemic agents act through an ionophoretic inward calcium transport mechanism (Couturier and Malaisse, 1980).

While it would not be possible to discuss in depth the mechanism of action of all of the agents which promote insulin secretion, some consideration must be given to the action of glucagon because of its important physiological role as a putative mediator of insulin release. While the stimulatory action of glucagon on insulin secretion is proportional to the extracellular calcium concentration, release is not blocked by verapamil and, like theophylline and cyclic AMP, is not associated with an increase in calcium uptake into islet tumor cells (Fleischer et al., 1981); this suggests that glucagon induces a redistribution of intracellular calcium rather than an influx from external sources. Since glucagon stimulates β-cell adenylate cyclase, at least part of the inhibitory action of calcium deprivation on glucagon-induced insulin release may be exerted through an impairment of the adenylate cyclase-cyclic AMP system (see Chapter 4).

There are at least two endogenous inhibitors of insulin secretion, namely catecholamines and somatostatin. In addition, prostaglandins and other arachidonic acid metabolites have also been proposed as endogenous inhibitors of insulin secretion, although the evidence so far in support of this claim is still inconclusive. Interest in these inhibitors within the scope of our discussion relates to their possible effects on calcium metabolism in the β cell. As we have seen, insulin secretagogues are associated with alterations in calcium handling by the β cell, so it is reasonable to suspect that inhibitors of insulin secretion interfere with calcium metabolism. However, clonidine, an α-adrenergic agonist, which is one of the most powerful inhibitors of insulin release, fails to affect glucose-stimulated ^{45}Ca net uptake but profoundly depresses the glucose-induced increase in ^{45}Ca efflux, suggesting that α-adrenergic agonists may affect the rate at which calcium is sequestered by subcellular organelles (Leclercq-Meyer et al., 1980). This postulate is not unique, for catecholamines alter calcium handling by subcellular organelles of salivary glands. The implications of such findings, if true, are far-reaching, for they reinforce the concept that the intracellular organelles of the β cell play an important physiological role in regulating calcium metabolism and hence directly participate in the secretory process.

Somatostatin, which was first identified in hypothalamic extracts and seen to dramatically inhibit growth-hormone secretion (Vale et al., 1977), was later found to impair insulin and glucagon release; so this hypothalamic hormone exerts a powerful inhibitory effect upon the function of the β cell, as well as that of the α cell of the mammalian pancreas. The δ cell, described as an element of a paracrine system, releases somatostatin which subsequently reaches the α and β cells. While early studies suggested that somatostatin exerts its inhibitory actions by blocking calcium uptake into the β cell, later findings demonstrated that somatostatin also effectively inhibits insulin secretion brought about by glucagon and phosphodiesterase inhibitors, even in the absence of extracellular calcium (Mandarino et al., 1980). These findings suggest that somatostatin does not block calcium entry into the cell but interferes with some calcium-dependent process associated with the translocation of intracellular calcium. Alternatively, somatostatin may possibly interfere with some aspect of cyclic AMP action not associated with calcium handling in the β cell.

The prostaglandins and other arachidonic acid metabolites must also be considered in this resume because of the available evidence which hints that these ubiquitous substances play a modulatory role in insulin secretion. Numerous *in vitro* and *in vivo* studies have attempted to elucidate the role of prostaglandins in modulating insulin release; however, the results have been conflicting, since some investigations report that prostaglandins increase insulin release in isolated rat islets and perfused pancreas, whereas others find inhibition or no effect on insulin release. Administration of drugs that inhibit endogenous prostaglandin synthesis can augment, or in some cases depress glucose-induced insulin release. Several studies in man, utilizing prostaglandin infusions, salicylate infusions, or aspirin ingestion suggest that prostaglandins inhibit glucose-induced insulin release (Robertson, 1979; Topol and Brodows, 1980). PGE, in particular, has been implicated as a negative modulator of insulin secretion, since not only does the administration of this prostaglandin blunt glucose-evoked insulin secretion, but inhibition of PGE synthesis in cultured islet cells by salicylate is accompanied by an augmented secretory response to glucose (Robertson and Chen, 1977; Metz *et al.*, 1981).

There are many questions that such experiments raise, such as the physiological validity of intravenous infusion of prostaglandins, since they are purported to act as intracellular messengers. Nevertheless, such studies cannot be ignored, not only because of their mechanistic implications, but also because of their potential pathophysiological significance with regard to the etiology and treatment of diabetes mellitus. The finding that infusion of sodium salicylate partially restores the acute insulin response to glucose in adult onset diabetes (Robertson, 1979) suggests that prostaglandins may play a role in the defective insulin response in diabetes mellitus. The critical role of calcium in this system relates to the calcium requirement for the enzyme phospholipase A_2 that provides the substrate precursor arachidonic acid for the synthesis of prostaglandins, as well as to the requirement of this cation for permitting arachidonic metabolites to express their effects. This matter will be dealt with in greater detail in the final chapter.

Any technical difficulties associated with specific investigations of the β cell are infinitely magnified by studies that confine themselves to the glucagon-secreting α cells, since they represent approximately 20% of the total cell population in the islets of Langerhans. So it is not surprising that the indefatigable investigators who have undertaken studies concerned with glucagon secretion have only succeeded in accumulating the most rudimentary of findings. Pioneering studies utilizing a variety of preparations, including incubated pieces of pancreas, *in situ* perfused rat pancreas, and a monolayer culture of fetal rat pancreas, found an enhancing effect of calcium deprivation upon glucagon release (see

Lundquist *et al.*, 1976). Conflicting data subsequently appeared in the literature when it was reported that arginine-induced glucagon release from the perfused rat pancreas was inhibited during calcium deprivation, while the ionophore A23187 increased glucagon release (Gerich *et al.*, 1974; Ashby and Speake, 1975). Some resolution of this problem was attained by the findings that calcium was able to stimulate glucagon release from the perfused pancreas in a stepwise manner as the extra-cellular calcium concentration was increased, but at high glucose concentrations calcium-stimulated glucagon release was suppressed at all calcium concentrations (Lundquist *et al.*, 1976) (Fig. 19). Moreover, low—but not high—glucose concentrations enhance ^{45}Ca uptake in islets rich in α cells (Berggren *et al.*, 1979). It, therefore, appears that calcium exerts a stimulatory effect on the secretory mechanism of the α cell, but it also plays a permissive role in allowing glucose to exert its normal inhibitory effect on glucagon secretion.

The additional fact that the paradoxical increase in glucagon secretion observed by some investigators during calcium deprivation is not universally observed in all preparations (Iversen and Hermansen, 1977) provides corroboration for the proposal that glucagon release fits the profile of endocrine processes that require calcium for secretion. Her-

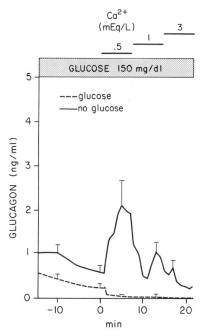

FIGURE 19. Interaction of calcium and glucose on glucagon secretion. Calcium was added in sequential steps to a glucose-free solution perfusing the rat pancreas. A low concentration of calcium (0.5 mEq/liter) elicited a large, transient release of glucagon, followed by weaker responses to higher calcium concentrations. By contrast, exposure to calcium in the presence of glucose caused glucagon secretion to diminish to undetectable levels (redrawn from Lundquist *et al.*, 1976).

mansen and Iversen (1978) cleverly used manganese to reinforce this concept by demonstrating that the manganese inhibition of glucagon release seen at both low and high glucose concentrations was reduced with increasing calcium concentrations, suggesting a competitive antagonism by manganese of calcium. While manganese inhibits glucagon release by competitively blocking calcium influx into the α cell, it also produces a calcium-dependent stimulation of glucagon release following the inhibitory phase. This facilitatory effect of manganese on glucagon secretion is not observed with either magnesium or verapamil, so that manganese, like magnesium, not only blocks the calcium channel, but permeates the slow calcium channel and displaces calcium from intracellular stores; on the other hand, manganese, like strontium and barium, may also directly activate the intracellular calcium receptor. So despite the inhibitory action of glucose, calcium-induced glucagon release is controlled by the same calcium-dependent exocytotic mechanism that so ubiquitously exists in other secretory organs (Esterhuizen and Howell, 1970), rather than belonging to the rare secretory systems, such as the parathyroid gland, which are inhibited by calcium.

In order to focus on the nature of calcium's role in the various secretory processes which exist in the islets of Langerhans, we have considered each cell type individually, while aware that intraislet interactions exist among the α, β, and δ cells. This is most clearly illustrated by the ability of somatostatin to inhibit both glucagon and insulin secretion, and of both glucagon and glucose to stimulate somatostatin release. Furthermore, the suppression of glucagon secretion by glucose might be expected to lead to a decrease in somatostatin secretion. It is, therefore, difficult to disentangle the direct effects of calcium on secretion of one or another of these hormones, particularly somatostatin. This complication notwithstanding, Hermansen and his colleagues (1979) demonstrated that acetylcholine, glucose, isoproterenol, and arginine accelerate somatostatin output from the isolated perfused canine pancreas, with a pattern of secretion that is very similar to that observed with insulin secretion. A stepwise increase in the calcium concentration also elevates somatostatin secretion. The stimulatory effect of calcium is more pronounced at high than at low glucose concentrations, although calcium acts as a secretagogue even in the absence of glucose.

Sodium also appears to play a role in the secretory function of the pancreatic δ cell, since veratridine causes a typical biphasic pattern of somatostatin secretion that is blocked by tetrodotoxin (Hermansen, 1980). Ouabain and potassium deprivation also augment somatostatin secretion, and this fact further supports a role of sodium in the mechanism of release of somatostatin. The stimulatory actions of veratridine, sodium deprivation, and inhibitors of Na^+-K^+ pump are all blocked by calcium-deprivation (Hermansen, 1980; Hermansen et al., 1980), suggesting that the changes in extracellular sodium that influence somatostatin release are mediated by altering the calcium handling of the cell through the sodium–calcium countertransport mechanism.

The fact that the stimulating action of veratridine on somatostatin release is blocked by omission of extracellular calcium contrasts with the stimulating action of veratridine on insulin release that is not blocked by omission of calcium from the extracellular medium (Lowe et al., 1976). These contrasting findings imply not only that the calcium-dependent mechanism of somatostatin secretion is not identical to that of insulin release (Kanatsuka et al., 1981) but also that the effects on somatostatin release revealed by these investigations are not principally secondary to effects on insulin secretion. But while somatostatin is located in secretory granules of δ cells and is extruded by calcium-dependent exocytosis into the pericapillary space (Sihusdziarra et al., 1978), more definitive information concerning the nature of the mechanism of somatostatin secretion, as well as glucagon secretion, must await the development of techniques for separating the various cell types of the islet to probe each of these secretory systems in isolation.

2. Anterior Pituitary Hormones

The integration of the nervous and endocrine systems is no more clearly illustrated than by a consideration of the mechanism of release of adenohypophysial hormones. The landmark monograph of Harris (1955) provided the impetus for establishing that secretion of adenohypophysial hormones is under the control of hypothalamic releasing factors, which are released into the hypophysial–portal circulation and ultimately reach the cells of the adenohypophysis. This "master gland" controls a variety of functions in the body and a different cell type is assigned the task of producing each specific hormone. Immunocytochemical procedures have identified somatotrophs, mammotrophs, thyrotrophs, and corticotrophs that secrete growth hormone, prolactin, thyrotropin and corticotropin, respectively (Farquhar et al., 1975). Gonadotrophs sequester both follicle-stimulating hormone (FSH) and luteinizing hormone (LH), and there is a controversy as to whether these two gonadotropic hormones occur in the same cell. Somatotrophs are the predominant cell type, although mammotrophs have the largest secretory granules and may be the predominant cell type in the pituitary glands of lactating females.

The first scheme for the secretion of an anterior pituitary hormone was proposed in 1966 for the mammotroph by Marilyn Farquhar (Smith and Farquhar, 1966), and conforms to the model established by Palade for the exocrine pancreas, with the synthesis of protein on attached polysomes and the concentration of protein and formation of secretory granules in the Golgi zone. There is also persuasive morphological evidence for exocytosis as the mode of secretion by each of the various cell types (Vila-Porcile and Olivier, 1978).

The cellular heterogeneity of the mammalian adenohypophysis offers a major obstacle to the study of this important gland. Consequently, research devoted to a study of the mechanisms of synthesis and release of pituitary hormones has been hindered by the lack of suitable homogeneous preparations for studying these mechanisms and thus in certain situations has led to spurious findings and contradictory interpretations. Indeed, perhaps more than for any other gland, the multiplicity of the hormonal activities of the adenohypophysis requires a correct identification of the cells under scrutiny. And as this resume evolves, it will be seen that only after isolated homogeneous cell preparations were developed could an accurate assessment be made of the concomitant cellular events associated with the selective stimulation of a particular cell type.

Before considering each of the individual cell types, some general statements can be made that apply to all of the cells of the adenohypophysis, particularly with regard to the role of calcium. As is the case with most secretory organs, the initial work devoted to investigating the role of calcium in adenohypophysial secretion followed the pattern of determining the effect of removal of extracellular calcium on the ability of high potassium or crude hypothalamic extracts to release hormones. The earliest such reports were those of Vale and Guillemin (1967), who showed that calcium ions were necessary for both thyrotropin-releasing factor and high potassium to stimulate the secretion of thyrotropin *in vitro*. These studies were followed by a series of similar investigations showing the ability of various secretagogues to release luteinizing hormone (LH), follicle-stimulating hormone (FSH), corticotropin, growth hormone, and prolactin (Geschwind, 1971; Kraicer, 1975). Some investigators found that a calcium-chelating agent such as EGTA was required to abolish the response, and others also observed that the response to high potassium was more sensitive to calcium deprivation than the response to the hypothalamic extract. Such findings prompted the obvious conclusion that the mechanism of secretion from the anterior pituitary gland conformed to the Douglas model of stimulus–secretion coupling.

Milligan and Kraicer (Kraicer, 1975) were the first to consider the possibility that an alteration in the cellular distribution of calcium in the adenohypophysis serves an important role in stimulus–secretion coupling. These conclusions were based upon the findings that only a brief wash in calcium-free medium was required to reduce the effects of high potassium, but that longer periods of calcium deprivation were needed to decrease the secretory response to crude hypothalamic extract or phosphodiesterase inhibitors. This concept was supported by findings that ^{45}Ca uptake was increased by high potassium, but that theophylline, cyclic AMP, and hypothalamic extract did not increase calcium uptake, nor were their actions blocked by verapamil (Moriarty, 1977).

These early findings, while clearly establishing a role for calcium in the mechanism of secretion of anterior pituitary hormones, were unable to define the source of the "trigger" calcium and more definitive studies had to await the development of purified populations of specific cell types. However, despite the limitations in methodology, the early studies indicated that the action of calcium was primarily exerted on the secretory response and not on other cellular functions such as hormone biosynthesis (Geschwind, 1969). Later studies, for the most part, bear out this conclusion (Bourne and Baldwin, 1980).

a. Gonadotropin

A major breakthrough in the development of our knowledge of cellular mechanisms associated with pituitary function came with the advent of techniques for the preparation and purification of gonadotrope cell cultures. A critical role played by extracellular calcium in this preparation is indicated by the findings that a brief wash of pituitary cell cultures with a low calcium medium results in a complete loss of gonadotroph responsiveness to gonadotropin-releasing hormone (GN-RH),[*] and D-600 blocks GN-RH-stimulated LH release (Conn *et al.*, 1981). The action of calcium in the mechanism of LH release appears to be similar to its actions in other secretory organs, wherein magnesium cannot substitute for calcium but barium acts as a secretagogue even in the absence of the primary stimulus; and lanthanum, manganese, and cobalt are inhibitory. Calcium also appears to be a sufficient stimulus in that either ionophore A23187 or liposomes, conveying calcium into pituitary cell, increase LH release.

While direct evidence for the existence of voltage-dependent calcium channels in gonadotropes is still lacking, the finding that veratridine increases gonadotropin release by a mechanism that is blocked by tetrodotoxin or EGTA supports the existence of membrane channels (Conn

[*] Also designated LH-RH—luteinizing hormone-releasing hormone.

and Rogers, 1980). However, the inability of tetrodotoxin or sodium deprivation to depress the secretory response to GN-RH suggests that GN-RH-stimulated calcium mobilization is not mediated through an action on the sodium channel; receptor-coupled calcium channels could provide the mechanism for the regulation of calcium fluxes in response to GN-RH (Fig. 20).

Yet, one must be circumspect about concluding that the physiological stimulation of LH release involves transmembrane calcium flux. For example, in hog pituitary cells in culture, GN-RH, high potassium, and ionophore A23187 all induce a biphasic pattern of LH release; however, GN-RH is much less dependent on the availability of extracellular calcium than high potassium or A23187, and this fact again prompts speculation that GN-RH mobilizes calcium from membrane or other cellular stores (Hopkins and Walker, 1978). In a perifused system (Bourne and Baldwin, 1980) the biphasic release in response to GN-RH stimulation exhibits a complete dependency on extracellular calcium only during the first phase of the secretory response. The second phase, which occurs 60–90 min after the initial acute phase, is thought to represent the secretion of both stored and newly synthesized hormone, and thus the newly synthesized LH could represent the extracellular calcium-independent component.

While some insight into the mechanism by which the anterior pituitary gland handles calcium may be gleaned by stimulating these cells in the presence of radioactive calcium and carrying out subcellular fractionations, such an experiment is logistically difficult because of the small amount of tissue one can derive from this gland. Another approach to this problem involves a study of the cyclic nucleotides. While some studies reveal that cyclic AMP formation is increased during stimulation by GN-RH (Labrie *et al.*, 1978; Walker and Hopkins, 1978), as more

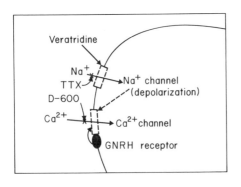

FIGURE 20. Proposed mechanism of action of GN-RH. Occupancy of the GN-RH receptor leads directly to calcium mobilization and luteinizing hormone secretion without involvement of voltage-dependent sodium channels. By contrast, veratridine activates tetrodotoxin-sensitive channels bringing about sodium entry and depolarization. As a consequence, the voltage-dependent calcium channels are activated, allowing calcium to enter the cell and evoke hormone secretion (Conn *et al.*, 1981).

specific stimuli were developed and homogeneous cell preparations became available, a dissociation between altered cyclic AMP levels and LH release could be demonstrated (Naor and Catt, 1980; Conn et al., 1981). Low doses of GN-RH stimulate LH release in the absence of measurable elevations of cyclic AMP levels; while cholera toxin increases cyclic AMP formation, in certain systems it is unable to stimulate LH release. Moreover, effects of cyclic AMP or its derivatives on gonadotrophin release have been variable, in contrast to the consistent stimulatory effects of prostaglandins, for example.

During the last few years, a role for cyclic GMP in the mechanism of gonadotropin release has become much more appealing in light of the fact that GN-RH stimulates cyclic GMP production in vitro (Rigler et al., 1978; Snyder et al., 1980). While a dissociation between an increase in cyclic GMP levels and LH release can be demonstrated, the finding that extracellular calcium is a prerequisite for GN-RH action on both cyclic GMP production and LH release (Naor et al., 1978) makes it tempting to consider that at least a portion of the actions of calcium on the secretory response of the anterior pituitary gland is expressed through the guanylate cyclase–cyclic GMP system. But, simply stated, the question of whether cyclic nucleotides play any specific role in the mechanism of stimulated LH release is still unanswered.

b. Growth Hormone

Growth hormone secretion from the mammalian anterior pituitary gland is chiefly under hypothalamic control mediated by growth hormone-releasing hormone (GH-RH) and somatostatin. Despite the development of a purified preparation of somatotrophs (Sheppard et al., 1980), the nagging problem remains of whether extracellular or intracellular calcium plays the primary role in the mechanism of growth hormone secretion, because while lowering the extracellular calcium blocks the release of growth hormone induced by a number of secretagogues, including high potassium, ionophore A23187, cyclic AMP or its derivatives, and by prostaglandins (particularly those of the E series) (Hertelendy et al., 1978; Sheppard et al., 1980), such experiments do not rule out a critical role for an alteration in the intracellular distribution of calcium.

While radiotracer analysis or morphological data have not provided us with much insight into the role of calcium in growth hormone release, studies concerned with the possible interactions of calcium with other putative mediators have revealed that the physiological mechanism governing GH-RH-induced growth hormone secretion may involve a cyclic

AMP-dependent step—which may also include calcium and prostaglandins. The answer to the question whether GH-RH acts directly on the cell membrane to increase membrane permeability to calcium, or via cyclic AMP to mobilize intracellular calcium will require the isolation and purification of this particular hypothalamic peptide and thorough investigation as to its mode of action on purified somatotrophs. But the possible role of prostaglandins in this sequence cannot be ignored in light of their potent stimulation actions on growth hormone secretion (Hedge, 1977; Hertelendy *et al.*, 1978) and the fact that they promote growth hormone release by a mechanism that requires external calcium and involves cyclic AMP (Kraicer and Spence, 1981). We will see that in the adrenal cortex, prostaglandins and cyclic AMP or its derivatives mobilize intracellular calcium; while the responses to PGE and dibutyryl cyclic AMP on growth hormone release are reduced under calcium-free conditions, their stimulating actions on growth hormone release are not as dependent on extracellular calcium as those of high potassium or ionophore A23187, suggesting that as in the adrenal cortex, prostaglandins and cyclic AMP stimulate growth hormone release by mobilizing a cellular pool of calcium.

One might look to somatostatin to aid in unravelling the complex interrelationships involved in growth hormone release. Somatostatin does not act like a calcium channel blocker since its inhibitory actions on calcium fluxes do not parallel those of verapamil (Schofield and Bicknell, 1978). Moreover, the inhibitory effect of somatostatin appears unaffected by alterations in the extracellular calcium concentration. The fact that somatostatin inhibits the release of growth hormone by agents that apparently trigger release by different mechanisms, such as high potassium and dibutyryl cyclic AMP (Hayaski-Kimura and Takahashi, 1979), suggests that somatostatin inhibits secretion by an action at a late step in the secretory process common to most secretagogues, such as the formation of cyclic AMP or the expression of its actions (Wollheim *et al.*, 1977; Bent-Hansen *et al.*, 1979). Since we obviously do not know all of the steps involved in the mechanism of growth hormone release, it would not be profitable to belabor consideration of the various possibilities by which somatostatin might exert its effects, particularly since the actions of somatostatin on the physiological stimulus GH-RH have yet to be analyzed.

c. Corticotropin

The presence of voltage-dependent calcium channels appear likely in the corticotrope, although there is no direct evidence that physiological secretagogues activate these channels. Still, potassium stimulates calcium uptake into pituitary tissue and pituitary cells in culture and elicits corticotropin secretion; calcium deprivation blocks this response to high potassium, as does cobalt and somatostatin (Kraicer, 1975; Richardson and Schonbrunn, 1981). By contrast, a purified hypothalamic extract was unable to stimulate ^{45}Ca uptake, although it provoked corticotropin secretion (Eto et al., 1974). The secretory response to hypothalamic factor is also depressed by somatostatin, in a manner which suggests that its inhibitory action is directed towards calcium (Richardson and Schonbrunn, 1981). However, the present controversy regarding the nature of the inhibitory action of somatostatin precludes any deductions as to the manner in which hypothalamic factors alter calcium handling by the corticotrope.

d. Thyrotropin

Thyrotropin release from the pituitary is controlled by hypothalamic thyrotropin-releasing hormone (TRH), which, like GN-RH and somatostatin, has been isolated and characterized (cf. Vale et al., 1977). The fact that the structure of TRH has been determined has accelerated progress on investigations concerned with the physiological mechanism of thyrotropin secretion. Michael Schrey and his associates (Schrey et al., 1978) utilized a perifused isolated cell system to study the control of thyrotropin secretion. This preparation represents a decided advantage over other such systems since it enables precise monitoring of transient changes in hormone secretion and provides a closer simulation of conditions existing in vivo. The continuous turnover of medium also minimizes the effects of accumulation of secretory and metabolic products and the feedback effects that they might otherwise exert. When such a system is utilized, extracellular calcium deprivation decreases the response to TRH while the responses to phosphodiesterase inhibitors are still capable of provoking a rise in thyrotropin secretion. The response to TRH is also curtailed by D-600 and is associated with an increased efflux of ^{45}Ca. Such results suggest that TRH utilizes extracellular or a superficially bound pool of membrane calcium.

Although the conventional neuron and neurosecretory fibers such as those of the posterior pituitary are the most obvious examples of calcium entry as a consequence of action potentials, we have also seen that electrical activity is a property of endocrine cells of the pancreas and neuroendocrine cells of the adrenal medulla. An even more widespread involvement of such electrical events in secretory phenomena is revealed by the demonstration of action potentials in normal and in a clonal strain of functional pituitary cells (Taraskevich and Douglas, 1977; Ozawa and Kimura, 1979). TRH increases spike frequency in such cells, and electrical activity can be obtained under sodium-deprived conditions but is suppressed by D600 and cobalt, indicating an important calcium component. Various experimental manipulations that alter the secretory response produce similar effects on the electrical discharges. For example, strontium, which is known to substitute for calcium in other secretory processes, replaces calcium in its ability to generate action potentials to depolarizing pulses. Barium, which depolarizes adrenal chromaffin cells, and like strontium, can substitute for calcium as an inward current carrier (Fatt and Ginsborg, 1958), depolarizes clonal pituitary cells and initiates secretion (Ozawa and Miyazaki, 1979; Gard et al., 1981).

TRH releases not only thyrotropin but also prolactin. Taraskevich and Douglas (1980), using teleost fish in which the prolactin cells are grouped in a distinct lobe, clearly demonstrated the presence of a calcium component of TRH-induced electrical activity by blocking potassium efflux with tetraethylammonium, resulting in a large regenerative potential that is reversibly suppressed by the addition of manganese. Additionally, catecholamines such as dopamine and norepinephrine, which are inhibitors of prolactin and melanophore-stimulating hormone secretion, also inhibit spontaneous firing of pituitary cells (Taraskevich and Douglas, 1978). Physiological modulators of certain of the anterior pituitary cells thus control secretion by affecting electrical activity in their target cells to regulate calcium influx. Whether this increase in calcium entry is directly responsible for activating the secretory apparatus remains an unanswered question.

The existence of electrical activity in endocrine calls may seem paradoxical, although the argument has been advanced by Pearse that endocrine cells arise from neuroectodermal origin as do the adrenal chromaffin cells (see Douglas, 1978), and may be regarded as a component of a third (endocrine) division of the nervous system that includes somatic and autonomic divisions (Pearse and Takor, 1976). But regardless of the validity of this concept, it is clear that cells of the anterior pituitary possess ion channels through which calcium ions enter the cell and in some way participate in the secretory process.

While it is clear that the ability to generate action potentials is a widespread phenomenon in the endocrine system (Williams, 1981), the action potentials of endocrine cells do not all share basically similar properties. For example, the action potentials generated by medullary chromaffin cells are mainly the result of a sodium current (Brandt et al., 1976), whereas the action potentials of the pancreatic β cell are predominantly related to an inward calcium current (Matthews, 1977), and the action potentials of cells of the anterior pituitary gland appear to have both a sodium and a calcium component (Taraskevich and Douglas,

1978; Ozawa and Miyazaki, 1979). The physiological implications of these apparent differences are not obvious, but they all reflect the association of ion movements with the primary stimulus and concomitant with these ion movements is the entry of calcium into the cell to trigger the secretory response. In the case of the thyrotroph, these inward calcium movements are too small to be measured by tracer analysis (Fleckman et al., 1981).

The role of cyclic AMP in the mechanism of action of TRH is unresolved. While cyclic AMP analogues or phosphodiesterase inhibitors promote TSH release (Nakano et al., 1976), small increases in intracellular levels of cyclic AMP in response to TRH administration have been observed by some, but not all, investigators (Barnes et al., 1978; Labrie et al., 1979). A role for cyclic AMP in TRH-induced thyrotropin release would imply a cyclic AMP-dependent mobilization of intracellular calcium stores in addition to the activation of calcium channels in the cell membrane. Alternatively, cyclic AMP-dependent phosphorylation reactions may play an important role in the secretory response (see Chapters 3 and 4).

e. Prolactin

TRH stimulates prolactin release from clonal pituitary cells by a calcium-dependent process (Tashjian et al., 1978; Tam and Dannies, 1980). The properties of this particular system closely resemble those of other systems in that strontium and barium are capable of substituting for calcium in triggering hormone secretion (Ozawa and Miyazaki, 1979), magnesium cannot replace calcium in sustaining TRH-induced prolactin release, and cobalt acts as an inhibitor (Gautvik et al., 1980). Calcium appears to be a sufficient stimulus because the ionophore A23187 and the readdition of calcium after calcium deprivation provoke prolactin secretion from rat pituitary cells (Gautvik et al., 1980; Tam and Dannies, 1980). The TRH-induced ^{45}Ca efflux from clonal cells represents calcium mobilized from a cellular pool (Gershengorn, 1980; Gershengorn et al., 1981). Nevertheless, the fact that prolactin release can be elicited by high K^+ (Nakano et al., 1976), taken together with the demonstration of a TRH-sensitive calcium component of action potentials, suggests that calcium influx somehow participates in the events associated with the secretory response.

An important role for cyclic AMP in the mechanism of prolactin secretion is indicated not only by the ability of TRH to induce an increase in cyclic AMP formation in isolated cells but also by the ability of dopamine to produce parallel inhibition of cyclic AMP synthesis and prolactin release (Barnes *et al.*, 1978). An increase in ^{45}Ca release from pre-labeled cells induced by TRH occurs at a time when prolactin release and cyclic AMP formation are also elevated (Gautvik *et al.*, 1980), suggesting that TRH exerts its effects through a cooperative action on calcium mobilization and cyclic AMP formation.

While the available evidence provides us with some knowledge of the mechanism of release of anterior pituitary hormones, it is apparent that further advances in the development of homogeneous cell preparations are required before more definitive information becomes available to enable us to dissect out the various factors associated with calcium-dependent hormone secretion in this particular system. Additionally, identification and purification of various physiological stimulants and inhibitors are mandatory in order to better understand the physiological mechanisms of secretion in this system; for while the use of such agents as ionophore A23187 and high potassium provides us with information regarding the significance of elevated intracellular calcium concentration in the secretory process, they do not afford insight into the physiological mechanism of secretion. It is thus incumbent upon the investigator to employ physiological stimulants whenever possible to study physiological mechanisms; in the case of the anterior pituitary, this has not always been possible. But with the inroads presently being made in this particular area, when a volume similar to this one is published some years from now it will almost certainly provide us with much more definitive insight into the role of calcium in the release of anterior pituitary hormones.

3. Hypothalamic-Pituitary Releasing Hormones and Gut Peptides

Progress toward elucidating the nature of the biochemical events associated with secretory activity of the hypothalamic peptides has been hampered by the complex control that the hypothalamus exerts on pituitary hormone synthesis and release, involving various stimulatory as well as inhibitory components. Without complete knowledge of the physiological mediators of the release of a given peptide, it is difficult to employ the appropriate stimulating agents to elucidate cellular mechanisms. Moreover, the central nervous system is a diffuse network of neural elements that in certain respects defies attempts to isolate various segments of the system and study them in isolation. However, the development of radioimmunoassay techniques and the application of immunochemical studies have demonstrated that these hypothalamic peptides are concentrated in secretory granules in nerve terminals in the hypothalamus and are released in response to stimulation from higher centers by a variety of neurotransmitters including biogenic amines, as well as other putative central neurotransmitters (Reichlin, 1978).

The hypothalamic factors, which are generally recognized as regulators of adenohypophysial activity, are not exclusively localized to the central nervous system but are also distributed in the peripheral nervous system and in the gut-enteropancreatic system (Fujita and Kobayashi, 1977). The relative ubiquity of these peptides obviously presents a problem in classifying them, for not only are these substances released into the blood stream to be delivered to distant target organs—and thus conform to the classification of hormones as originally formulated by Bayliss and Starling (1904)—but they may also have a localized range of extracellular movements. As a consequence, the concept of paracrine—rather than endocrine—secretion would seem more appropriate in certain situations, extending even to the point where a functional distinction between these peptides and conventional neurotransmitters becomes clouded (Guillemin, 1978).

Various preparations have been employed to study the mechanism of release of hypothalamic releasing factors. For example, superfused or incubated rat hypothalami or isolated hypothalamic synaptosomes are capable of releasing corticotropin-releasing factor, gonadotropin-releasing hormone (luteinizing hormone-releasing hormone), somatostatin, and thyrotropin-releasing hormone in response to high potassium or veratridine by a calcium-dependent mechanism (Bennett and Edwardson, 1975; Schaeffer et al., 1977; Warbert et al., 1977; Bigdeli and Snyder, 1978). Not only is secretion induced by depolarizing agents related to the extracellular calcium concentration (Drouva et al., 1981), but it is blocked by verapamil, D-600, magnesium, and manganese (Bigdeli and Snyder, 1978; Drouva et al., 1981); but not by tetrodotoxin (Hartter and Ramirez, 1980). These data support the vast array of existing evidence in other systems that neurons are caused to secrete by depolarization of the neuronal membrane promoting calcium influx through "slow" calcium channels which triggers the discharge of product. While there is evidence from studies utilizing the intact animal that prostaglandins may participate in the process regulating the release of hypothalamic hormones (Hedge, 1977), it is still premature to seriously implicate the prostaglandins as physiological modulators of secretion in this particular system. Nevertheless, prostaglandins also stimulate GN-RH release from perifused or incubated rat hypothalami by a mechanism dependent on the presence of extracellular calcium (Gallardo and Ramirez, 1977; Bigdeli and Snyder, 1978).

There are other substances that are found in the nervous system and in the gastrointestinal tract that may be released by exocytosis to serve as neurotransmitters or modulators (Kobayashi, 1979); one such substance is neurotensin, which is present not only in the hypothalamus, but also in the pituitary gland and gastrointestinal tract (Bloom and Polak, 1978). Neurotensin is released from rat brain tissue by a calcium-dependent mechanism in response to high potassium (Iversen *et al.*, 1978a). The opioid peptides such as enkephalins and endorphins have also been isolated from brain slices and synaptosomal preparations. The pattern is the same for their release in that enkephalin output from rabbit striatal synaptosomes or globus pallidus in response to high potassium is blocked by calcium deprivation (Henderson *et al.*, 1978; Iversen *et al.*, 1978b), thereby defining a possible neurotransmitter role for these opioid peptides (Fig. 21). The source of the enkephalins is neuronal. O. H. Viveros and his associates, using the adrenal medulla as a paradigm for studying neuronal function, have demonstrated that opiate-like peptides

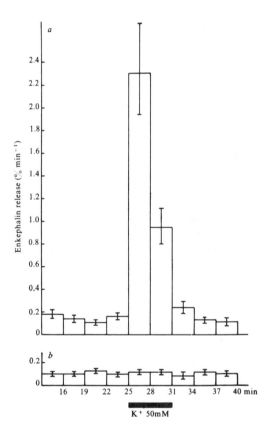

FIGURE 21. Potassium-induced release of enkephalin immunoreactivity from slices of rat globus pallidus superfused with (a) normal or (b) calcium free medium. Results are expressed as the percentage total tissue enkephalin released per minute based upon measurements of the total enkephalin content at the end of the superfusion (Iversen *et al.*, 1978b).

are stored in medullary chromaffin granules together with catechol-amines, carrier proteins, and adenine nucleotides, and are released in response to acetylcholine by a calcium-dependent mechanism (Viveros *et al.*, 1979). This important discovery suggests that these peptides are released into the peripheral circulation to exert important neuroendo-crine functions outside the central nervous system, perhaps involving the functional activity of secretory cells.

Vasoactive intestinal polypeptide (VIP), a 28-residue peptide with structural similarity to secretin and glucagon, has been isolated from porcine small intestine (Said, 1978). VIP was originally considered a gut hormone candidate but immunocytochemical and radioimmunochemi-cal studies have demonstrated the presence of VIP in neurons of both the central and peripheral nervous systems; in fact, it has been localized to the synaptic vesicle fraction, distinct from adrenergic and cholinergic nerves, and can be isolated in synaptosomal preparations of cerebral cortex, hypothalamus, and striatum of rat brain. Stimulation of VIP release with high potassium from synaptosomes or superfused hypo-thalamic tissue is reduced by a low calcium or high magnesium medium (Giachetti *et al.*, 1977; Emson *et al.*, 1978), so VIP may be added to the expanding list of peptides that may function as neurotransmitters or modulators in both the central and peripheral nervous systems. A cal-cium-dependent release of VIP and cholecystokinin can also be dem-onstrated from the isolated duodenum (Kanno *et al.*, 1979), implying that a similar calcium-dependent mechanism governs peptide secretion both in the central nervous system and in the gut.

It has long been known that the nervous and endocrine systems are linked by the adrenal medullary chromaffin cells and the neurosecretory cells of the neurohypophysis. The chromaffin cell and the nerve terminal of the neurohypophysis are of neuronal ancestry and, like endocrine cells, possess secretory granules which discharge their hormones into the circulation. However, it is apparent from our narrative that it is becoming more and more difficult to categorize some cells as nerve cells rather than endocrine cells and *vice versa*, which has led to the devel-opment of the paraneuron concept (Fujita and Kobayashi, 1979). The importance of the paraneuron concept for the basis of our present dis-cussion is the recognition that the boundaries between neural and non-neural secretory cells are obscured. Viewing the gut endocrine cells, for example, as members of a large family of endocrine cells (paraneurons), which are closely related to neurons in structure, function, and metab-olism implies a continuum. Implicit in this concept is the idea that the secretory substances of all of these cells will be produced, transported, and packaged in granules by a common mechanism; and, most relevant for the topic under discussion here, that the granules will be released by a common calcium-dependent exocytotic mechanism.

The remarkable phenomenon of calcium being a mediator in a spectrum of secretory systems seemingly so divergent has previously been alluded to (Rubin, 1974a). It is now apparent that we must revise our perspective and view the action of calcium as a key element in systems that basically are structurally and functionally, if not embryologically, related. Those who have been struck by the commonality of calcium's action on the secretory process are probably not surprised that biologists are now discovering, by a somewhat different tack, that secretory systems are fundamentally similar after all.

4. Thyroid Hormone

In our extended sojourn we have encountered systems that utilize various means to elevate intracellular calcium for the purpose of augmenting secretory activity. However, the final few secretory systems that we will consider will reveal a somewhat clouded picture with regard to calcium as a modulator of the secretory process. This will be observed in the thyroid gland and adrenal cortex, where hormone secretion may not be exocytotic. The parathyroid gland and renin-secreting juxtaglomerular cells contain preformed secretory product in membrane-bound organelles, yet calcium appears to inhibit secretory activity. So anyone who hopes to develop a unitary theory of calcium action must integrate into this theory a body of established facts and generalizations that differ from the basic stimulus–secretion coupling model.

The thyroid gland, like several other endocrine organs, is controlled by the pituitary gland through the action of pituitary thyrotropin. Amine-containing systems are also present within the thyroid gland (Melander, 1977), including sympathetic nerves and mast cells that release norepinephrine, histamine, 5-hydroxytryptamine, and dopamine. These biogenic amines not only function as mediators of blood flow but also may exert in some species a stimulatory effect on thyrotropin-induced synthesis and release of thyroid hormone. The smallest functional unit of the thyroid gland is the follicle, which consists of a single layer of epithelial cells enclosing a cavity called the follicular lumen. The follicular cells possess the unique property of concentrating iodide several hundredfold for the purpose of synthesizing thyroid hormone. In no other endocrine organ do we find a situation where the active uptake of ion is directly linked to hormone synthesis. Glycoprotein subunits sequestered in vesicles are released into the follicule lumen by exocytosis

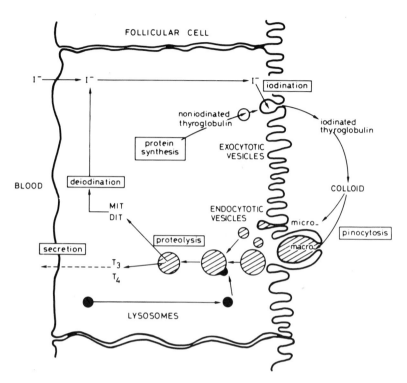

FIGURE 22. Iodine metabolism in the thyroid gland. Iodine is concentrated in the follicular cell and iodination of the protein matrix occurs prior to the discharge of thyroglobulin into the follicular lumen by exocytosis. Upon stimulation, the colloid droplets reenter the follicular cell by exocytosis, and fuse with lysomes, releasing thyroid hormone (T_3,T_4) (Van den Hove-Vandenbroucke, 1980).

upon exposure to thyrotropin and become incorporated into the colloid. Iodination occurs after synthesis of the protein matrix at the epithelial cell interface (Taurog, 1978; Ericson, 1981) (Fig. 22).

The thyroid follicular cell differs from other endocrine cells—but resembles pancreatic acinar cells—in that the portion of the cell membrane exposed to the primary stimulus differs from the one that is involved with the discharge of secretory product. Thyrotropin stimulation of the basal portion of the cell produces an engulfment of thyroglobulin by pseudopods and a reentry of the protein across the luminal membrane of the follicular cell by endocytosis. Within 10 min after the injection of thyrotropin there is an increase in the number of colloid droplets, which migrate slowly through the cytoplasm toward the base of the cell. Lysosomes move toward the apex to coalesce with the colloid droplets, and thyroglobulin is degraded to thyroid hormone (thyroxine and tri-iodiothyronine) that is extruded from the cell (Van den Hove-Vanden-broucke, 1980) (Fig. 22).

While existing evidence affirms that thyroid hormone is released by a nonexocytotic mechanism, certain colloid droplets migrating toward the basal region of the follicular cell increase in density and decrease in size. These mature droplets come to resemble the large dense granules of unstimulated secretory glands. Thyroid hormone may be discharged by way of these dense granules, although the lack of loss of lysosomal enzymes coincident with thyroid hormone discredits the possibility that secretion occurs by exocytosis (Wolff and Williams, 1973). Despite the possibility that the mechanism of secretion of thyroid hormone may differ from that of most other hormones, a role for microtubules and microfilaments has been advocated in the secretory process of the thyroid gland (Wolff and Williams, 1973).

In light of the seemingly unique mechanism of thyroid hormone secretion, it is not surprising that the role of calcium in thyrotropin-induced thyroid hormone secretion does not conform to the classical stimulus–secretion coupling model. While early experiments disclosed that thyrotropin decreases membrane resistance of thyroid cells, suggesting an alteration in membrane permeability (J. A. Williams, 1970), a concentration of extracellular potassium that would be expected to produce a 20–25 mV depolarization did not affect basal or thyrotropin-induced thyroid hormone release (Williams, 1972a). So while normal thyroidal tissue exhibits electrical activity, and medullary carcinoma cells of thyroid demonstrate all-or-nothing action potentials (Tischler et al., 1976) and may therefore be considered electrically excitable, the action of thyrotropin as a secretagogue does not directly involve calcium influx through voltage-dependent channels. Radiotracer analysis suggests that a translocation of cellular calcium may be a crucial step in activation of the thyroid cell (Rodesch et al., 1976), although calcium entry through receptor-operated channels cannot be excluded.

The significance of extracellular calcium in the action of thyrotropin was discredited by reports that stimulation by thyrotropin of intracellular colloid droplet formation and ^{131}I release (iodine and thyroid hormones) are not inhibited by calcium deprivation, even in the presence of calcium-chelating agents. Other thyrotropin effects including activation of glucose oxidation and iodide organification are diminished (Dumont et al., 1971; Williams, 1972b). Prolonged calcium and magnesium deprivation partially obtunds the secretory response (Williams, 1972b), but by removing both cations, a general disturbance of cell function was probably effected. The prolonged calcium deprivation required for blockade of secretion contrasts with the rapid effect of sodium deprivation to pro-duce depression of secretion (Williams, 1972a). The fact that ouabain is unable to stimulate, but in fact inhibits, thyrotropin-stimulated hormone release indicates that sodium accumulation is not an obligatory step, although taken together with the dependence of extracellular sodium it suggests that a functioning sodium–potassium transport system is required for secretion.

As calcium lack does little to impair the secretory response of the thyroid gland, experimental manipulations designed to augment cellular concentrations also fail to enhance thyroid hormone release. Thus, thy-roid hormone secretion and colloid droplet formation in resting or stim-ulated dog thyroid slices *in vitro* are not increased by the addition of calcium, barium, or ionophore A23187, but, in fact, elevation of the calcium concentration in the follicular cell actually inhibits thyrotropin-stimulated hormone secretion (Dumont *et al.*, 1977). In addition to ion-ophore A23187, other agents that increase cyclic GMP formation, such as acetylcholine, high potassium concentrations, and 5-hydroxytrypt-amine also depress secretion induced by thyrotropin (Dumont *et al.*, 1977). The implication here is that calcium acts as an inhibitor of thyroid hormone secretion, and perhaps part of this inhibitory effect is due to a calcium modulation of cyclic GMP levels.

Initial reports showed that many, if not all, of the actions of thy-rotropin on growth and function of the thyroid gland are controlled by the activity of the adenylate cyclase–cyclic AMP system (Dumont *et al.*, 1971). Thyrotropin stimulates adenylate cyclase activity of bovine and canine thyroid homogenates and increases cyclic AMP levels in thyroid glands. Moreover, some of the effects of thyrotropin in thyroid function can be duplicated by cyclic AMP or its derivatives (Dumont *et al.*, 1971). The thyrotropin-induced increase in cyclic AMP formation is blunted by the simultaneous administration of agents that elevate cyclic GMP levels (Dumont *et al.*, 1977). Calcium, perhaps in concert with cyclic GMP, is responsible for this inhibitory effect on the cyclic AMP response, possibly through the activation of phosphodiesterase (Champion and Jacquemin, 1978; Spaulding, 1979). Therefore, in the thyroid gland calcium may exert its cellular actions by controlling the tissue concentrations of cyclic nucleotides. Calcium may also allow cyclic AMP to express some of its cellular effects (Dumont *et al.*, 1971), although this aspect of calcium's action on thyroidal function is not well elucidated.

A final potential site of calcium action may involve the phospholi-pase–prostaglandin system, since prostaglandins may function as mod-ulators of thyroid hormone synthesis and release. Thyrotropin, ionop-hore A23187, and cholinergic and α-adrenergic agents increase prostaglandin levels in thyroidal tissue (Burke et al., 1973; Boeynaems et al., 1979; Takasu et al., 1981). In addition, enzymes for synthesizing various prostaglandins have been found in bovine thyroid and their activity is increased by thyrotropin (Friedman et al., 1976). Since it is well known that calcium is required for activation of phospholipase A_2 and the synthesis of prostaglandins in other tissues, calcium may have a similar action in the thyroid gland (Haye and Jacquemin, 1977; Boeynaems et al., 1979). However, the question as to whether thyrotropin is capable of stimulating phospholipase A_2 is a controversial one (Irvine et al., 1980).

The importance of prostaglandins in thyroidal function is underscored not only by the ability of PGE and prostacyclin to mimic most of the effects of thyrotropin (Field et al., 1971; Takasu et al., 1981), but also by the finding that indomethacin, an inhibitor of prostaglandin synthesis, abolishes the secretory response in rats to exogenous thyrotropin (Thompson et al., 1977). The actions of prostaglandins, like those of thyrotropin, may be expressed through calcium and cyclic AMP (Dekker and Field, 1970; Takasu et al., 1981). However, the activation of adenylate cyclase by thyrotropin and PGE occurs through different mechanisms (Wolff and Cook, 1973) and inhibitors of prostaglandin synthesis do not alter the cyclic AMP response to thyrotropin (see Boeynaems et al., 1979). So, there are sufficient grounds for suspecting that while prostaglandins may be important mediators of the action of thyrotropin, they may act at a step distal to cyclic AMP formation. Nevertheless, it is still not possible to provide a clear explanation of the complex interrelations among calcium, cyclic nucleotides, and the phospholipase–prostaglandin systems in the mechanism of action of thyrotropin.

Progress toward establishing whether thyrotropin action is associated with an alteration in calcium handling by the thyroid gland has been hampered by the absence of a suitable in vitro system for studying the cellular events associated with thyroid hormone secretion. The development of techniques for isolating thyroid cells as well as follicles (Rousset et al., 1980; Denef et al., 1980) should aid in obtaining a more precise assessment of the role of calcium in thyroid hormone secretion by providing a better kinetic view of its metabolism.

But due to the unique nature of the secretory process associated with thyroid hormone secretion, it would be rather surprising to discover ultimately that the action of calcium in the thyroid gland parallels its action in the adrenal medulla and endocrine pancreas, for example. Even so, intracellular calcium appears to play a very critical role in expressing the effects of thyrotropin on various metabolic processes in the thyroid through the mediation of other cellular messengers. Whether these actions of calcium are directly translated into heightened secretory activity seems problematic, but the question as to the specific role of calcium in the mechanism of thyroid hormone secretion still affords a challenging problem.

5. Steroid Hormones (Cortical and Gonadal)

a. Corticosteroids

As this narrative has unfolded, a clear picture has begun to emerge with regard to the critical role of calcium in exocytotic secretion. However, the thyroid gland, as we have seen, presents a divergent pattern, and the adrenal cortex manifests some unique features, facts that further temper our enthusiasm for viewing the secretory process in a unified manner.

Unlike other endocrine organs, the mammalian adrenal cortex is divided into three concentric zones. The zona glomerulosa is relatively independent of pituitary control and secretes mineralocorticoid hormones which regulate salt and water metabolism, while the inner two zones—the zona fasciculata and zona reticularis—depend on pituitary corticotropin for activity and secrete steroids that influence carbohydrate and protein metabolism. The goal of the regulatory mechanism responsible for corticoid secretion is not to maintain a fixed concentration of hormone in the circulation but to effect fluctuations in the concentration of steroids, which are a reflection of changes in the rate of pituitary corticotropin. The necessity for fluctuating—rather than relatively static—concentrations of steroid hormone in the circulation may in some way account for the fact that, unlike most other hormones, corticosteroids are not stored in secretory organelles but are synthesized upon demand by a complex series of reactions that brings about the conversion of cholesterol to corticosteroid, involving pregnenolone as a key intermediate (Samuels and Nelson, 1975). These reactions take place both in the mitochondria and endoplasmic reticulum, proceeding through a series of hydroxylation reactions involving cytochrome P-450.

While it is clear that the cortical cell lacks any appreciable storage of secretory product, the question has not been answered whether the newly synthesized hormone is incorporated in secretory granules prior to its discharge from the cell. Extremely low levels of corticosteroids are universally found in the "resting" or unstimulated adrenal cortex, and difficulties encountered in visualizing secretory granules in cortical cells have made it difficult to formulate a suitable model of steroid release (Malamed, 1975). In fact, the concept has long been espoused—without documentation—that steroids are synthesized in response to corticotropin and immediately released by simple diffusion. On the other hand, the view that an exocytotic mechanism underlies hormone release in steroid-producing cells—which has gained more and more support in recent years—rests with the detection of electron-dense granules, frequently located near the plasma membrane (Nussdorfer *et al.*, 1978). Moreover, corticotropin-elicited increase in the steroid content of the gland parallels the increase in the number of these granules (Gemmell *et al.*, 1977). Additionally, the corticotropin-induced steroid output by the adrenal gland is accompanied by an increase in "secretory" proteins in the perfusate (Rubin *et al.*, 1974), a fact that suggests that the mechanism of steroid hormone release shares certain basic features with the secretory mechanism of other endocrine glands where exocytosis is the mechanism of secretion. In the avian adrenal, evidence has also been presented for exocytotic release of steroid hormone on the basis of: (1) the presence of membrane-bound vesicles that appear to contain "lipid-like" material; (2) vesicles fusing with the plasma membrane and (3) the presence of coated pits, a common feature of exocytotic processes (Pearce *et al.*, 1977).

Although such data are in accord with the concept that steroid secretion is a process related to, or identical with exocytosis, additional documentation is needed to support this claim. It is necessary to isolate and identify a granular fraction from the cortex that contains steroid hormone, as well as other granular constituents. The specific identification of a granular protein and its concomitant secretion from the gland with the hormone would certainly add credence to a concept that is presently based only on rudimentary evidence.

In light of the link forged between calcium and exocytotic secretion, further insight into the mechanism of steroid release may be gleaned by investigations on the role of calcium in the steroidogenic process. The requirement for extracellular calcium in the steroidogenic response to corticotropin was described many years ago (Birmingham *et al.*, 1953) and it is now clear that calcium functions as the second messenger during corticotropin stimulation of steroidogenesis (Halkerston, 1975). However, the mechanism of action of calcium remains controversial because it is difficult to dissociate its effects as a second messenger from a variety of other functions it performs in the cortical cell. The crucial point to emphasize is that in the absence of extracellular calcium, corticotropin-induced steroid synthesis is abolished (Halkerston, 1975); since there is little performed hormone available, it is extremely difficult to dissociate the effects of calcium on steroid synthesis from any possible direct effect on the secretory process *per se*. The fact that calcium alone, under certain experimental conditions, mimics the effects of corticotropin in eliciting steroidogenesis (Podesta *et al.*, 1980) provides convincing evidence for a direct action of calcium on the machinery controlling the steroidogenic process, but, since there is little stored hormone in the gland, a direct effect of calcium on the secretory process cannot be ascertained. The fact that ionophore A23187 is at best only a weak steroidogenic agent (Schrey and Rubin, 1979) suggests that other mediators are also involved in the steroidogenic pathway.

Matthews provided insight into the mechanism of action of corticotropin and the possible role of calcium by analyzing the electrical activity associated with corticotropin-stimulated corticosteroid release. The findings that high potassium or calcium deprivation depolarizes the cortical cell but does not markedly affect the steroidogenic response to corticotropin make it improbable that the action of corticotropin is associated with the activation of voltage-dependent channels (Matthews and Saffran, 1968; 1973). To confirm this conclusion, the fact that replacement of sodium by choline or the addition of tetrodotoxin does not impair corticotropin-induced steroidogenesis suggests that sodium influx is probably not an important factor in the action of corticotropin. Corticotropin may still, however, elicit calcium influx independently of any change in the membrane potential, as evidenced by the ability of corticotropin to elicit tetrodotoxin-resistant action potentials under certain experimental conditions.

However, radiotracer analysis has provided little evidence for an increase in ^{45}Ca uptake in response to corticotropin. In perfused cat adrenal glands, corticotropin appears to bring about a mobilization of cellular calcium (Jaanus and Rubin, 1971), and, in rat adrenal glands, the corticotropin-induced enhancement of calcium uptake appears to be a consequence of the increase in steroidogenesis (Leier and Jungmann, 1973).

Since corticotropin does not express its protean actions by increasing membrane permeability to calcium, the effects of this cation may be confined to the surface of the cortical cell; alternatively, cellular messengers activated by corticotropin may mobilize intracellular calcium to trigger the requisite cellular responses. While calcium is not required for the binding of corticotropin to its receptors on the surface of the cortical cell, it may be required for the following functions: (1) activation of membrane-bound enzymes such as adenylate cyclase and phospholipase A_2, (2) hydroxylation reactions in the steroidogenic pathway, and (3) secretion of steroid.

An essential feature of the control mechanism regulating corticosteroid production and release is the adenylate cyclase–cyclic AMP system. In fact, one of the earliest recognized actions of cyclic AMP was to mediate the effect of corticotropin on steroidogenesis (Grahame-Smith *et al.*, 1967). While over the years a controversy has raged over the actual importance of cyclic AMP in the steroidogenic and trophic responses of the cortex (Schimmer, 1980), cyclic AMP appears to serve an important function, although only a small proportion of a cell's ability to produce cyclic AMP need be activated to increase steroidogenesis maximally (Schulster and Schwyzer, 1980).

The effects of calcium on the adenylate cyclase–cyclic AMP system are complex in that calcium in relatively high concentrations directly inhibits the catalytic activity of adenylate cyclase (Ontjes, 1980). However, in adrenocortical cells increasing calcium concentrations produce a progressive increase in corticotropin-stimulated cyclic AMP formation and steroid release, implicating calcium in the transmission of the signal between the corticotropin receptor and the adenylate cyclase enzymic subunit (Sayers *et al.*, 1972). The stimulatory effects of calcium appear to be exerted at the step where the enzyme is activated by guanyl nucleotides (GTP) (Mahaffee and Ontjes, 1980); this effect probably accounts for the role of calcium in "coupling" of hormone-receptor binding and adenylate cyclase activation. The fact that lanthanum—which does not permeate the cortical cell—inhibits both the cyclic AMP and the steroid response to corticotropin, but not the steroidogenic effect of dibutyryl cyclic AMP, reinforces the concept that calcium acts at superficial binding sites on the cell membrane to play an important role in the regulation of adenylate cyclase by corticotropin (Haksar *et al.*, 1976). While calcium may also regulate the transduction of information between the corticotropin receptor and guanylate cyclase to promote cyclic GMP formation (Perchellet and Sharma, 1979), the consensus of evidence consigns to cyclic GMP only an ancillary role, at best, in the control mechanisms governing steroid production and release (Laychock and Hardman, 1978).

The phospholipase–prostaglandin system may also be important in mediating the steroidogenic actions of corticotropin. Corticotropin activates a calcium-dependent phospholipase A_2 and prostaglandin synthesis in cat and rat adrenocortical cells (Rubin and Laychock, 1978a,b). The physiological significance of the prostaglandins in the steroidogenic response is still controversial, particularly in light of the fact that, in several systems, inhibitors of prostaglandin synthesis are unable to significantly inhibit the steroidogenic response to corticotropin (Schimmer, 1980), although in cat cortical cells both indomethacin (a cyclooxygenase inhibitor) and ETYA (a cyclooxygenase and lipoxygenase inhibitor) produce profound inhibition of corticotropin-induced steroid output at relatively low concentrations (Laychock and Rubin, 1976). While prostaglandins are potent stimulators of steroidogenesis in several mammalian species (Schimmer, 1980), the steroidogenic actions of exogenously administered prostaglandins and cyclic AMP or its derivatives are much less dependent on extracellular calcium than corticotropin (Rubin, 1981), implying that these putative intracellular mediators of steroidogenesis bypass several of the calcium-dependent pathways localized to the cell membrane, i.e. adenylate cyclase and phospholipase activation.

We therefore appear to have a rather complex compartmentalization of calcium involved in the steroidogenic process in that, on the one hand, superficially bound membranous calcium is crucial for the primary signals to be expressed by corticotropin–receptor interactions. These primary signals bring about the formation of the intracellular mediators such as cyclic AMP and prostaglandins; however, in order for the intracellular mediators to exert their effects, intracellular calcium pools are also mobilized (Fig. 23). The mobilization of cellular calcium may promote steroid hydroxylation and cholesterol side chain cleavage reactions by increasing the mitochondrial synthesis of pregnenolone (Simpson *et al.*, 1975; Farese and Prudente, 1978). The finding that ruthenium red, which impairs mitochondrial function by inhibiting calcium transport, blocks corticotropin- and cyclic AMP-induced steroidogenesis in adrenal tumor cells (Warner and Carchman, 1978) points to the mitochondria as a critical site of calcium action in the adrenal cortex.

b. Mineralocorticoids

While corticotropin is the major regulator of glucorticoid production in the adrenal fasciculata and reticularis cells, angiotensin II and high potassium and low sodium serve as regulators of aldosterone secretion by the glomerulosa cells. Corticotropin also promotes aldosterone secretion, although the cells of the zona glomerulosa are much less dependent on anterior pituitary control than those of the zona fasciculata or reticularis. Angiotensin and high potassium enhance aldosterone production and secretion through a cyclic AMP-independent mechanism that has a high requirement for extracellular calcium and is sensitive to inhibition by lanthanum or verapamil (Shima *et al.*, 1978; Fakunding *et al.*, 1979; Fakunding and Catt, 1980). The fact that the response to corticotropin or cyclic AMP is less sensitive to calcium depletion than that mediated by angiotensin or potassium suggests that the actions of corticotropin on the glomerulosa cell are mediated through a cyclic-AMP-regulated mechanism involving an intracellular mobilization of calcium (Fakunding *et al.*, 1979).

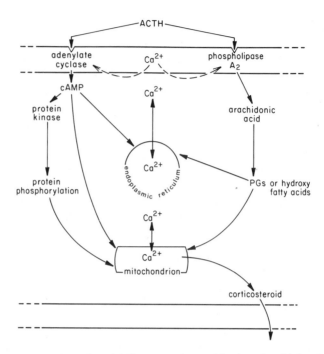

FIGURE 23. A hypothetical model illustrating the participation of multiple calcium pools in the mode of action of corticotropin (ACTH). When the hormone binds to receptors on the surface of the adrenocortical cell, an external pool of calcium required to couple receptor activation to stimulation of adenylate cyclase is mobilized. The calcium released from this superficial compartment also activates phospholipase A_2, which like adenylate cyclase, is localized to the plasma membrane. The resulting formation of cyclic AMP and prostaglandins or other arachidonic acid metabolites leads to the mobilization of an internal pool of cell calcium, perhaps localized to the mitochondria or endoplasmic reticulum to promote steroid synthesis and release (see text for further details).

Although angiotensin, potassium, and ouabain depolarize glomerulosa cells (Natke and Kabela, 1979), any proposed step in the secretory response that might involve depolarization cannot be primarily mediated by sodium entry into the glomerulosa cell since sodium deprivation does not diminish—but may even enhance—the response of the zona glomerulosa to a secretagogue (Kramer et al., 1979). Consequently, a transmembrane calcium influx mediated through receptor-operated membrane channels is more likely to be responsible for triggering angiotensin-induced aldosterone synthesis and secretion. The actions of angiotensin on smooth muscle contraction, which also manifest dependence on extracellular calcium (Freer, 1975), may be expressed through a similar mechanism. On the other hand, a small increase in the potassium concentration required to stimulate aldosterone secretion by glomerulosa cells is associated with a slowing of the washout of calcium from these cells, rather than with an increased calcium uptake, suggesting that, as in the perfused cat adrenal gland, stimulation of steroidogenesis involves a redistribution of calcium from one intracellular pool to another (Mackie et al., 1978). The electrophysiological and radiotracer analyses may not be in conflict, but may only reflect, once again, the mobilization of several pools of calcium.

c. Gonadal Steroids

A pivotal role for calcium in the mechanism of secretion of gonadal steroids has also been established, although these preliminary studies do not provide us with insight into the critical calcium pools involved in this process or into the mechanism by which calcium exerts its effects. Based largely upon the work of Gemmell et al. (1974) and Niswender and his associates (Sawyer et al., 1979), the notion that gonadal steroids are sequestered in secretory granules and released by an exocytotic mechanism can be seriously entertained. The additional finding that luteinizing-hormone stimulation of testosterone production by Leydig cells of rat testis requires external calcium fits the stimulus–secretion coupling model (Janszen et al., 1976). However, in contrast to the adrenal cortex, where the calcium requirement can be overcome by increasing the concentration of corticotropin—leading to the conclusion that calcium is involved in the transmission of the signal arising from corticotropin-receptor interactions—maximum luteinizing-hormone stimulation of testosterone production is absolutely dependent upon the presence of calcium. It has therefore been suggested that a principal effect of calcium in gonadal steroidogenesis is exerted on a late step in the sequence of events, perhaps involving the activation of protein kinase (Janszen et al., 1976). The additional finding that the administration of luteinizing hormone in vivo markedly enhances the stimulatory effect of calcium on pregnenolone production by isolated mitochondrial fractions indicates that luteinizing hormone increases pregnenolone production by increasing mitochondrial uptake of calcium (Van der Vusse et al., 1975, 1976).

In ovine corpora lutea, where calcium also plays a critical role in mediating the steroidogenic response (Higuchi et al., 1976), luteinizing hormone is unable to produce a detectable change in the membrane potential when progesterone release is enhanced. These data taken together with the finding that sodium deprivation increases progesterone secretion but does not affect the resting membrane potential support the conclusion that—analogous with findings obtained on the adrenal cortex—depolarization and calcium entry are not essential for progesterone release from luteal cells (Higuchi et al., 1976). To emphasize the apparent importance of intracellular calcium in these reactions, it may be mentioned that the stimulatory effects of sodium deprivation on steroid synthesis are ascribed to an enhanced calcium uptake into mitochondria resulting from a decrease in intracellular sodium (Kramer et al., 1979).

It is likely that luteinizing hormone acts through intracellular messengers to translocate cellular calcium rather than by directly activating voltage-dependent ion channels in the plasma membrane. An interplay of cellular calcium with cyclic AMP probably takes place, because in luteal and Leydig cells, stimulation of steroid production by gonadotropins is accompanied by coincident changes in the production of steroid and cyclic AMP bound to the regulator component of protein kinase (Podesta et al., 1978; Ling et al., 1980). On the other hand, the role of cyclic GMP in steroidogenesis is still uncertain since luteinizing hormone does not markedly alter cyclic GMP levels in concentrations that increase cyclic AMP and progesterone accumulation in luteal cells (Ling et al., 1980).

While a physiological role for prostaglandins in the reproductive system has not been unequivocally established, they have been identified in the testis and ovary, and one of the major changes essential to ovulation that occurs upon stimulation by luteinizing hormone is an increase in prostaglandin synthesis by the ovary (Clark et al., 1978). The luteolytic action of prostaglandins to terminate the ovarian cycle is also well documented. The question as to whether prostaglandins act as modulators (positive or negative) for controlling gonadal steroid production and release is an important one to answer, since investigations on adrenal steroidogenesis have provided intriguing evidence that the phospholipase–prostaglandin system, along with the adenylate cyclase–cyclic AMP system, cooperates with calcium in mediating the actions of the primary stimulus (Rubin and Laychock, 1978a,b; Rubin, 1981).

6. Parathyroid Hormone

Secretory systems exist in which calcium has a well-defined role, but as a negative modulator of secretion. The parathyroid gland is atypical among the exocytotic systems in that the release mechanism is impaired by physiological calcium concentrations. When ionic calcium decreases, the gland is released from this inhibition and more parathyroid hormone is discharged. This inhibitory effect of calcium, which is necessary for homeostatic regulation of calcium metabolism, seemingly represents a significant gap in our knowledge of the control mechanisms governing the secretory process. Yet, paradoxically, it may help us gain a better perspective on how this cation causes cells to secrete.

When marine vertebrates migrated onto land, the evolution of the parathyroid gland accompanied the change from a high calcium environment to one in which calcium is in relatively short supply. The hormone elaborated by the parathyroid gland functions in the body to maintain the concentration of free calcium in the extracellular fluid within the physiological range, via its direct action in promoting the efflux of calcium from bone and by increasing the reabsorption of calcium from the renal tubular fluid. This hormone also increases the conversion of 25-hydroxyvitamin D_3 to the much more potent 1,25-dihydroxyvitamin D_3 which stimulates intestinal calcium absorption. Extracellular ionic calcium in turn, by negative feedback control on the parathyroid gland, regulates the release of parathyroid hormone. The parathyroid gland has no known tropic hormone, with the possible exception of β-adrenergic agonists, and the decrease in ionized plasma calcium is a direct stimulus to parathyroid hormone release. When the calcium concentration in the blood that passes through the parathyroid gland drops below 2.5 mM, hormone secretion is enhanced.

The parathyroid gland has ultrastructural characteristics like those of other endocrine organs in that the hormone is synthesized in the endoplasmic reticulum and packaged in the Golgi system. Actually, parathyroid hormone is initially synthesized as a preprohormone, converted to prohormone, and upon transfer to the Golgi zone is converted to parathyroid hormone (Habener et al., 1977; Habener and Potts, 1978). Thus, parathyroid hormone is not directly synthesized, but like insulin, is formed by the proteolytic cleavage of a prohormone. Electron dense bodies, representing secretory granules that contain the storage form of the parathyroid hormone, can be followed from their formation in the Golgi apparatus to the discharge of their contents from the cell (Munger and Roth, 1963; Habener et al., 1979).

Not only can parathyroid hormone be isolated from the particulate fraction of the gland homogenate (MacGregor et al., 1973), but protein distinct from parathyroid hormone can be detected as a component of parathyroid secretion (Kemper et al., 1974). This secretory protein (designated SP-I) is contained within membranous structures of the gland, as is parathyroid hormone (Morrissey et al., 1980), and its discharge from the cell is, likewise, obtunded by calcium (Morrissey and Cohn, 1979a). Thus SP-I is likely to play a role in the transport and binding of parathyroid hormone in a manner similar to the chromogranins in the adrenal medulla and the neurophysins in the neurohypophysis. Extending this analogy, the concomitant release of secretory protein and hormone can be construed as biochemical evidence for exocytosis as the mode of secretion by the parathyroid gland. But despite the clear association of calcium and exocytotic release in other secretory systems, the processes associated with parathyroid hormone production storage, secretion, and even intraglandular destruction are complex and involve a number of individual steps, several of which might be controlled by calcium.

The first evidence that calcium could regulate parathyroid hormone secretion was provided in 1942 by Patt and Luckhardt when they perfused calcium-deficient blood through an isolated dog thyroid-parathyroid gland; when this perfusate was subsequently injected into a second dog, an increase in blood calcium was observed that was similar to that obtained with crude parathyroid extract. Direct measurement of parathyroid hormone by bioassay and radioimmunoassay both in *in vivo* and *in vitro* model systems confirmed the control of parathyroid hormone secretion by extracellular calcium. Louis Sherwood and his colleagues demonstrated in the perfused cow, goat, and sheep thyroid–parathyroid systems that there is an inverse relationship between the blood calcium concentration and parathyroid hormone secretion, and that infusions of EDTA lead to a marked increase in parathyroid hormone secretion within five minutes (Sherwood *et al.*, 1966, 1968). However, if medium calcium is reduced to zero, then hormone output is diminished (Ramp *et al.*, 1979).

While *in situ* preparations have provided valuable information about the factors that control the dynamics of parathyroid secretion, *in vitro* systems were required for analyzing the cellular events associated with the secretory response. Bovine slice preparations, glands maintained in organ culture, and dispersed cell preparations have all been used to demonstrate that the amount of parathyroid hormone released into the medium is inversely proportional to the extracellular calcium concentration (Targovnik *et al.*, 1971; Williams *et al.*, 1973; Brown *et al.*, 1976). The unique signal transmitted by calcium to the secretory apparatus of the parathyroid cell is also exemplified by the inhibitory action of ionophore A23187 (Fig. 24), and by the fact that high levels of magnesium, a well-known calcium antagonist, mimic the effect of calcium in inhibiting parathyroid hormone release, although it is approximately two times less potent in this regard (Brown *et al.*, 1980; Habener and Potts, 1976). The relatively rapid effect of calcium deprivation in triggering parathyroid hormone secretion denotes a direct action on the secretory apparatus, and the ability of ionophore A23187 to decrease parathyroid hormone secretion without affecting synthesis supports this concept

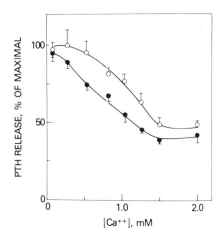

FIGURE 24. Inhibitory effect of ionophore A23187 on calcium-regulated hormone release by isolated bovine parathyroid cells. Cells were incubated for 2 hours with varying calcium concentrations with (●) or without (○) 1 μM ionophore A23187 (Brown et al., 1980).

(Habener et al., 1977). Calcium is therefore recognized as the principal regulator of parathyroid hormone secretion, but, paradoxically, the action of calcium is inhibitory.

Calcium also serves to regulate intracellular levels of parathyroid hormone. The complex pattern of parathyroid hormone synthesis, involving the formation of two prohormones, makes it difficult to assess the precise effect of calcium, but apparently hypocalcemia enhances the amount of hormone found in the parathyroid cell not by a direct effect on synthesis but by a reduced degradation of the parathyroid hormone or its prohormone(s) (MacGregor and Cohn, 1978). It would be simplistic to conclude that calcium impedes the release of parathyroid hormone by reducing the total concentration of hormone in the gland because there are at least two hormone and secretory protein pools that may be individually recruitable, one consisting of more recently synthesized protein and the other of older "storage" protein (Morrissey et al., 1980). It is apparent from such findings that the effects of calcium on the biosynthesis and release of parathyroid hormone have many ramifications and clearly cannot be considered from only one aspect.

In searching for additional clues as to the nature of this rather unique action of calcium, we naturally look to other putative mediators of hormone secretion. Parathyroid hormone is released from dispersed bovine parathyroid cells and slices upon exposure to β-adrenergic agonists, cyclic AMP or its derivatives, and prostaglandins of the E series (Abe and Sherwood, 1972; Brown et al., 1978; Gardner et al., 1978). The ability of β-adrenergic agonists and prostaglandin E to regulate parathyroid hormone release is probably mediated through cyclic AMP; a close relationship exists in parathyroid cells between parathyroid hormone release and cyclic AMP, extending over a wide range of cyclic nucleotide concentrations (Brown *et al.*, 1978). This prompts speculation that calcium does not act directly on the exocytotic process but indirectly by altering the intracellular cyclic AMP levels either by inhibiting adenylate cyclase with an attendant fall in cyclic AMP (Matsuzaki and Dumont, 1972), or by activating phosphodiesterase through an interaction with calcium-dependent regulator protein (Brown, 1980). A calmodulin-regulated phosphodiesterase has been demonstrated in bovine parathyroid cells, and this enzyme complex may contribute to the calcium-induced reduction in intracellular cyclic AMP content as well as parathyroid hormone release (Brown, 1980).

However, calcium can also modulate the secretory response to cyclic AMP, as well as the intracellular concentrations of this cyclic nucleotide (Brown *et al.*, 1978). So it is necessary to postulate an additional site of calcium action, distal to the formation of cyclic AMP. While it is tempting to believe that cyclic AMP might be the mediator through which calcium exerts its modulatory action on parathyroid hormone secretion, it again may be simplistic to view this system in such a manner, since the effects of calcium deprivation on parathyroid hormone secretion are not completely mimicked by cyclic AMP. Thus, while hypocalcemia mobilizes both new and old pools of hormone, isoproterenol and dibutyryl cyclic AMP bring about the release solely of older hormone (Morrissey and Cohn, 1979a,b). Calcium therefore does not exclusively express its actions through this cyclic nucleotide.

Prostaglandins of the E series also increase cyclic AMP and parathyroid hormone release, and the dose–response relationships of PGE_2 to cyclic AMP and parathyroid hormone stimulation are very similar, thereby coupling the actions of these two proposed mediators (Gardner *et al.*, 1978). However, the physiological importance of prostaglandins in the normal regulation of parathyroid hormone secretion has not been established, and we are still lacking certain critical pieces of the jigsaw puzzle that will enable us to formulate a model that encompasses all of the existing evidence. These anticipated results will probably reveal that calcium also controls parathyroid hormone release by regulating the prostaglandin content of the parathyroid cell.

However, another approach worthy of consideration relates to the possibility that the parathyroid cell does not possess the requisite buffer systems for maintaining a low cytosolic level of calcium, enabling calcium to rise to a level in the cell that would elicit inhibitory effects (Glick and Mockel, 1980). This theory derives support from the knowledge that some extracellular calcium is required for parathyroid hormone secretion to occur (Ramp *et al.*, 1979). It then follows from our foregoing discussion that when the intracellular calcium buffering systems of a secretory cell are overwhelmed by an excessive amount of ionic calcium, this cation may become an inhibitor rather than a stimulator of activity. Future studies on the ability of intracellular organelles of the parathyroid gland to sequester calcium should provide us with some valuable information along these lines and lead us closer to our ultimate goal in gaining a fundamental understanding concerning the mode of action of calcium in regulating the secretory apparatus.

7. Calcitonin

Apart from thyroid hormone, a second hormone is secreted from the thyroid gland whose function is to act in concert with the parathyroid gland to regulate the calcium concentration of the extracellular fluid. In 1961 Copp and his associates showed that the perfusion of hypercalcemic blood through the thyroid–parathyroid complex of dogs rapidly reduced the calcium concentration in the systemic circulation (see Copp, 1969). The experimental observations, which were not the result of a decrease in parathyroid hormone release, were subsequently confirmed in goats and pigs (Munson, 1976). Copp and his associates named the hormone calcitonin because of its presumed role in regulating the calcium concentration in body fluids. But in 1963, Hirsch and Munson (Munson, 1976) succeeded in extracting the hormone from rat and dog thyroid. This work led to the isolation and chemical characterization of the porcine hormone, a polypeptide which acts primarily on bone to nullify the demineralizing action of parathyroid hormone.

The successful demonstration by immunofluorescence of calcitonin in the parafollicular cells of the thyroid gland, together with the observation that the cells were not invariably parafollicular, led to the proposal that they be called C cells (C for calcitonin) (Pearse, 1976). These cells—spawned by the neural crest—migrate to the ultimobranchial glands of the mammalian embryo, and subsequently invade the thyroid and parathyroid glands. The main characteristics of these C cells are poorly developed rough endoplasmic reticulum and specific secretory granules. Although exocytosis has not been described in these cells, the evidence that the prominent cytoplasmic granules decrease in number after elevation of blood calcium (Ekholm and Ericson, 1968) favors a mechanism of secretion which resembles that found in most other secretory systems. Moreover, since the C cells, like the medullary chromaffin cells, are derived from the neural crest, their fundamental mechanisms of secretion would be expected to be basically similar.

The thyroid gland is adapted to secrete calcitonin in response to acute hypercalcemia, implying a direct effect of calcium on preformed hormone. Procedures used to monitor mammalian calcitonin secretion have included acute incubation of porcine thyroid tissue slices, whole glands incubated *in vitro,* monolayer cultures of C cells from human medullary thyroid cancer tissue, as well as the intact perfused gland. An increase in the calcium concentration is an adequate stimulus for all of these preparations (Care *et al.,* 1971; Bell and Queener, 1974; Pento *et al.,* 1974; Cooper, 1975; Roos *et al.,* 1975). The rate of calcitonin secretion increases when the calcium concentration of the fluid perfusing the thyroid is increased, and there is a high correlation between the calcium concentration in the thyroid venous plasma and the rate of calcitonin release (Care *et al.,* 1968; Hirsch and Munson, 1969). Apart from the direct effects of calcium, gut peptides, including cholecystokinin and gastrin, are also capable of stimulating calcitonin secretion (Munson, 1976; Deftos *et al.,* 1978). Adrenergic drugs, particularly those which stimulate β receptors, are also potent secretagogues as well as cyclic AMP or its derivatives and prostaglandins (Munson, 1976; Vora *et al.,* 1978). But due to the dearth of available methods for preparing homogeneous mammalian C cell preparations, the data are inadequate to discern what roles, if any, cyclic AMP and/or prostaglandins play in the mechanism of calcitonin secretion.

In light of the neuronal ancestry of the parafollicular cells, it is not surprising that high potassium stimulates calcitonin secretion from the thyroid gland perfused *in situ,* as well as from tumor cells in culture (Cooper, 1975; Gagel *et al.,* 1980). The additional finding that potassium-induced calcitonin release is calcium-dependent connotes a pattern resembling the archetypal model of stimulus–secretion coupling. This conclusion is supported by findings that cations related to calcium, including magnesium, barium, and strontium, also modify calcitonin secretion (cf. Deftos *et al.,* 1978). While high concentrations of magnesium stimulate calcitonin release, when hypermagnesemia is superimposed on hypercalcemia a decrease in the secretion rate ensues (Care *et al.,* 1971), a result demonstrating that magnesium also acts as a calcium antagonist in this endocrine system. Verapamil, likewise, reduces the release of calcitonin from the rat thyroid (Ramp *et al.,* 1979), and this result supports the contention that calcium entry into the C cell is a prelude to calcitonin release. These findings, taken together with the knowledge that barium and strontium stimulate calcitonin secretion (Pento, 1974; Cooper, 1975), are concordant with the proposal that the mechanism of calcitonin secretion conforms to the stimulus–secretion coupling model.

A relatively homogeneous population of calcitonin-producing cells found in avian and fish ultimobranchial glands, which persists as a separate gland in submammalian vertebrates, has facilitated exploration of the mechanism of calcitonin production and secretion. For example, trout ultimobranchial glands treated with proteolytic enzymes yield enriched fractions of viable C cells, which can be maintained in culture and respond briskly to various peptide hormones. However, in contrast to the potent stimulatory effect of calcium in human C cells, calcium elicits only a slight biphasic response in trout cells (Roos and Deftos, 1976), suggesting that there may be differences in regulation of C-cell function in higher and lower vertebrates.

These differences may be a reflection of the inability of calcium to permeate the invertebrate C cell, thereby creating the need for a trophic hormone, neurotransmitter, or autacoid to facilitate calcium influx. Calcium alone is a sufficient stimulus to trigger secretion in such secretory organs as the adrenal medulla and endocrine pancreas when a prior period of calcium deprivation promotes its accumulation within the cell. Consequently, the exquisite sensitivity of the mammalian C cells to small changes in the plasma calcium concentration may be ascribed to an increase in the amount of calcium that enters the cell to reestablish equilibrium conditions. Such speculation is supported by the fact that a prior period of hypocalcemia—producing augmented membrane permeability—causes an even higher rate of calcitonin secretion upon a given hypercalcemic challenge (Munson, 1976).

The findings that glucagon, pentagastrin, and the closely related cholecystokinin stimulate calcitonin secretion in certain experimental animals, in tumor cell cultures, and in the Zollinger–Ellison syndrome (gastrin-secreting tumors), taken together with a postprandial increase in calcitonin secretion in experimental animals that is not related to changes in the calcium concentration (cf. Deftos et al., 1978), forge a functional link between the gastrointestinal tract and calcitonin secretion. Knowledge that patients with gastrinomas respond to a calcium challenge with a brisk increase in gastric secretion implies that at least part of the effect of high calcium on calcitonin secretion may be exerted through the stimulated release of gastrointestinal peptides (Deftos et al., 1978; Gagel et al., 1980). The additional finding that somatostatin inhibits calcitonin secretion in various mammalian preparations both in vivo and in vitro (Hargis et al., 1978; Linehan et al., 1979), not only reinforces the association between gut peptides and calcitonin release (cf. Munson, 1976), but also substantiates the proposal that the mechanism of calcitonin release conforms in terms of its basic properties to the stimulus–secretion coupling model, with calcium providing the link between the membrane events triggered by the primary stimulus and the discharge of calcitonin from the C cell.

Yet, it is clear from the evidence obtained using *in vitro* systems that calcium alone is capable of exerting a direct action on the C cell. Moreover, the increase in calcitonin secretion induced in humans is correlated with changes in serum calcium levels, but not with changes in peptide (gastrin) levels (Austin *et al.*, 1979). The inescapable conclusion is, therefore, that calcium is the principal physiological regulator of calcitonin secretion, at least in man.

To sum up, since the primary function of calcitonin and parathyroid hormone is to maintain the extracellular calcium concentration within rather narrow limits, the rate of secretion of these hormones is regulated in a feedback manner by the concentration of ionized calcium impinging on these cells. While the calcium-sensitive receptor within the parathyroid cell may be somewhat unique, the calcitonin-secreting parafollicular cells of the thyroid gland seem to behave in a relatively orthodox manner, with the exception that these cells are more readily able to utilize calcium as a sufficient stimulus, without the interposition of a primary signal.

While systematic studies of calcitonin biosynthesis in mammals have been few because of the relative lack of homogeneous mammalian C-cell systems, it is clear that calcium acts directly on the secretory process and not through an indirect action on calcitonin biosynthesis. A direct effect of calcium on calcitonin release is indicated by the rapidity of the release response. This response is associated with a decrease in the calcitonin content of the thyroid gland, implying that in C cells, which are equipped for short term periods of activity, synthesis does not keep up with demand (Munson, 1976). By contrast, a more important role for calcium exists in the formation and degradation of parathyroid hormone, which teleologically is crucial for sustaining the more prolonged actions of the parathyroid gland on calcium homeostasis. Further insight into the cellular mechanisms by which calcium expresses its positive and negative control over these two endocrine organs awaits the advent of a greater volume of experimental data derived from homogeneous cell populations.

8. Renin

We now come to the final system which we will consider, the renin-secreting juxtaglomerular cells of the kidney, which in light of the fact that they are modified myoepithelial cells closely related to smooth muscle cells, might connote an action of calcium on renin release that parallels the property of this cation to induce muscle contraction. However, the existing evidence in this regard is conflicting; in fact, calcium may act as a negative modulator of renin secretion through a mechanism more akin to that which mediates muscular relaxation.

Renin was originally described as a pressor substance of renal origin, but is actually the enzyme catalyzing the first step in the formation of another vasoactive hormone, angiotensin II. The juxtaglomerular cells, which synthesize, store, and secrete renin are usually found in the media of the renal afferent arterioles, just adjacent to the glomerulus. They have well-developed endoplasmic reticulum and Golgi membranes, and possess cytological characteristics that are consistent with an endocrine function. A characteristic ultrastructural feature of the juxtaglomerular cells is the presence of membrane-bound granules in the cytosol, although certain of these cells possess smaller protogranules and others appear devoid of granules (Barajas, 1979). Studies utilizing gradient centrifugation confirm that renin is stored in these granules, possibly in an inactive form (Morimoto *et al.*, 1975; Morris and Johnston, 1976). The relation of the granules to renin secretion, however, is still unclear. There is no definitive evidence for exocytotic release, and it has been suggested that renin release occurs by a mechanism similar to that proposed for the thyroid gland involving fusion of lysosomes with the secretory granule (Morris and Johnston, 1976); however, exocytotic release is still a viable possibility.

A prerequisite for a subsequent discussion of the cellular events associated with this secretory process is an understanding of the physiological factors controlling renin release, which include renal perfusion pressure, renal sympathetic tone and plasma catecholamine levels, and sodium load and/or concentration at the macula densa (Davis and Freeman, 1976). Renin secretion is stimulated by a decrease in renal perfusion pressure and inhibited by an increase in renal perfusion pressure; these effects are mediated through an intrarenal baroreceptor in the afferent arterioles. Sympathetic nerves innervate the juxtaglomerular cells of vertebrates, and morphological evidence exists for a close functional relationship between the sympathetic nervous system and renin secretion mediated through a β-1 adrenoceptor mechanism (Kopp *et al.*, 1980) that operates in the absence of any functional or hemodynamic changes in the kidney (Keeton and Campbell, 1981). Thus, catecholamines released from sympathetic nerve endings or from the adrenal medulla promote renin secretion through the direct activation of β-adrenoceptors on the juxtaglomerular cell. Renin release is also inversely related to the sodium load in the renal tubules. This mechanism is controlled by the macula densa region of the distal tubule, which is in close contact with the juxtaglomerular cells.

So in attempting to delineate the role of calcium in renin secretion, there are complicating factors which could hamper our progress along these lines. For example, since calcium can also inhibit sodium reabsorption, alter catecholamine release, and affect vascular tone, it may be difficult to assess the direct role of calcium in renin secretion. As a consequence, it is necessary to compare the effects of calcium on renin release in *in vivo* models with effects on isolated *in vitro* systems, such as cortical slices, cell suspensions, and isolated glomeruli, where influences of glomerular filtration and tubular transport of ions can be circumvented. To explore the effects of calcium ions on renin secretion without the influences of changes in systemic hemodynamics, renal nerve activity, or circulating hormones, isolated perfused kidneys have also been utilized. In addition to employing various types of preparations, the effect of calcium has been investigated on both "basal" and stimulated renin secretion, and a clear distinction must be made between the effects of calcium on these two parameters.

Studies employing *in vivo* models have revealed that calcium infusions can either increase, decrease, or cause no change in basal renin secretion (Keeton and Campbell, 1981); on the other hand, an increase in renin secretion was found when isolated perfused kidneys, isolated superfused glomeruli and renal cortical slices were exposed to a calcium-deprived medium (Van Dongen and Peart, 1974; Baumbach and Leyssac, 1977; Park and Malvin, 1978). The additional findings that the calcium ionophore A23187 is more effective as a renin secretagogue in calcium-deficient media (Baumbach and Leyssac, 1977; Fynn *et al.*, 1977), and that magnesium, a well-known inhibitor of calcium, enhances renin secretion by an action that is nullified by calcium (Churchill and Lyons, 1976; Wilcox, 1978; Ettienne and Fray, 1979) support the supposition that calcium exerts an inhibitory action on renin release.

This is a plausible mechanism to account for these results particularly when additional findings are considered which bear on this problem. Thus, renin release is abolished when cortical slices are incubated in high potassium medium which depolarizes juxtaglomerular cells (Fishman, 1976) and this inhibitory action of potassium is reduced by the removal of calcium from the medium or by the addition of D-600 (Park and Malvin, 1978; Churchill, 1980) (Fig. 25). Such data suggest that depolarization opens voltage-dependent calcium channels of the juxtaglomerular cell and that the entry of calcium causes a *decrease* in

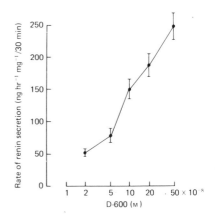

FIGURE 25. Stimulatory effect of D-600 on renin release from rat renal cortical slices incubated in high potassium. High potassium (60 mM) reduces renin secretion to near control levels, and, in the presence of increasing concentrations of D-600, the secretion rate is restored to a value not significantly different from the uninhibited rate of basal secretion (Churchill, 1980).

renin release. Moreover, ouabain, an inhibitor of the sodium–potassium pump and a secretagogue in most other secretory systems diminishes basal renin secretion, as does potassium deprivation (Churchill and Churchill, 1980).

Since the inhibition of secretion produced by these experimental manipulations depends upon the presence of calcium in the medium, an attractive postulate meriting consideration is that inhibition of the sodium pump causes an intracellular accumulation of sodium, leading to an increase in cellular calcium either by the inhibition of calcium efflux or the increase in calcium influx (Churchill, 1979). The resulting increase in intracellular calcium culminates in the reduction of renin secretion. A link between the sodium pump and renin secretion is strengthened by the finding that pharmacologic agents which stimulate ATPase activity enhance renin secretion (Churchill *et al.*, 1979).

While these data offer strong support for the concept that, as in the parathyroid gland, the entry of calcium into the juxtaglomerular cell exerts an inhibitory rather than stimulatory effect on the discharge of secretory product, this is not the complete story. Apart from its inhibitory action, calcium can act as a sufficient stimulus for renin secretion when restored to the medium after a period of calcium deprivation (Chen and Poisner, 1976; Lester and Rubin, 1977). Moreover, ionophore A23187 stimulates renin secretion from isolated feline glomeruli if the ionophore and calcium concentrations are sufficiently low and only inhibits it when the concentrations of stimulant are relatively high (Harada *et al.*, 1979). On the basis of such studies, it has been proposed that an intracellular calcium receptor similar to that found in other secretory cells is also present in the juxtaglomerular cell.

The question of the significance of spontaneous release as opposed to evoked release has not been considered in depth up to now; however, equating spontaneous release with stimulated release invokes a dubious premise, particularly since spontaneous release may not occur by exocytosis, as has been demonstrated in the glucagon-secreting α cells of the pancreas (Carpentier *et al.*, 1977). Secretory organs, for the most part, are only sporadically active, responding when the occasion arises to the appropriate stimulus and then returning to a resting state of activity. Consequently, analysis of the role of calcium in the physiological mechanism of secretion of a given secretory organ can only be legitimately assessed when physiological stimuli are employed to enhance secretory activity.

Accordingly, the stimulant action of catecholamines on renin secretion does not depend upon the presence of extracellular calcium (Lester and Rubin, 1977). Neither verapamil nor manganese alters the response to isoproterenol in isolated perfused rat kidney (Logan and Chatzilias, 1980), and D-600 is unable to block the first phase of isoproterenol-induced renin release from the perfused cat kidney, but blocks a second phase which is thought to involve the mobilization or synthesis of renin (Lester and Rubin, 1977). The ability of catecholamines to hyperpolarize the juxtaglomerular cell (Fishman, 1976) is also consistent with the lack of an inwardly-directed calcium current and with the inability of channel blockers to affect the response to catecholamines. It is thus agreed that a net influx of calcium into the juxtaglomerular cell is not an obligatory requirement for eliciting renin release to β-adrenergic receptor stimulation and, in fact, may inhibit renin secretion (Churchill and Churchill, 1980).

Since recognition of the signal generated by β-adrenergic receptor involves cyclic AMP as a mediator (Robison *et al.*, 1971), it is mandatory to consider the role of cyclic AMP in catecholamine-induced renin secretion. While the evidence that has accumulated on this general subject mainly relates to the stimulant actions of cyclic AMP or its derivatives and phosphodiesterase inhibitors, due to the lack of an adequate isolated juxtaglomerular cell system required to appropriately explore this problem, a role for cyclic AMP in renin secretion has not been definitively established (Davis and Freeman, 1976; Keeton and Campbell, 1981).

A larger volume of definitive evidence actually exists for a mediator role for the prostaglandins in the mechanism of renin release (Franco-Saenz *et al.*, 1980a; Keeton and Campbell, 1981). Intrarenal infusion of prostaglandins or the prostaglandin precursor arachidonic acid stimulates renin release in dogs, rabbits, rats, and humans through a direct effect on the juxtaglomerular cell (Osborne et al., 1978). Nonsteroidal antiinflammatory drugs such as indomethacin, which block prostaglandin synthesis, also block the increase in renin secretion caused by isoproterenol, further supporting a role for prostaglandins in sympathetic-mediated renin secretion. Moreover, in Bartter's syndrome, which is associated with an abnormally high production of prostaglandins, there is an increase in renin secretion and hyperaldosteronism, and many of the abnormalities associated with this syndrome can be corrected by nonsteroidal antiinflammatory agents. This dependence of excess renin secretion on prostaglandin production can be interpreted as strong evidence in favor of a pivotal role for prostaglandins in the regulation of renin secretion.

Of all of the prostaglandins investigated, PGE_2 and PGI_2 (prostacyclin) are regarded as the most likely candidates for mediating renin secretion; and the finding that PGI_2 elevates renin release by cortical slices (Whorton et al., 1977) affirms that prostaglandins act directly on the juxtaglomerular cell. Whether the juxtaglomerular cell is capable of synthesizing PGI_2 remains an open question. But juxtaglomerular cells are modified vascular elements, and PGI_2 is produced by vascular tissue; it would, therefore, seem likely that the juxtaglomerular cells have the capacity to synthesize prostaglandins.

A possible interrelationship between the prostaglandins and cyclic AMP is indicated by the fact the theophylline plus subthreshold doses of PGE_2 increase renin secretion from a superfused system of cortical slices (Franco-Saenz et al., 1980b). The renin-releasing action of PGE_2 may be mediated through a specific prostaglandin receptor in the juxtaglomerular cell that activates adenylate cyclase, enabling cyclic AMP to trigger renin secretion. Prostaglandins are capable of stimulating cyclic AMP formation in other secretory organs, so that the mechanism by which the prostaglandins enhance renin secretion may well involve cyclic AMP, particularly since cyclic AMP is an adequate stimulus for renin secretion in certain test preparations (Franco-Saenz et al., 1980b).

One site of calcium action in this proposed sequence of events is in the formation of renal prostaglandins (Zenser and Davis, 1978), a reaction catalyzed by the activation of a calcium-dependent phospholipase A_2 to increase the available substrate (arachidonic acid) (Zusman and Keiser, 1977; Zenser et al., 1980). So as we have seen in the adrenal cortex (Fig. 23), one locus of calcium action may be at the cell surface to control phospholipase A_2 and adenylate cyclase activities, which may be key elements in the mechanism of renin secretion by catalyzing the formation of prostaglandins and cyclic AMP, respectively.

In the parotid gland and adrenal cortex, the actions of cyclic AMP and prostaglandins are expressed through the mobilization of cellular pools of calcium. Whether a similar scenario ensues in the juxtaglomerular cell remains to be answered. The finding that catecholamine-induced renin release is associated with calcium efflux from the perfused cat kidney even after prolonged calcium deprivation (Harada and Rubin, 1978) provides support for the concept that the process of catecholamine-induced renin secretion involves the mobilization of calcium from a cellular site, perhaps provoked by cyclic AMP or the prostaglandins. This postulate, if true, consigns the juxtaglomerular cell to the camp of secretory systems that utilize intracellular calcium to activate secretion.

On the other hand, acknowledging the significant volume of evidence supporting the concept that calcium acts as a negative modulator of renin release, the catecholamine-induced calcium efflux may reflect the stimulation of the sodium–potassium pump with the resulting decrease in the intracellular sodium concentration bringing about net calcium extrusion via sodium–calcium exchange. This *decrease* in the intracellular calcium concentration may be responsible for triggering renin secretion. Since the juxtaglomerular cell can be described as a smooth muscle cell modified for secretory activity, one may envision a similar mechanism operative to effect catecholamine-induced relaxation of smooth muscle (Scheid *et al.*, 1979), although a cyclic-AMP-mediated increase in calcium sequestration, rather than a net efflux of calcium, may be the basis for this mechanism (Mueller and Van Breemen, 1979). The relationship between secretion and contraction becomes even more apparent when one considers the very close parallels between the role of calcium in stimulus–secretion coupling and in excitation–contraction coupling—a topic which we will consider in the following chapter.

The question of whether the disconcerting differences regarding the role of calcium in renin release reflect species variability (rat vs. cat), experimental variability, or inherent limitations of the experimental preparations still must be dealt with, but it is clear that as more homogeneous cell preparations have been developed for probing secretory systems, a better understanding of the complex interactions of various putative mediators, including calcium, has been obtained. Consequently, the cellular mechanisms regulating renin secretion at the cellular level will only be clearly defined after isolated cell preparations have been more effectively utilized. But for the time being at least, consideration must be committed to the proposition that in the juxtaglomerular cell, calcium acts as a negative modulator of secretion just as it does in the parathyroid gland (Fray, 1980).

J. PERSPECTIVE

We have now completed our comprehensive survey of the secretory systems, and a pattern has emerged that implicates calcium as a critical messenger in the mechanism involved in the discharge of product from the secretory cell. An increase in the calcium concentration of a cell specialized for secretion inevitably leads, in most cases, to the discharge of secretory product. The accumulation of calcium within the cell may be brought about by diverse mechanisms involving calcium entry from the extracellular medium through voltage-dependent or receptor-operated channels in the plasma membrane. The mobilization of calcium bound to the cell membrane may also be involved in triggering the secretory response. Calcium may also be released from cellular stores through the auspices of cellular messengers. Alternatively, intracellular calcium may be released by the entry of calcium from the cell membrane or the extracellular medium. Even within a given secretory cell, the duration of stimulation or the type of receptor activated dictates the

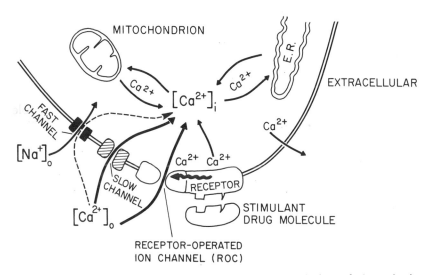

FIGURE 26. A schematic representation of the redistribution of calcium during activation of the secretory process. Following stimulation, calcium enters excitable cells through voltage-dependent "slow" channels and inexcitable cells through receptor-operated channels (ROC). Calcium may also be mobilized from the plasma membrane, or with the aid of intracellular mediators from the mitochondria or the endoplasmic reticulum (ER). The free calcium concentration is restored to prestimulation levels by sequestration into intracellular organelles or by extrusion from the cell.

manner in which calcium is mobilized. But, despite the nuances that exist with regard to the manner in which calcium is handled, a common thread runs through each system linking each of them to the action of calcium (Fig. 26).

Attempts to define a secretory system in which calcium does not serve any role in the control mechanisms governing secretion have generally proven futile with the possible exception of the thyroid gland, which seemingly possesses a rather unique secretory mechanism. In the adrenal cortex there is also no convincing evidence that calcium plays a direct role in the machinery directly involved in the extrusion of corticoids, although it is clear that calcium plays a critical role in the actions of corticotropin in bringing about the synthesis and possibly the mobilization of hormone. But since it is not possible to dissociate the effects of calcium on corticosteroid synthesis from those on secretion, a strong claim for the participation of calcium in the process regulating steroid release cannot be made.

With regard to the so-called "inhibitory" role of calcium in parathyroid hormone and renin secretion, we may attribute some fundamental difference in the properties of the intracellular calcium receptor that triggers their discharge. It may also be constructive to consider that the manner in which calcium is handled by the cell enables the free intracellular calcium to attain levels that exert inhibitory effects. However, such speculation will have to be scrutinized more carefully with more sophisticated methodologies than are presently being utilized.

But, as mentioned in Chapter 1, the fact that exceptions exist to a fundamental concept should not cause us to summarily reject it. Apart from certain exceptions, the evidence that we have considered for these many pages validates the theory that calcium is the critical coupling agent in linking the membrane events associated with activation of the cell by the primary stimulus with the cellular events that culminate in the discharge of secretory product. The final two chapters of this volume will be devoted to a consideration of the cellular events by which calcium is involved in mediating the secretory response.

CHAPTER 3

Nature of the Role of
Calcium in Secretion

A. MEMBRANE FUSION

The large body of evidence that we have considered links the pivotal
action of calcium on the secretory process to exocytosis, which is a
fusion–fission response involving the interaction of the plasma membrane and the secretory granule membrane. Membrane fusion is not
confined to secretory cells but is a fundamental process in biological
systems, involving the interaction of membranes of two different cells
or membranes of the same cell. Cell fusion is observed in the presence
of calcium, barium, or strontium, but not magnesium (Poste and Allison,
1973), strengthening the relationship between cell fusion and secretion
and implying that elucidation of the basic and fundamental action of
calcium in secretion might be sought in a better understanding of membrane fusion phenomena. While Palade (1959) was the first to recognize
the importance of membrane fusion in the activation of the secretory
process, DelCastillo and Katz proposed that there are specific and com-
plementary sites on the synaptic vesicles and axonal membrane such
that only when the two sites interact can membrane fusion and release
of transmitter occur (Katz, 1969).

The mechanisms of membrane–membrane recognition and subsequent fusion and fission are complex. Recognition or "receptor" sites,
which may be either lipid or protein in nature, must be located asym-
metrically on interacting membranes. The existence of lipid asymmetry,
with phosphatidylserine, for example, on the inner aspect of the plasma

membrane while neutral lipids such as phosphatidylcholine are localized on the outer aspect, therefore, may have important implications for fusion (Chap *et al.*, 1977; Rothman and Lenard, 1977). Another interesting point relates to the selectivity of only certain membranes to fuse rather than a generalized fusion of all cellular membranes; insight into this particular aspect of the problem may provide a better understanding of the mechanism of membrane fusion as it applies to exocytosis.

Early proposals for a general mechanism of membrane fusion such as the induction of membrane instability via removal of calcium and the hydrolysis of ATP from membranes (Poste and Allison, 1973) afford some insight into the mechanism of cell fusion, but do not provide detailed information at the molecular level. The most specific evidence concerning the molecular mechanism of membrane fusion derives from the work of Papahadjopoulos and his associates on the fusion of model membranes (Papahadjopoulos, 1978; Papahadjopoulos *et al.*, 1978). Using liposomes containing various phospholipids so that the membrane composition and ionic environment can be controlled, the key event in the fusion of model membranes was determined to be the interaction of calcium and acidic phospholipids. Calcium-induced lateral phase separation of acidic phospholipids within a mixed lipid membrane creates an unstable, highly permeable membrane susceptible to fusion with other similarly perturbed membranes. According to this view then, the fusion susceptibility of a given membrane will be determined by: (1) the concentration of calcium in the vicinity of the membrane and (2) the content and distribution of acidic phospholipids within the lipid bilayer. While the use of artificial membranes has afforded us greater insight into the mechanism of membrane fusion, its relevance to exocytotic secretion remains problematic, because these models are incomplete from a biological standpoint and the mechanism involved in the fusion of artificial membranes may not necessarily be relevant to exocytosis *in vivo*.

When considering exocytotic secretion, several key points ought to be taken into account:

1. Two distinct events are associated with exocytosis, i.e. mobilization of the secretory granules to the cell membrane and the actual fusion–fission process itself.
2. The membrane component of the secretory granules are distinct from the membrane components of the plasma membrane and any proposed mechanism involving membrane fusion must acknowledge the heterogeneity of the two membranes which fuse.

3. Lipid and protein membrane components comprise key elements in the fusion–fission process.

4. Calcium and ATP also play critical roles and without either of these two critical components membrane fusion would not occur.

So the molecular basis of action of calcium on the secretory process can be viewed from a number of different perspectives, none of which is mutually exclusive. The perspectives encompass ATP-dependent events possibly involving the cytoskeletal system in delivering the granules to the cell membrane, and the interaction of calcium with specific cellular proteins or with lipid-metabolizing enzymes. A common theme underlying our impending discussion is that the basic mechanism regulating the secretory process may involve contractile events corresponding to those operating in muscle.

B. ENERGY METABOLISM AND SECRETION

When considering the role of calcium in the molecular mechanism of stimulus-evoked secretion, it is also necessary to include ATP as an essential factor. In fact, in certain secretory systems, such as the mast cell and endocrine pancreas, ATP is one of many agonists which stimulates calcium-dependent secretion by an action on the cell surface (Loubatieres-Mariani et al., 1979; Cockcroft and Gomperts, 1979). This effect of ATP, though occurring at low concentrations (micromolar range), does not directly bear on the findings of an energy requirement for exocytosis. The ability of the secretory cell to produce metabolic energy, which can be supplied by either anaerobic glycolysis or oxidative metabolism, closely correlates with the activity of calcium as a secretagogue, and a quantitative relation between the calcium and cellular ATP concentrations is critical for triggering hormone release (Rubin, 1970, 1974a) (Fig. 27).

An energy-requiring chemical reaction would necessitate the splitting of ATP. As a consequence calcium-magnesium ATPase has been sought and found in the membrane of secretory organelles from a variety of cells (Douglas, 1974a; Michaelson and Ophir, 1980), and secretory responses have been correlated with tissue ATP levels (Johansen and Chakravarty, 1972; Holmsen, 1975). Moreover, ATP stimulates release of secretory product from a variety of isolated secretory vesicles (Oberg and Robinovitch, 1980). This releasing activity involves granule lysis. These experiments, taken together with other findings, prompted the

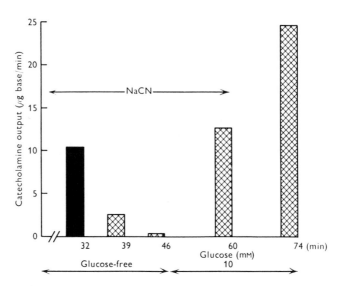

FIGURE 27. The effect of inhibition of energy metabolism on the secretory response of the perfused cat adrenal gland to calcium. Adrenal glands perfused with glucose-free, calcium-free, high potassium solution plus sodium cyanide (0.2 mM) were exposed to calcium for 2-min periods. Control responses were subsequently obtained after glucose was added and cyanide removed from the perfusion medium. The vertical columns depict catecholamine outputs obtained in response to calcium (■ 0.5 mM) and (▩ 3.0 mM). (Rubin, 1970).

proposal that ATP causes the granules to lyse by serving as a substrate for an inwardly directed proton-pumping ATPase, which secondarily generates an osmotic ion gradient across the granule membrane (Pollard *et al.*, 1979). According to this view, ATP may serve *in vivo* to prime secretory granules for the fusion–fission response by maintaining an osmotic ion gradient across the secretory granule membrane.

Although ATP and ATPase induce release of product from isolated granule preparations *in vitro*, calcium does not produce a consistent and/ or marked effect on the release of secretory product from various isolated granule preparations, suggesting that the primary locus of the calcium-ATP interaction is not the secretory granule (cf. Rubin, 1974a). On the other hand, calcium does promote the fusion of secretory vesicles with one another (Thorn *et al.*, 1978). Exocytosis not only involves interaction of the plasma and granule membranes, but in certain secretory cells, including the mast cell, salivary acinar and adenohypophysial cells, vesicles fuse with one another to form channels which open to the cell surface for discharging secretory product (Douglas 1974a; Meldolesi *et al.*, 1978). Hence, the process of vesicle aggregation has been presented as a valid model for the initial interaction that occurs between vesicles and the plasma membrane during exocytosis. Isolated adrenal chromaffin granules aggregate in the presence of high levels of calcium. This finding, taken together with the discovery that a calcium-binding protein called synexin promotes calcium-dependent fusion of chromaffin granule membranes, prompted speculation that synexin may be the intracellular receptor for calcium in the process of exocytosis (Pollard *et al.*, 1979, 1981). The location of synexin-binding sites on isolated chromaffin granules and on the inner aspect of plasma membrane fragments of isolated medullary chromaffin cells satisfies a critical requirement for a protein presumed to be involved in the regulation of membrane recognition and fusion during exocytosis (Scott *et al.*, 1982). This theory, however, suffers from some weaknesses, particularly that strontium and barium, which are able to replace calcium in mediating catecholamine release from the adrenal medulla, do not support synexin-mediated granule aggregation.

Reconstituted *in vitro* secretory systems comprised of secretory granules and plasmalemma may provide a better model system for studying exocytotic events. Davis and Lazarus (1976) developed a system of mouse islet granules and cod islet plasma membranes that exhibits calcium-dependent glucose-induced insulin release. ATP also acted as a secretagogue but the most striking finding was that glucose, ATP and calcium together acted in a cooperative fashion, releasing most of the stored insulin. Validation of such findings and the use of similar models in other secretory systems will clearly focus on the critical factors in exocytosis (see also Milutinovic *et al.*, 1977).

While energy is clearly required for certain steps in stimulus–secretion coupling, as well as being needed to maintain the general function of the cell, the critical question is whether hydrolysis of ATP is required for membrane fusion itself or for some prior step in the sequence. An ATPase stimulated by calcium was selectively localized by cytochemical methods at exocytotic sites in *Paramecia* (Plattner, 1978; Plattner *et al.*, 1980). Extending the analysis to mutants revealed that attachment of vesicles does not require ATPase but discharge of product does, implying that local ATP hydrolysis directly promotes fusion by "destabilizing" membranes. This type of analysis, if it can be extrapolated to exocytotic events occurring in higher organisms, should help to elucidate the role of ATP in the secretory process and considerably enhance our understanding of the cellular mechanism of calcium's action in secretion.

Before leaving the discussion of the role of metabolic energy in the secretory process, it must be mentioned that whereas metabolic energy is required for stimulated secretion, spontaneous secretion is enhanced during inhibition of metabolism (Rubin, 1969; Rahamimoff, 1976). This seemingly paradoxical effect is readily interpretable on the basis of the release of calcium from ATP-dependent intracellular binding sites. The resulting increase in the free intracellular calcium concentration will, of course, promote spontaneous release. The interesting but unanswered question is whether this release is also exocytotic.

C. MICROTUBULE, MICROFILAMENT SYSTEM

A wide variety of eukaryotic cells contain filamentous structures (a) microfilaments, which are mainly composed of actin-like elements, and (b) microtubules, which are hollow tubes consisting mainly of the protein tubulin. These proteins are thought to mediate various types of cell movement such as protoplasmic streaming, as well as active movements of cell organelles, including secretory granules. The motility phenomena whether they be cell shape changes, muscle contraction, or secretion all appear to have features in common (Allison, 1973; Hitchcock, 1977).

Microtubules are prevalent wherever streaming occurs, but whether they provide directionality and a supportive function for the movement of intracellular organelles has still not been established. The purported role of the microtubules in secretory activity is based mainly upon pharmacological evidence in that agents such as colchicine and vinblastine, which bind to tubulin and disrupt microtubules, inhibit the release response. Lacy and Malaisse are particularly strong proponents of the theory that secretory granules are attached to microtubules and that calcium triggers contraction or a change in the physical conformation of the microtubule system which results in granule discharge (Lacy and Malaisse, 1975; Malaisse et al., 1975). Although the response of secretory cells seems to be governed by the state of polymerization of the microtubules, it appears that colchicine inhibits the secretory process mainly by disrupting the intracellular transport of protein rather than by interfering with the discharge step (Chambaut-Guerin et al., 1978). Moreover, the inconsistent and variable effects of microtubule inhibitors on the secretory response do not support the concept that the microtubules play a direct role in extruding secretory product (Pecot-Dechavassine, 1976). The suggestion that calcium regulates secretory activity by an effect on microtubule assembly seems unlikely because low concentrations of calcium depolymerize microtubules *in vitro* (see Butcher, 1978). On the other hand, supraphysiological concentrations of calcium are required to promote microtubule assembly. It is therefore more reasonable to consider that microtubules serve to translocate the secretory granule from its site of packaging in the Golgi apparatus to the periphery of the cell (Patzelt *et al.*, 1977). Alternatively, rather than exerting a dynamic effect on cell motility, microtubules may function primarily as skeletal structures, preserving the structural integrity of the cell.

The association of actin with the microfilaments has given rise to the possible involvement of this component of the cytoskeletal system in the secretory process because of the possible functional relationship between secretion and contraction. However, the effects of microfilament-active drugs on secretory cells are extremely variable. Cytochalasin B, a fungal metabolite which disrupts microfilaments, may inhibit or enhance the secretory response, or may even have no effect, depending upon the system studied (Allison and Davies, 1973; Nemeth and Douglas, 1978). Moreover, the specificity of action of these drugs has been open to question for many years, and several of them are general metabolic inhibitors and have profound effects on the overall functionality of the secretory cell (Allison and Davies, 1973; Meldolesi *et al.*, 1978). Analysis of distribution patterns of microfilaments suggests other possible roles in cellular function. Actin-containing filaments are usually distributed in areas just beneath or associated with the plasma membrane (Clark and Spudich, 1977; Gröschel-Stewart, 1980), so that the microfilament system may be involved in the regulation of cell surface receptor distribution and mobility.

In conclusion, it is difficult to reconcile all of the published data with the postulate that the microtubules provide a framework to direct the granules towards the plasma membrane, and the microfilaments, powered by an actomyosin-induced contraction, provide the motive force for granule extrusion. The microtubules and microfilaments more likely exert their effects on mechanisms involved with events that are a prelude to secretion such as the transmission of the signal from the receptor or the packaging and mobilization of granules.

D.   SECRETION AS A CONTRACTILE EVENT

The term stimulus–secretion coupling was originally proposed to emphasize the correlation between the action of calcium on secretory process and its action on excitation–contraction coupling of muscle.

These analogies have been written about extensively, and the striking parallelisms between the processes regulating secretion and those regulating contraction have given rise to the concept that a common mechanism is involved in the activation of both of these fundamental biological processes (Douglas, 1968; Rubin, 1974a). In addition to the fact that a common denominator in secretion and contraction appears to be an intracellular accumulation or translocation of calcium ions, the properties of both calcium-dependent activation mechanisms are quite similar in that of all other cations tested only strontium and barium are able to stimulate directly both processes, whereas magnesium inhibits the action of calcium in both systems. Further analogies can be made with regard to the metabolic events involved in these two processes. Contraction, like secretion, requires both ATP and calcium ions, and splitting of ATP by ATPase is required in both processes. The most valid conclusion that emerges is that a similar mechanism exists for triggering these two processes which involves the interaction of calcium, ATP, and ATPase.

Alan Poisner (1973), an early proponent of the concept of secretion as a contractile event, proposed that ATP released from the plasma membrane serves as a substrate for ATPase in the secretory granule membrane and initiates a contractile event just as it does in the muscle fiber. The findings that ATP, but not calcium, releases secretory product from isolated secretory granules are consistent with this concept. The adherence of the granule to the plasma membrane in the presence of calcium and ATP is believed to initiate the contractile event.

This model of exocytosis involves the presence of contractile proteins in the secretory granule membrane as well as in the plasma membrane. Consistent with this theory is the finding that contractile proteins, which were once thought to be localized exclusively in muscle, have been isolated from secretory cells and even localized to the secretory granule (Trifaro, 1977). The oft-quoted postulate that exocytosis is a contractile event involving fusion between vesicular myosin and synaptic membrane actin (Berl et al., 1973) is deemed highly unlikely (Trifaro, 1977), particularly since actin, not myosin, has been localized to secretory vesicle membranes (Meyer and Burger, 1979). Tropomyosin and troponin are also present in brain and adrenal medulla, although the relationship of troponin to the structurally similar calcium regulatory protein, calmodulin, remains to be determined (Trifaro, 1977).

While calcium may regulate exocytosis by acting on the troponin–tropomyosin system in the cytoplasm of a secretory cell, the caveat here is that in the animal kingdom, two distinctly different calcium regulatory systems exist for triggering muscle contraction (Lehman,

1976). In skeletal muscle, troponin C controls calcium sensitivity of the contractile system in that calcium binding to troponin C alters its conformation, thereby modifying the position of tropomyosin; this allows actin to interact with myosin resulting in ATP hydrolysis and muscle contraction. In smooth muscle and nonmuscle cells troponin is not present and calcium binds directly to myosin to stimulate contraction. Thus different systems have evolved for regulating contraction, and as yet we have no idea how the biochemical effects of calcium on the secretory apparatus compare with those on muscle contraction.

The contractile proteins of nonmuscle cells possess physico-chemical properties that resemble those found in muscle; and nonmuscle actins will activate the Mg-ATPase of muscle myosin (Huxley, 1978). If secretion involves a sliding filament mechanism akin to muscle contraction, calcium may act to dissociate cross-linked actin filaments, which after interacting with myosin, are free to slide, causing contractile activity (Huxley, 1976). Apart from the actin–myosin interactions, polymerization–depolymerization of the contractile proteins and cytoplasmic gel formation might also be responsible for structural changes associated with intracellular movement. Nonmuscle actin, when polymerized in the presence of actin-binding proteins, forms gels which may alter the properties of the cytoplasm (Taylor *et al.*, 1978). The dissolution of actin-related gels by calcium raises the intriguing possibility that calcium, by affecting actin–actin interactions, causes solation of the cytoplasm which promotes the interaction of the secretory granule with the plasma membrane.

Yet, there are still insufficient grounds for tacitly regarding the secretory cell as a modified contractile element. As a matter of fact, the properties of nonmuscle actin differ from those of muscle actin in certain basic respects. For example, nonmuscle actins are not always organized into filaments, as are muscle actins, but exist in a nonfilamentous form which only polymerizes into filaments during cellular activity (Hatano and Oosawa, 1978). We, therefore, cannot exclude the possibility that the function of the actomyosin system in secretory cells may simply be related to cell architecture or to specialized functions as might exist in the platelet or neutrophil (i.e., clot retraction, chemotaxis), rather than to the secretory process itself. A direct action of calcium on the actomyosin system of secretory cells hinges on the attainment of convincing evidence that actomyosin functions in the generation of the force required in the various expressions of the exocytotic process, whether by directly promoting calcium-dependent membrane fusion, or by participating in the intracellular movements of vesicles.

E. CALCIUM-BINDING PROTEINS

In our survey of secretory systems we have alluded to the intra-cellular calcium "receptor" which appears to be universally present in secretory tissue and is responsible for triggering the secretory response. But because of the dearth of information regarding the identity and chemical nature of this receptor, the cellular mechanism underlying the action of calcium remains unknown. The pronouncement by Kretsinger (1975) that the processes in which calcium functions as a second mes-senger are mediated by calcium-binding proteins was upheld by the discovery of calmodulin, a ubiquitous calcium binding protein that reg-ulates some calcium-mediated cellular processes. During the late 1960's and early 1970's, Cheung and Kakiuchi (see Cheung, 1979) independ-ently discovered that during the purification of brain phosphodiesterase, enzyme activity was lost unless calcium and another protein were also present. On the basis of these studies, a calcium-binding protein was identified as an activator of phosphodiesterase.

However, phosphodiesterase is not unique in its dependence on this protein factor, because this calcium-binding protein was found to be a multifunctional enzyme regulator, controlling the activity of such enzymes as calcium-ATPase, adenylate cyclase, phospholipase A_2, and myosin light chain kinase. The potential physiological significance of calmodulin is underscored by the knowledge that it is found in almost every plant and animal cell, and that the structure of calmodulin is almost perfectly conserved throughout evolution, suggesting a funda-mental and crucial role for this protein. The diverse, yet specific, actions of calcium can be related to the degree and manner in which the cation binds to the receptor proteins, since the cell possesses multiple acceptor proteins and each protein has multiple calcium-binding sites (Klee *et al.*, 1980; Means and Dedman, 1980). The scheme for calcium–calmodulin interaction is provided in Fig 28.

Recognition of the potential importance of calmodulin in secretory phenomena is afforded by the findings that it has been identified in the neurohypophysis, pancreatic β cells, and exocrine pancreas (Thorn *et*

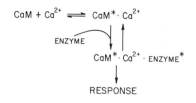

FIGURE 28. Scheme for the interaction of cal-cium and calmodulin. The binding of calcium to calmodulin (CaM) induces a conformational change in the protein activator. The calcium–calmodulin complex then interacts with the target enzyme to modify activity.

al., 1978; Sugden *et al.*, 1979; Bartelt and Scheele, 1980). In the pancreatic β cell where cyclic AMP is purported to be an important modulator of glucose-induced insulin secretion, calcium alone decreases adenylate cyclase-activity in islet homogenates, but together with calmodulin, it increases enzyme activity (Sharp *et al*, 1980). Moreover, the phenothiazine antipsychotic agent, trifluoperazine, which binds to this regulator protein in the presence of calcium, preventing stimulation of enzyme activity by added calmodulin, impairs glucose– and potassium–induced insulin secretion (Sugden *et al.*, 1979; Schubart *et al.*, 1980a) and blocks TRH induced thyrotropin release (Fleischer *et al.*, 1981). But this drug also inhibits potassium-induced ^{45}Ca uptake, indicating that it can affect calcium movement as well as calcium action (Fleckman *et al.*, 1981).

If, for the sake of argument, we assume that calmodulin is an intracellular calcium receptor in secretory cells, the question arises of the mechanism by which the interaction of calcium with its protein receptor triggers the secretory response. The ability of the calcium–calmodulin complex to enhance adenylate cyclase and phosphodiesterase activites provides a potentially important link between calcium and cyclic AMP, and we will address this important point in greater detail in the next chapter. Also, the ability of calmodulin to promote prostaglandin synthesis in blood platelets may stem from an action on the enzymes involved in prostaglandin metabolism, including phospholipase A_2 (Wong *et al.*, 1980).

Another aspect of this interesting and rapidly emerging area of investigation was spawned by the recognition that calcium-dependent phosphorylation of certain specific membrane proteins, through the activation of protein kinases that require calmodulin as a co-enzyme, can affect a number of biological processes in the cell, including the secretory process (Schulman and Greengard, 1978). Calcium-dependent neurotransmitter release evoked from synaptosomes either by depolarization or ionophore A23187 involves a calmodulin-mediated activation of a calcium-dependent protein kinase that phosphorylates proteins in the vesicle membrane (DeLorenzo *et al.*, 1979; Michaelson and Avissar, 1979). The increase in acetylcholine release and phosphorylation occur on a similar time scale; strontium and barium, which increase transmitter release, also enhance the phosphorylation mechanism, while magnesium is devoid of phosphorylating activity. Cytosol prepared from insulinoma cells also contain proteins that are phosphorylated by a mechanism involving calcium and calmodulin (Schubart *et al.*, 1980b). Such studies implicate calcium-dependent phosphorylation as an important component of calcium-induced secretion, although they do not necessarily imply a causal relationship.

It is within the context of contractile phenomena that calmodulin assumes particular importance, since calmodulin is chemically similar, although not identical, to troponin C, reaffirming a possible association of secretory activity with contractile events. The interaction of actin and myosin from blood platelets is similar to that of smooth muscle (see Chapter 2, Section F). Thus, protein-kinase-induced phosphorylation increases actomyosin ATPase activity leading to contraction of the actomyosin system (Lebowitz and Cooke, 1978). Since calmodulin mediates the calcium stimulation of platelet myosin light chain kinase (Adelstein et al., 1980), it is tempting to speculate on these grounds that an increase in cellular calcium and the increase in phosphorylation produced by calcium binding to calmodulin mediates an actomyosin-induced contraction to trigger the secretory process. Alternatively, a calmodulin-dependent phosphorylation of tubulin may be responsible for mediating calcium's action on the secretory process (Dedman et al., 1979; Burke and DeLorenzo, 1981). The finding that protein phosphorylation precedes the release reaction implies some functional correlation between these two events (Lyons and Shaw, 1980). However, phosphorylation could also be related to other cellular functions, as for example shape change and/or aggregation of the blood platelet.

This hypothetical mechanism that invokes a calcium–calmodulin-induced phosphorylation of protein to activate contractile proteins in initiating secretion, while attractive and worthy of careful analysis, must be balanced against the complicating variables including the dichotomy between calcium–calmodulin- and cyclic-AMP-induced phosphorylations; so the interaction of alternate phosphorylating mechanisms must be taken into account. Moreover, activation of protein kinases can be induced not only by cyclic nucleotides and the calcium–calmodulin complex but also by phospholipids (Takai et al., 1979; Kuo et al., 1980); this latter calcium-dependent activation of protein kinase occurs independently of calmodulin, so that realistically one cannot consider calmodulin as the calcium receptor to the exclusion of other biochemical or cellular mechanisms (Fig. 29). Nevertheless, the anticipated revelations involving calmodulin should contribute important pieces to the puzzle and allows us to obtain a more complete picture of the events associated with calcium-activated secretion. Additionally, the need for calmodulin in calcium transport processes as revealed by its regulation of calcium-ATPase gives rise to the supposition that calmodulin may not only control the action of calcium, but also the means by which its effects are terminated (Pershadsingh et al., 1980; Conigrave et al., 1981) (Table 2).

While calmodulin obviously has immense potential importance with regard to the cellular actions of calcium, we cannot ignore the possible

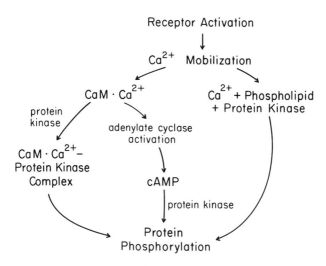

FIGURE 29. Mechanisms for calcium-induced activation of protein phosphorylation. Stimulus-induced calcium mobilization leads to the activation of protein kinase and protein phosphorylation through (a) calcium-calmodulin complexation, (b) the activation of the adenylate cyclase-cyclic AMP system, and (c) phospholipid-induced stimulation of protein kinase.

roles of other calcium-binding proteins in the processes associated with secretion. We have already considered the intriguing studies involving the actions of synexin in membrane fusion phenomena taking place in adrenomedullary chromaffin cells, and it is not unreasonable to entertain the possibility that still other as yet undiscovered calcium-binding proteins will be identified in secretory cells. The role of calcium in the process of secretion should therefore be viewed as being closely linked to specific cellular proteins. Perhaps equally important to consider as another modality for expressing the actions of calcium are the phospholipids of the cell; they will be dealt with in the succeeding section.

**TABLE 2.   Possible Sites of the
Regulation of Secretion by Calmodulin**

1. Cyclic nucleotide metabolism
2. Arachidonic acid metabolism
3. Activation of the actomyosin system
   (regulation of myosin light chain kinase)
4. Control of microtubule assembly
5. Calcium pump (Ca^{2+}-Mg-ATPase)

F. PHOSPHOLIPID METABOLISM

The growing awareness of the potential importance of arachidonic acid metabolism in eliciting cell activation has fostered studies that implicate phospholipid turnover as a potential locus of calcium action in the secretory process. Although the turnover of phospholipids, particularly phosphatidylinositol, may be important in the events that occur between receptor activation and the secretory process, possibly involving calcium mobilization, alterations in phospholipids produced by phospholipases and acylating enzymes in the cell membrane may also contribute directly to exocytosis. The idea that phospholipases regulate secretory events is appealing because of the multiple ways in which the actions of these enzymes, by regulating arachidonic acid metabolism, can be envisioned as forging a link between the triggering events and the subsequent intracellular changes that accompany the secretory response. The basic element that makes phospholipases such an intriguing candidate for a critical role in the calcium-dependent events associated with secretion rests with the finding that phospholipases, particularly phospholipase A_2, require calcium for activity (Rubin and Laychock, 1978b).

Phospholipase A_2 is a ubiquitous enzyme found in the membranous fractions of most cells; the nonlysosomal hormone-activated phospholipase can be distinguished from lysosomal phospholipase A_2 by an alkaline pH optimum and the absolute dependence on the presence of calcium (Brockerhoff and Jensen, 1974; Rubin and Laychock, 1978a; Jesse and Franson, 1979). The activation of a number of secretory organs, including the blood platelet, thyroid gland, adrenal cortex, neutrophils, and the adenohypophysis, all involve the stimulation of a calcium-dependent phospholipase A_2, resulting in the activation of arachidonic acid turnover (Bills *et al.*, 1976; Feinstein *et al.*, 1977; Haye and Jacquemin, 1977; Schrey and Rubin, 1979; Rubin *et al.*, 1981a; Naor and Catt, 1981) (Fig. 30). The concept that the availability of free calcium is a pivotal step in the generation of free fatty acid from phospholipid is supported by the finding that ionophore A23187 stimulates phospholipase A_2 activity in several secretory cells, including the blood platelet, adrenocortical cell, and neutrophil (Pickett *et al.*, 1977; Feinstein *et al.*, 1977; Schrey and Rubin, 1979; Rubin *et al.*, 1981b). Additionally, local anesthetics, which impair calcium mobilization in platelets and stimulus-induced platelet release reaction, inhibit phospholipase A_2 activity, either by directly inhibiting calcium-dependent enzyme activity or by preventing stimulus-induced mobilization of intracellular caclium which is required for phospholipase A_2 activity (Vanderhoek and Feinstein, 1979).

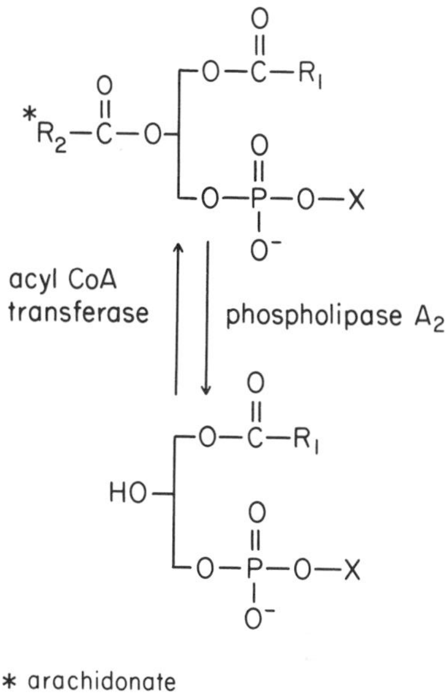

$$
\begin{array}{c}
\overset{\displaystyle O}{\overset{\|}{R_2{-}C{-}O}}\!-\!\!\begin{bmatrix}
\;{-}O{-}\overset{\displaystyle O}{\overset{\|}{C}}{-}R_I\\[2mm]
\;{-}O{-}\overset{\displaystyle O}{\overset{\|}{P}}{-}O{-}X\\[1mm]
\qquad\;\;|\\[0mm]
\qquad\;\;O^-
\end{bmatrix}
\end{array}
$$

acyl CoA
transferase   phospholipase A_2

$$
HO\!-\!\!\begin{bmatrix}
\;{-}O{-}\overset{\displaystyle O}{\overset{\|}{C}}{-}R_I\\[2mm]
\;{-}O{-}\overset{\displaystyle O}{\overset{\|}{P}}{-}O{-}X\\[1mm]
\qquad\;\;|\\[0mm]
\qquad\;\;O^-
\end{bmatrix}
$$

* arachidonate

FIGURE 30. The release of arachidonic acid from phospholipid (deacylation) and its reincorporation into lysophospholipid (reacylation) by a sequence catalyzed by phospholipase A_2 and acyl CoA transferase, respectively.

One product of phospholipase activation, the lysophospholipids, has been of particular interest in the mechanism of membrane fusion reactions since they possess fusogenic properties (Lucy, 1978). It has been proposed that lysophosphatidylcholine, for example, might be involved in the secretion of catecholamines and histamine. Many years ago Hogberg and Unväs (1957) described the disruptive action of phospholipase on mast cells and suggested that mast cell secretagogues activate a "lytic enzyme attached to the mast cell membrane." Products of this lipolytic reaction, lysophosphatidylcholine and lysophosphatidylserine, potentiate anaphylactic histamine release from mast cells (Sydbom and Uvnäs, 1976; Martin and Lagunoff, 1979).

A possible role for endogenous lysophosphatidylcholine in the secretory process has been considered, due to its relatively high concentration within adrenomedullary chromaffin granule membranes (Winkler et al., 1974). However, the lysophosphatidylcholine content of these granules does not change during secretory activity. Moreover, other types of secretory vesicles do not possess such a reletively high concentration of lysophosphatidylcholine (Winkler et al., 1974), and when high concentrations are detected, they are attributed to postmortem hydrolysis of phospholipids (Arthur and Sheltawy, 1980). We can, therefore, assume that lysophosphatidylcholine is not directly involved in the process of membrane fusion; however, as we will see later, it may modulate the activity of adenylate and guanylate cyclase (Shier *et al.*, 1976; Zwiller *et al.*, 1976; Houslay and Palmer, 1979).

Viewed from another perspective, the rapid reacylation of the lysophospholipid brings about the partitioning of free fatty acids into different domains of membrane phospholipids that may selectively perturb particular lipid regions of the membrane (Klausner *et al.*, 1980). At the same time, reacylation prevents the potentially cytotoxic lysophospholipid from rising to levels that are detrimental to the cell. The existence of phospholipases and acylating enzymes in the cell membrane could thus provide the basis for the formation of specific phospholipids with acyl derivatives of different chain lengths and degrees of saturation, reactions that may be important for regulating such membrane functions as the control of ionic permeability or even exocytotic secretion (Lands and Crawford, 1976) (Fig. 30).

It is, therefore, necessary that any consideration of the mechanism of calcium-dependent exocytosis should include the possibility that calcium-induced activation of phospholipase A_2 sets into motion a train of events that leads to granule and plasma membrane fusion and the subsequent discharge of secretory product. The secretory response of certain cells may be mediated by arachidonic acid metabolites which mobilize internal calcium stores, or perhaps enhance calcium entry into the cell. This possibility is most convincingly represented by the neutrophil and blood platelet, where functional activity, including the release response, is closely correlated with arachidonic acid metabolism and the alteration of calcium availability (Feinstein, 1978a; Naccache *et al.*, 1981). Calcium pools localized at or near the plasma membrane may regulate phospholipase A_2 and adenylate cyclase activities, while intracellular pools of calcium may participate more directly in the intracellular mechanisms responsible for exocytotic release (see Figs. 14 and 23).

While a calcium-dependent increase in phospholipid turnover is an important component of the cellular events associated with activation of the secretory process, other pathways in addition to the activation of phospholipase A_2 must also be taken into account. Phospholipase C-mediated reactions, involving the cleavage of polar head groups from phospholipids to produce diacylglycerol and phosphatidic acid by subsequent rephosphorylation, may also be pivotal biochemical events responsible for membrane fusion (Billah *et al.*, 1979; Hawthorne and Pickard, 1979; Bell and Majerus, 1980 (Fig. 31). The importance of phospholipase C as a potential site of calcium action cannot be ignored not only because of the ability of calcium to activate this enzyme in certain instances but also because of its presence in a wide spectrum of tissues (Hostetler and Hall, 1980) and because of its selectivity for phos-

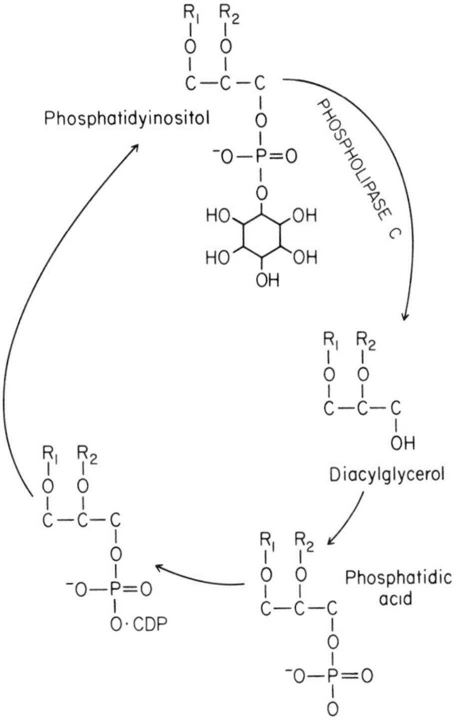

FIGURE 31. The phospholipase C-mediated reaction responsible for phosphatidylinositol metabolism. Activation of phospholipase C degrades phosphatidylinositol to 1,2 diacylglycerol and inositol 1,2-cyclic phosphate, 1,2 diacylglycerol is phosphorylated to phosphatidic acid which then combines with CTP (cytosine triphosphate) to form diacylglycerol-CDP, which reacts with inositol to reform phosphatidylinositol.

phatidylinositol as a substrate (Irvine and Dawson, 1978; Billah et al., 1980). One difficulty in assigning an important physiological role for phospholipase C is that it is localized mainly in the cell cytosolic fraction despite the fact that the substrate for its hydrolysis is localized to the cytoplasmic surface of the plasma membrane. Also a calcium requirement for phospholipase C is, in no way, as clearly demonstrable as that for phospholipase A_2 (Billah et al., 1980; Cockcroft et al., 1980).

Nevertheless, both phospholipase C and phospholipase A_2 must be considered in any theory that relates the action of calcium on the secretory process to turnover of membrane phospholipids (Der and Sun, 1981). These phospholipid effects are, in all probability, a reflection of a complex interaction of several enzymatic reactions. Not only may the effects of phospholipases C and A be linked (Billah et al., 1979), but phospholipid methylation reactions have been coupled to calcium influx and the release of arachidonic acid during activation of secretion in mast cells (Hirata and Axelrod, 1980). Such evidence, while again focusing on phospholipase A_2 as a potentially important vehicle for regulating calcium-induced exocytosis, must be evaluated in relation to other changes in phospholipid turnover.

While based upon our present state of knowledge it would seem premature and somewhat presumptuous to formulate a unitary theory to explain the action of calcium on the exocytotic process, it is clear from the foregoing discussion that both proteins and lipids occupy strategic positions in this scheme. Calcium binding to regulatory proteins and the hydrolysis of ATP, along with calcium-dependent turnover of membrane phospholipids, are also key elements in this sequence. But because phospholipids also control the activity of membrane enzymes, including ATPase (Coleman, 1973; Sandermann, 1978) a sequence of events highlighted by the activation of phospholipase assumes an even more critical posture. The redistribution of calcium during activation of the secretory process may not only promote the interaction of ATPase with its substrate ATP by a direct effect of the cation on the contractile machinery, but the accompanying stimulation of phospholipase activity may also modify ATPase activity through concomitant alterations in phospholipid metabolism. In fact, new insights may be reached by viewing phospholipid turnover as a critical factor controlling the activity level of a variety of membranous enzymes that, in one way or another, participate in the secretory process.

CHAPTER 4

Interaction of Calcium with Other Putative Mediators

A. ANALYSIS OF THE ROLE OF CYCLIC AMP IN SECRETION

1. Cyclic AMP as a Second Messenger

Our survey of the secretory systems established that cyclic AMP exerts an influence on a number of secretory systems, although no unitary theory can be promulgated regarding the role of cyclic AMP. In some systems cyclic AMP augments secretory activity. This positive modulatory role of cyclic AMP is nowhere more apparent than in the rat parotid gland where secretion evoked by β-adrenergic receptor activation proceeds through a cyclic-AMP-mediated mechanism (Butcher and Putney, 1980). In fact, in this system the role of cyclic AMP in β-adrenoreceptor-mediated amylase release has been more clearly established than the role for calcium.

On the other hand, in other systems, most notably the blood platelet, mast cell, and basophil leukocyte, compounds that increase cyclic AMP depress secretion and cyclic AMP acts as a negative modulator of secretion. There are also calcium-dependent secretory processes in which cyclic AMP does not appear to play any direct role; the most notable example of such a system is the adrenal medulla where cyclic AMP does not participate in the mechanism of catecholamine release, although it may function earlier in the secretory cycle perhaps in the synthesis and/or packaging of the hormone (Jaanus and Rubin, 1974; Kumakura et al., 1979). There is also evidence for and against the involvement of cyclic AMP in neurotransmitter release (Dretcher *et al.*, 1976; Duncan and Statham, 1977; Weiner, 1979). Many endogenous substances modulate neurotransmitter metabolism, particularly in sympathetic nerve endings (Westfall, 1977) and these agents may regulate transmitter metabolism and/or release through the adenylate cyclase–cyclic AMP system. However, there seems to be no consistent relationship between the actions of these substances on nervous tissue adenylate cyclase and the effects that they exert on norepinephrine release evoked by nerve stimulation. Isoproterenol, which is known to increase adenylate cyclase activity through activation of β adrenoreceptors, enhances nerve-stimulated norepinephrine release, while prostaglandins of the E series, which also activate adenylate cyclase, are negative modulators of norepinephrine secretion (Hedqvist, 1977; Weiner, 1979).

In other secretory systems such as the adrenal cortex and endocrine pancreas, cyclic AMP appears to participate in the sequence of events that culminates in the discharge of hormones. However, in these two endocrine systems, cyclic AMP serves as a positive modulator rather than an obligatory mediator of the release process, in that the response of these cells does not always bear a direct relation to tissue cyclic AMP levels (Charles *et al.*, 1975; Schimmer, 1980). By contrast, the protean action of corticotropin in the adrenal cortex is obligatorily dependent on calcium (Halkerston, 1975), and glucose-induced insulin secretion is correlated with calcium levels within the β cell (Malaisse *et al.*, 1978). On the other hand, we cannot ignore secretory systems such as the parathyroid gland and the renin-secreting juxtaglomerular cells, in which calcium appears to act as a negative modulator of secretion. In these systems cyclic AMP and arachidonic acid metabolites may play more critical roles in the activation of secretion.

Clearly, the most meaningful view of the interactions of calcium and cyclic AMP is the concept of monodirectional and bidirectional systems that was originally proposed by Michael Berridge (1975). In monodirectional systems, which include the adrenal medulla and the conventional neuron, endocrine and exocrine pancreas, adrenal cortex, and the salivary gland, cell activation is regulated in one direction only, and recovery is mediated by the removal of the stimulant. In such instances, cyclic AMP either is not required or augments the signal activated by calcium. In bidirectional systems, which include the blood platelet and mast cell, the positive modulatory actions of calcium are inhibited by activation of the cyclic AMP system. The basis for this calcium–cyclic AMP interaction will be discussed forthwith. But simply stated, in bidirectional systems stimulation results in a decline in the calcium level brought about by the negative feedback of cyclic AMP, whereas in monodirectional systems cyclic AMP serves to enhance the intracellular levels of ionized calcium.

But to view calcium–cyclic AMP interactions in such general terms oversimplifies what is a much more complex multiplicity of interactions, because not only can cyclic AMP modulate calcium homeostasis in the cell—although more convincing evidence to support this concept is sorely needed—calcium in turn exerts primary and pivotal effects on cyclic AMP metabolism. These points at which calcium and cyclic AMP systems interact will now be considered in further detail.

2. Calcium Regulation of Adenylate Cyclase

Calcium is intimately involved in the formation of cyclic AMP since a reduction of extracellular calcium can impede the capacity of the primary stimulus to augment cyclic AMP levels (Sayers et al., 1972; Harfield and Tenenhouse, 1973). Hormonally-stimulated adenylate cyclase consists of at least three components: a receptor, a regulator protein that binds GTP, and a catalytic unit; by enhancing GTP binding to the regulator protein, calcium exerts its facilitatory action (Mahaffee and Ontjes, 1980). It is important to point out that the positive modulatory action of calcium on adenylate cyclase does not require the interaction of the primary stimulus with a receptor, since stimulation of cyclic AMP formation in nervous tissue induced by depolarizing agents such as high potassium also requires calcium (Shimizu et al., 1970). This supports the important concept that the mobilization of calcium is sufficient to promote the interaction of GTP with the enzyme complex (Fig. 32).

Calcium is also a powerful inhibitor of basal adenylate cyclase activity in a variety of secretory systems and reduces the ability of secretagogues to induce enzyme activation (Burke, 1970; Rodan and Feinstein, 1976; Capito and Hedeskov, 1977; Renckens et al., 1980). The physiological implications of such in vitro studies are apparent. The rise in cytoplasmic calcium during activation of the secretory process, by inhibiting adenylate cyclase, could serve to restore cyclic AMP to prestimulation levels and help to terminate the secretory response. The inhibitory action of calcium is the result of a competition with the activator manganese at a second site, probably located on the catalytic subunit (Mahaffee and Ontjes, 1980). Consequently, in secretory cells, adenylate cyclase is an important locus of calcium action, which can be expressed either by activation or inhibition, with concomitant alterations in cyclic AMP production. Nevertheless, it must be emphasized that the primary action of calcium on secretion is exerted at a step distal to cyclic AMP

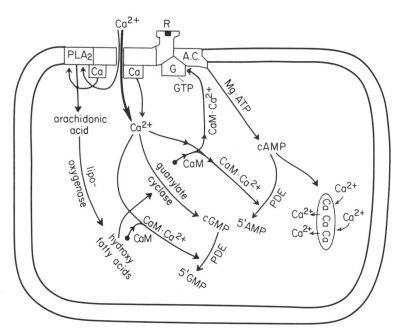

FIGURE 32. The interaction of calcium, cyclic nucleotides, and arachidonic acid metabolites. Agonist-induced stimulation causes calcium to accumulate within the cytosol by either entering through voltage-dependent or receptor-operated channels or being released from membrane-binding sites. Calcium may directly activate guanylate cyclase to promote cyclic GMP formation or by binding to calmodulin (CaM) may activate adenylate cyclase (A.C.), and thereby augment cyclic AMP levels. Cyclic AMP and/or cyclic GMP may serve as positive modulators of the secretory process by mobilizing calcium from intracellular binding sites. The calcium–calmodulin complex also leads to the activation of phosphodiesterase (PDE), thereby promoting the degradation of the cyclic nucleotides. Calcium-dependent activation of phospholipase A_2 (PLA$_2$) enhances the synthesis of hydroxy fatty acids by way of the lipoxygenase pathway, perhaps resulting in the activation of guanylate cyclase and/or further increases in calcium permeability (for further details see text).

synthesis, since they are instances where calcium deprivation blunts the secretory response without impairing the cyclic AMP response (Gautvik *et al.*, 1980; Spence *et al.*, 1980).

3. Calcium Regulation of Phosphodiesterase

Calcium is also involved in the degradation of cyclic nucleotide, because calcium stimulates phosphodiesterase activity. Multiple forms of phosphodiesterase exist in the intact cell, although the precise subcellular localization of any of these enzymes is not known. The mode of action of calcium in activating certain forms of phosphodiesterase has been elucidated, and it is clear now that calcium exerts its action by binding to the protein activator, which has been recognized as the multifunctional calcium-dependent regulator termed calmodulin. Protein activator together with calcium produces optimal stimulation of enzyme activity. The fact that the calcium concentration that causes half maximal activation of the enzyme in the presence of activator is within the presumed physiological range of the cytoplasmic calcium concentration testifies to the relevance of these findings to the intact cell (Wells and Hardman, 1977).

Thus, calcium is an important regulator of cyclic AMP homeostasis by controlling cyclic AMP formation and at the same time by activating the machinery involved in regulating cyclic AMP degradation (Fig. 32). The calcium-binding protein confers a calcium-dependent activation of adenylate cyclase and phosphodiesterase, resulting in an increase in cyclic AMP production by stimulating adenylate cyclase activity; the concomitant activation of phosphodiesterase results in a decrease in cyclic AMP levels. Such a coordinated mechanism for controlling cyclic AMP levels provides for a rapid activation as well as termination of the response. So whatever the role of cyclic AMP in the secretory cycle, the ubiquitous actions of calcium are closely linked to the metabolism of this cyclic nucleotide.

4. Cyclic AMP Regulation of Calcium Metabolism

In addition to the effects that calcium exerts on cyclic AMP metabolism, our survey of the secretory process has provided indirect evidence that cyclic AMP functions as a regulator of calcium homeostasis in the secretory cell. Mitochondria possess an efficient system for sequestering cytoplasmic calcium and are a principle site of calcium deposition in the secretory cell. The possibility that hormones influence calcium transport by mitochondria was initially considered by Howard Rasmussen (1966) some years ago. Subsequently, cyclic AMP was advocated as an important messenger in the control of cell calcium since it was thought to induce a release of calcium from isolated mitochondria (Borle, 1975). However, convincing confirmatory evidence to support this postulate has been difficult to muster. Part of the problem resides with our insufficient knowledge about the mechanism of calcium cycling in mitochondria; much more needs to be learned about this process before definitive statements concerning its physiological regulators can be espoused (Bygrave, 1978).

Although evidence favoring the endoplasmic reticulum as the locus of calcium mobilization is even less convincing than that favoring the mitochondria, the possibility that the endoplasmic reticulum of the cell acts as an important calcium-buffering system should receive serious consideration, particularly in the case of neural or other electrically excitable tissue. Of possible relevance to the termination of the secretory process are the studies of Arnold Katz and his associates (Tada et al., 1978) on the cyclic-AMP-dependent stimulation of calcium uptake into cardiac sarcoplasmic reticulum, which is mediated by the activation of cyclic AMP-dependent protein kinase. The phosphorylation of a protein called phospholamban determines the rate of turnover of the calcium-dependent ATPase as this molecule translocates calcium from outside to inside the sarcoplasmic reticulum. A cyclic-AMP-dependent calcium uptake mechanism has also been demonstrated in a microsomal preparation from blood platelets (Käser-Glanzmann et al., 1978), as well as from a preparation of adrenocortical microsomes (Laychock et al., 1978), suggesting a similar process at work in secretory cells to terminate the secretory response. But postulating the existence of a cyclic-AMP-dependent calcium uptake mechanism into the endoplasmic reticulum implies that cyclic AMP is involved in both the initiation and termination of the secretory process, with cyclic AMP perhaps facilitating release of calcium from mitochondria and stimulating calcium uptake into endoplasmic reticulum. Such a complex sequence of events would have to be well-coordinated by interlinked processes (Fig. 32).

Prostaglandins have also been implicated as mediators of calcium metabolism in the secretory cells (Rubin and Laychock, 1978b, Rubin, 1981), so whether cyclic AMP and/or metabolites of arachidonic acid function as primary mediators of calcium metabolism in secretory cells remains an unanswered question. But despite the controversy regarding the identity of the intracellular mediators that act in concert with calcium, their indispensable role is clearly indicated in secretory organs that utilize cellular calcium such as the rat parotid gland and blood platelet. These messengers serve to transmit the primary signal from the cell membrane to the intracellular stores of calcium, that is, if the source of cellular calcium is other than the plasma membrane.

Before leaving the topic of cyclic AMP regulation of calcium metabolism, we must also consider the possible role of cyclic AMP at the level of the cell membrane as a regulator of the activity of the calcium channel. Here again, we cannot make any general statements regarding the effects of cyclic AMP on the calcium channels in secretory organs because the evidence favoring a role for cyclic AMP in activating calcium gating at nerve endings is not convincing (Skirboll et al., 1977); while it appears that cyclic AMP controls the activity of the calcium channel in mast cells, it is not clear as to whether it acts to facilitate or prevent calcium entry (Foreman *et al.*, 1976, 1980). However, a cyclic-AMP-dependent phosphorylation reaction may be a critical step in the calcium-dependent facilitation of transmitter release by inactivating the potassium channel, which normally serves to switch off the voltage-dependent calcium channel (Castellucci *et al.*, 1980).

Obviously, it is difficult to reconcile all of the published results with a scheme that would explain the actions of cyclic AMP on calcium homeostasis in secretory cells. In fact, there should even be some reservation for the unequivocal acceptance of a significant and/or universal role for cyclic AMP as a regulator of calcium metabolism. The uncertainty regarding the physiological significance of the adenylate cyclase–cyclic AMP system in the secretory response mandates that other putative mediators, i.e. arachidonic acid metabolites, be given equal consideration as possible modulators of calcium metabolism. This idea draws intriguing support from studies on the platelet and neutrophil which have been considered and will be further amplified in Section C of this chapter.

5. Cyclic AMP Action at the Molecular Level

Since it has not been irrefutably established that cyclic AMP plays a primary and direct role in activation of secretion, it would not be profitable to attempt a detailed discussion of the possible biochemical mechanisms through which cyclic AMP may exert its effects on secretory systems. But even the most cursory consideration of the molecular basis of cyclic AMP action must include protein kinase. The situation in this regard becomes clouded because of the existence of cyclic-AMP-, cyclic GMP-, and calcium-activated protein kinases, each of which plays a specific physiological role owing to its respective apparent subcellular compartmentalization (Nishizuka *et al.*, 1979). It is, therefore, not unlikely that there would be a multiplicity of phosphorylation reactions associated with the activation, as well as the termination, of the secretory response.

Keeping these complexities in mind, and considering the contingency that we are not able to fit cyclic AMP and calcium into a general scheme, there are nevertheless certain models that have been proposed which are worthy of consideration and may provide the necessary impetus for achieving a fundamental understanding of the role of cyclic AMP in the secretory process. Rasmussen and Nagata (1970) offered a scheme in which cyclic AMP-activated phosphorylation of microtubules in the cytoskeleton converts the microtubules from a calcium-insensitive to a calcium-sensitive state allowing a calcium–cytoskeleton interaction that leads to the discharge of secretory product. Cyclic AMP may also prime the secretory process by phosphorylating the membrane of the secretory granule through the activation of a specific protein kinase. Granule movement would then become sensitive to calcium (Berridge, 1975).

Both of these theories impart a primary action to cyclic AMP to prime the calcium-sensitive system. However, in certain secretory systems, introducing calcium into the cell either by ionophoresis or by the administration of the ionophore A23187 elicits an explosive release of secretory product in the absence of any discernible increase in cyclic AMP production. The inescapable conclusion that derives from such experiments is that calcium plays a primary and direct role in the secretory process. Any role that cyclic AMP plays in such systems can only be a secondary one and will probably be determined by the manner in which calcium is handled by each particular secretory cell.

B. ANALYSIS OF THE ROLE OF CYCLIC GMP IN SECRETION

1. Cyclic GMP as a Second Messenger

While much is known about the mechanism by which adenylate cyclase is activated, as well as the possible physiological role of cyclic AMP, the mechanism by which hormones activate guanylate cyclase in intact cells, as well as the physiological role of cyclic GMP, is still obscure. Nevertheless, certain parallelisms exist between cyclic GMP and cyclic AMP. The apparent parallels concern the forms of protein kinase and phosphodiesterase that are selective for one or another of the two cyclic nucleotides and that mediate their respective actions, as well as their rates of degradation. On the other hand, the characteristics of guanylate cyclase clearly distinguish it from adenylate cyclase. While adenylate cyclase is a particulate enzyme which is susceptible to stimulation by hormones in broken cell preparations, a major portion of guanylate cyclase resides in the soluble fraction after homogenization (Goldberg *et al.*, 1973) and physiological agonists are unable to activate guanylate cyclase in cell-free systems (Goldberg *et al.*, 1975). Moreover, certain pharmacological agents clearly distinguish the activation of guanylate cyclase from adenylate cyclase in that azide and nitro compounds, including sodium nitroprusside and nitrosamines, increase guanylate cyclase activity without altering cyclic AMP levels, whereas fluoride increases adenylate cyclase activity but does not increase cyclic GMP levels.

Part of the problem in elucidating the role of cyclic GMP in secretory activity relates to the fact that researchers have employed the Sutherland criteria for establishing cyclic AMP as a second messenger (Robison *et al.*, 1971) in attempting to interpret the actions of cyclic GMP, when in fact the actions of cyclic GMP in certain tissues are even antagonistic to those mediated through an increase in cyclic AMP. Goldberg and his colleagues (1973) originally advocated a "dualism" between these two nucleotides through the opposing actions of cyclic GMP and cyclic AMP. Such a dualism of action is most clearly illustrated by the control of cardiac function in which the positive inotropic action of catecholamines is purported to be expressed by an increase in cyclic AMP levels mediated through β-adrenergic receptor stimulation, and the negative inotropic action of acetylcholine is mediated through an increase in cyclic GMP formation provoked by activation of muscarinic receptors.

However, as seen in the thyroid gland, attempts to verify this hypothesis have commonly met with frustration and failure. The problem with verifying the hypothesis stems mainly from the fact that the original notion that cyclic GMP mediates certain effects opposing those of cyclic AMP is no longer tenable. The outcome at times even favors the proposition that these two cyclic nucleotides act cooperatively. Most notably, cyclic GMP, or a less readily hydrolyzed analogue such as 8-bromo-cyclic GMP, does not elicit effects that oppose those of cyclic AMP, but in many cases mimics those of cyclic AMP. Examples of this can be found in a number of secretory cells (Ignarro and George, 1974; Butcher *et al.*, 1976; Gardner, 1979). Another complication arises from the fact that although endogenous levels of cyclic GMP rise in response to many physiological stimuli of secretory cells (Christophe *et al.*, 1976; Snyder *et al.*, 1980; Laychock, 1981)—although this is not always the case (see for example Laychock and Hardman, 1978)—endogenous cyclic GMP concentrations are approximately one order of magnitude less than cyclic AMP levels. Still, the concentrations of cyclic GMP needed for producing an increase in secretion by exogenous cyclic GMP are equivalent to those of cyclic AMP and are orders of magnitude greater than endogenous levels of cyclic GMP. These artificially high concentrations of cyclic GMP may elicit effects either by increasing cyclic AMP levels by inhibiting phosphodiesterase or by stimulating cyclic-AMP-dependent protein kinase.

Another strategy employed to probe the physiological significance of cyclic GMP relates to the use of pharmacological agents that elevate cyclic GMP levels in the absence of any increase in cyclic AMP levels.

Sodium nitroprusside and ascorbic acid increase cyclic GMP levels in various secretory cells including the adrenal cortex, platelets, rat pancreatic acini and islets (Laychock and Hardman, 1978; Schoepflin et al., 1977; Butcher and Putney, 1980; Laychock, 1981). But these agents do not universally augment the secretory response, although in rat pancreatic islets the increase in insulin release induced by sodium nitroprusside parallels the increase in cyclic GMP levels (Laychock, 1981). Since in certain secretory tissues an increase in cyclic GMP may be a sufficient stimulus for activating the secretory response (Laychock, 1981), while in others it is apparently not (Laychock and Hardman, 1978), we are forced to conclude that either the increase in cyclic GMP levels elicited pharmacologically does not enable the cyclic nucleotide to gain access to the appropriate protein kinase (Lincoln and Keely, 1980), or that the elevation of cellular cyclic GMP levels is in and of itself not always adequate to trigger the secretory response (Butcher and Putney, 1980). Apart from these data, a direct role for cyclic GMP in the physiological mechanism governing secretion is discredited by the inability of phosphodiesterase inhibitors to potentiate the secretory response despite an enhancement of tissue cyclic GMP levels (Heisler and Lambert, 1978).

There are not many clues as to the biological role of cyclic GMP in secretory cells. Whatever the actions of this nucleotide, its ubiquitous distribution connotes an activity that transcends the specialized function of secretory cells. However, a less diverse role for cyclic GMP in the regulation of cell function than cyclic AMP is inferred from our knowledge that the cyclic-GMP-dependent protein kinase has a more stringent substrate specificity than the cyclic-AMP-dependent kinase (Lincoln and Corbin, 1978). Cellular cyclic GMP may render cellular components more sensitive to the alterations induced by the primary stimulus, as for example, supporting a more effective interaction between calcium and its intracellular "receptor." Alternatively, it is just as reasonable to consider that cyclic GMP may be involved at the other end of the spectrum in the "turning off" of some of the events induced by the primary stimulus (Murad et al., 1979). But as we will see, whatever the action of cyclic GMP in secretory cells, it also bears an important relation to the principal messenger, calcium.

2. Calcium and Cyclic GMP Interactions

The guanylate cyclase–cyclic GMP system shares with the adenylate cyclase–cyclic AMP system the basic property of being regulated by calcium. Tissues deprived of calcium fail to exhibit an increase in cyclic

GMP accumulation in response to various secretagogues which mediate their effects through receptor activation (Murad et al., 1979). The fact that an increase in cyclic GMP levels can also be elicited by the calcium ionophore A23187 and membrane depolarization (Smith and Ignarro, 1975; Ferrendelli et al., 1976; Nestler et al., 1978) suggests that increases in cyclic GMP levels are mediated through a mobilization of calcium, rather than through stimulation of a receptor-mediated mechanism directly coupled to guanylate cyclase. On the other hand, the ability of nitrosamines and sodium azide to increase cyclic GMP differs, in that their actions can be fully expressed in the absence of extracellular calcium (DeRubertis and Craven, 1976).

An important clue to furthering our understanding of the control mechanisms involved with regulating guanylate cyclase activity may be provided by the knowledge that the physiological agonists of cyclic GMP that require extracellular calcium fail to activate guanylate cyclase in broken cell preparations, implying that the increase in tissue cyclic GMP levels produced by these agents not only depends upon the mobilization of calcium, but may also involve one or another cellular intermediate. The fact that fatty acids and their metabolites, as well as phospholipase A_2 and lysophosphatidylcholine, are capable of activating soluble guanylate cyclase from a variety of tissues suggests that the increase in cyclic GMP levels may be a consequence of phospholipase A_2 activation (Shier et al., 1976; Aunis et al., 1978; Goldberg et al., 1978). This hypothesis explains the indirect effects of agents on cyclic GMP levels that are dependent on calcium by postulating the locus of calcium action at the level of the calcium-dependent phospholipase A_2. We will discuss this important concept in more detail with regard to the putative role of arachidonic acid metabolites in secretion, but accumulating evidence favors a pivotal role for the release and subsequent oxygenation of arachidonic acid in the calcium-dependent modulation of tissue cyclic GMP (Goldberg and Haddox, 1977; Murad et al., 1979) (Fig. 32).

Testimony that the relationship between calcium and cyclic GMP represents an important aspect of calcium's action on the secretory process is provided by the knowledge that barium and strontium substitute for calcium in maintaining the depolarization-induced increase in cyclic GMP levels in neural tissue; furthermore, magnesium has no effect and in high concentrations attenuates the increase in cyclic GMP stimulation (Ferrendelli et al., 1976). The parallel between the ability of various cations to substitute for calcium in maintaining both neurotransmitter release and cyclic GMP levels avers that, at least in excitable cells, stimulation of cyclic GMP is closely linked to some calcium-dependent mechanism mediating the secretory response.

While calcium seems to be critical for regulating cyclic GMP production in secretory cells, it may also be an important factor in regulating its degradation. The increased availability of calcium allows only a transient accumulation of cyclic AMP because calmodulin governs both the synthesis and degradation of cyclic AMP (Fig. 32). However, cytoplasmic phosphodiesterase also catalyzes the hydrolysis of cyclic GMP; in fact, at a physiological concentration of substrates, the rate of cyclic GMP hydrolysis is greater than that of cyclic AMP (Cheung, 1979). The increase in cellular calcium availability could therefore result in an increase in cyclic AMP and a concomitant decrease in cyclic GMP. But this brings us full-circle to the "dualism" theory of cyclic AMP and cyclic GMP interactions, which emphasizes the complex interplay that exists between the second messengers.

C. ANALYSIS OF THE ROLE OF PROSTAGLANDINS IN SECRETION

1. Prostaglandins as Second Messengers

While we have reserved our discussion of the family of compounds termed the prostaglandins to complete the existing picture, this in no way implies that this group occupies a secondary or less important role in mediating the secretory process than do the cyclic nucleotides. As a matter of fact, if the reader refers back to Chapter 2, he will find that in many systems including the blood platelet, adenohypophysis, adrenal cortex and the renin-secreting juxtaglomerular cells, there is convincing evidence favoring prostaglandins as modulators of the secretory response. But again, no clear picture emerges regarding a specific and well-delineated action of prostaglandins on secretory activity, or the particular prostaglandin(s) directly involved in the process. However, like the cyclic nucleotides, the ubiquity with which prostaglandins are found in biological tissues implies that they perform some critical and general function in controlling cellular activity.

The first prostaglandins discovered were the more stable species such as PGE and $PGF_{2\alpha}$. The major pathway involved in the formation of these prostaglandins is catalyzed by a membranous cyclooxygenase complex that converts arachidonic acid to unstable intermediates called endoperoxides (PGG_2 and PGH_2), which then undergo enzymic transformation to unstable products, thromboxane A_2 and prostacyclin (PGI_2). In 1979, another major route of arachidonic acid metabolism, the lipoxygenase pathway, was discovered which gives rise to the leukotrienes and other hydroxy fatty acids (Fig. 33). Although little is known about the possible physiological relevance of lipoxygenase products, researchers working in this particular area are optimistic about ultimately demonstrating significance of these substances in regulating various cellular functions, including secretion.

a. The primary stimulus should be able to augment the metabolism of arachidonic acid.

b. Prostaglandins or arachidonic acid (or its metabolites) should mimic the physiological effect of the primary stimulus.

c. The administration of drugs which interfere with metabolism should alter the response to the primary stimulus.

We have seen that prostaglandins and hydroxy acids are synthesized and released in response to secretagogues in such secretory tissues as the thyroid gland, blood platelet, neutrophil, and exocrine and endocrine pancreas (Burke *et al.*, 1973; Gorman, 1979; Bokoch and Reed, 1980; Marshall *et al.*, 1980; Kelly and Laychock, 1981). Our own investigations using the adrenal gland as a model for hormonal stimulus–secretion phenomena have attested to the role of prostaglandins as cellular intermediates in corticotropin-induced steroid production and release. In the intact perfused cat adrenal gland, the rise in prostaglandin release produced by corticotropin precedes the enhancement of steroid production and release (Laychock *et al.*, 1977). This *in situ* observation was supported by the response to corticotropin of dispersed cat adrenocortical cells, which show an increase in PGE and PGF synthesis and release upon exposure to the tropic hormone (Rubin and Laychock, 1978a).

Prostaglandins and arachidonic acid (or its metabolites) are capable of stimulating the secretory process; however, since the concentrations used to stimulate secretion are generally within the micromolar range, whereas endogenous concentrations are in the picomolar to nanomolar range, caution must be exercised in interpreting the secretory response to exogenous prostaglandins and arachidonic acid metabolites within the scope of physiological mechanisms. An alternative approach to this problem is to assess functional activity after inhibition of arachidonic acid metabolism. However, this type of study is complicated by the fact that the end products of arachidonic acid metabolism, which differ in their biological properties, vary from tissue to tissue. Thus, the variable effects often observed with the use of these inhibitors may be due to the differential inhibition of one or another of the end-products. Further complicating interpretations of effects of cyclooxygenase inhibitors is

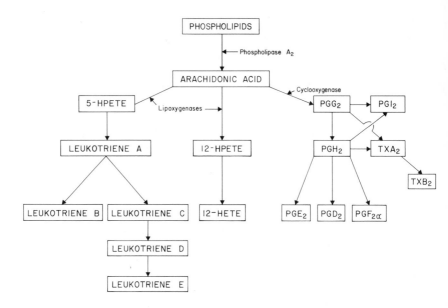

FIGURE 33. A summary of the known pathways of arachidonic acid metabolism. Arachidonic acid is liberated from phospholipids through the action of phospholipase A_2 and metabolized by membrane-bound cyclooxygenase to cyclic endoperoxides (PGG$_2$ and PGH$_2$) and then to thromboxane A_2, prostacyclin, and/or the stable prostaglandins. Arachidonic acid may also be transformed by the action of cytoplasmic lipoxygenase enzymes to hydroxy fatty acids.

their potential for altering cellular events distinct from prostaglandin synthesis (Flower, 1974). Indomethacin, for example, has been recognized as a calcium antagonist (Northover, 1977), as well as an inhibitor of phosphodiesterase (Flower, 1974). Although the realization of the potential importance of the lipoxygenase pathway is only in a rudimentary stage, the use of inhibitors of this pathway, such as ETYA, have indicated that this pathway may indeed prove to be of great significance in spawning metabolites that serve as mediators of the secretory process. However, all of the existing data are difficult to interpret in light of our as yet incomplete understanding of the specific roles of one or another of the metabolites of arachidonic acid in secretory activity. So the generalization that arachidonic acid metabolites are obligatory intermediates in secretion will require considerable scrutiny.

2. Calcium and Prostaglandin Production

If prostaglandins and other arachidonic acid metabolites function as intermediates in the sequence of events associated with secretion, then this role must somehow be linked to the action of calcium. The interaction between calcium and arachidonic acid metabolism at one level involves phospholipase A_2, since this calcium-requiring enzyme appears to be a key factor regulating arachidonic acid metabolism by providing free precursor fatty acid (Rubin and Laychock, 1978b).

We have seen that the redistribution of calcium within the secretory cell is sufficient to increase the availability of free fatty acid or the synthesis of prostaglandins and other arachidonic acid metabolites (see Section F, Chapter 3). This calcium-induced turnover of membrane phospholipids may also modify the activity of enzymes that participate in the secretory process. Thus, arachidonic acid and other unsaturated fatty acids may elevate cyclic GMP levels (Goldberg and Haddox, 1977), or may stimulate a calcium-activated ATPase serving to translocate calcium (Seiler and Hasselbach, 1971; Repke *et al.*, 1976). Moreover, the regulation of adenylate cyclase activity appears to depend upon the nature of the lipid components within the cell membrane (Ross and Gilman, 1980). It should be emphasized that the important control mechanism for altering the structure and function of the cell membrane not only involves the breakdown (deacylation) of phospholipids but also their reacylation by fatty acids (Lands and Crawford, 1976). Acyl transferase reactions—which have been detected in membrane fractions of various tissues endowed with secretory activity (Elsbach *et al.*, 1972; Corbin and Sun, 1978; Sun, 1979)—constitute a potentially important vehicle for regulating the molecular species of membrane phospholipids and thus for determining membrane function. So it follows, then, that a cascade of potentially important reactions may be triggered by the calcium-dependent activation of phospholipase A_2 (Fig 30).

We have also alluded to the potential significance of other phospholipases, particularly phospholipase C, which in certain systems can be activated by calcium (see Chapter 3, Section F). Evidence obtained from studies of the blood platelet favors the view that a phosphatidylinositol-specific phospholipase C may be an important regulator of arachidonic acid metabolism through a reaction sequence involving diglyceride lipase (Rittenhouse-Simmons, 1979; Billah *et al.*, 1980). Thus the evidence emerging from the ever-increasing number of studies indicates that not only does the action of secretagogues involve the calcium-dependent turnover of cyclic nucleotides, but that an early and possibly pivotal control mechanism governing the secretory process involves the activation of arachidonic acid metabolism mediated by the calcium-activated phospholipases.

3. Calcium and Prostaglandin Action

Elucidation of the physiological role of prostaglandins is made difficult both by the proliferating number of arachidonic acid metabolites that have been identified and by their diverse actions. The most striking illustration of the protean actions of prostaglandins derives from their effects on smooth muscle, where their ability to either contract or relax a given preparation will depend upon both the smooth muscle and the prostaglandin employed. In an attempt to reconcile the antagonistic effects of prostaglandins on smooth muscle, Strong and Bohr (1967) suggested that the stimulant effect of prostaglandins are due to the enhanced cationic permeability of the membrane or to decreased binding to intracellular organelles. The inhibitory effects of prostaglandins, on the other hand, were ascribed to a decrease in free cytoplasmic calcium levels as a result of enhanced intracellular binding to organelles, such as the mitochondria.

In secretory cells, metabolites of arachidonic acid can either mimic the actions of the primary stimulus or serve as a negative modulator. Extending the analogy with smooth muscle would lead us to conclude that the putative mediators exert their effects on secretion by altering calcium permeability. The manner in which prostaglandins mediate release phenomena may be ascribed to their ionophoretic properties (Carsten and Miller, 1977; Serhan et al., 1980), although prostaglandins are generally not potent ionophores (Weissmann et al., 1980). In fact, a stronger case might be made for the contention that phosphatidic acid, which is generated by the phospholipase-C-mediated reaction, is a physiological calcium ionophore (Putney, 1981). Nevertheless, arachidonic acid metabolites of the lipoxygenase pathway are capable of enhancing calcium availability in the neutrophil (Naccache et al., 1981) (Fig. 13), and some argument can be made for the concept that prostaglandins modulate calcium activity in adrenocortical cells by mobilizing an active form of calcium from intracellular binding sites (Rubin, 1981).

Another important variable to consider is that the ability of prostaglandins to exert their actions on calcium metabolism might not only be related to a direct action on calcium availability but to effects expressed through cyclic AMP, since prostaglandins and prostacyclin are capable of stimulating adenylate cyclase–cyclic AMP system in a number of secretory organs. Therefore, at least in certain secretory systems, prostaglandins or other arachidonic acid metabolites may act as second messengers, with cyclic AMP acting as a third messenger (Kuehl et al., 1974). The potential for prostaglandins to mobilize calcium by promoting cyclic AMP formation sets these arachidonic acid metabolites apart from ionophorous antibiotics (e.g. A23187), which are unable to elevate cyclic AMP levels.

The complex, but coordinated, series of interactions between calcium, prostaglandins, and cyclic nucleotides can be most readily illustrated by employing the blood platelet as a model. In this bidirectionally-regulated secretory system, primary stimuli such as thrombin stimulate thromboxane production, which triggers the release reaction as well as aggregation by mobilizing cellular calcium. On the other hand, the production of prostaglandins of the E series, as well as prostacyclin, inhibits aggregation and the release reaction. These regulatory actions of the prostaglandins appear to be controlled through cyclic AMP since PGE and prostacyclin, which increase cyclic AMP levels inhibit the release reaction, while stimulators of the release reaction such as thromboxane A_2 decrease cellular cyclic AMP levels (Feinstein, 1978a; Gorman, 1979). The calcium mobilization induced by thromboxane to trigger the release reaction produces an inhibition of adenylate cyclase resulting in a decrease in cyclic AMP levels. The same stimulation of cyclooxygenase that promotes the formation of prostaglandins of the E series as well as prostacyclin results in an elevation of cyclic AMP levels that brings about an enhanced calcium uptake into intracellular organelles to terminate activity (Feinstein, 1978a; Gorman, 1979). Calcium also plays another key role in this sequence because the coordinated series of events is triggered by the activation of a calcium-dependent phospholipase to bring about the increased availability of arachidonate. A similar—though less complex—mechanism probably operates in sympathetic nerves, wherein prostaglandins serve as negative modulators of catecholamine release by altering the availability of free calcium within the nerve ending (Hedqvist, 1977).

In the neutrophil and adrenocortical cell, there is also particularly strong evidence that arachidonic acid metabolites act as mediators of the secretory process. The neutrophil represents a system where products of the lipoxygenase pathway appear to predominate quantitatively and functionally, although these newly recognized substances have not been characterized sufficiently to designate specific candidates as mediators of the secretory response. But the fact that exogenous arachidonic acid or its lipoxygenase metabolites mimic many of the effects of a primary secretagogue (Bokoch and Reed; 1981; Sha'afi et al., 1981) reinforces the supposition that stimulation of phospholipase A_2 and the resulting metabolism of arachidonic acid is an important and possibly even critical component of the mechanisms governing lysosomal enzyme release, as well as chemotaxis and aggregation in this cell type.

Investigations along these lines have also delineated a role for arachidonic acid metabolism in the events associated with corticotropin-induced corticosteroid production and release (Rubin and Laychock, 1978a; Schrey and Rubin 1979). Prostacyclin has proven to be the most interesting of the prostaglandins in this test system, not only because of its potency, but because it elevates cyclic AMP levels and mobilizes a cellular pool of calcium (Rubin, 1981). Whether other prostaglandins act to enhance calcium sequestration in the adrenocortical cell, as in the blood platelet for example, has not been determined; however, the complexity of the interactions among the various intracellular mediators is highlighted by the knowledge that in the adrenal cortex and juxtaglomerular cells, prostacyclin serves as a positive modulator, while in the blood platelet it functions as a negative modulator. But any theory that implicates arachidonic acid metabolites as regulators of calcium metabolism at either extra- or intracellular loci will remain speculative at least until it has been ascertained whether these substances primarily serve as intracellular messengers or as blood-borne mediators with primary effects on the plasma membrane. The fact that thromboxane A_2 is produced by the platelet while prostacyclin is formed in the vascular endothelial cell suggests that, at least in the platelet, they act at several cellular loci.

It is apparent from our discussion that a wide variety of biochemical mechanisms exists for controlling the secretory apparatus and that the role of intracellular mediators varies from one cell type—or one species—to another. But whatever the role of cyclic nucleotides and/or metabolites of arachidonic acid, calcium must be considered the one "absolute" in the apparent variations on a central theme. Arachidonate metabolites, like cyclic nucleotides, participate in the secretory process; however, their role is subservient to that of calcium, not only because calcium controls their metabolism, but because they serve either as positive or negative modulators of the secretory process by exerting effects on calcium metabolism in the cell. Due to the multitude of calcium-dependent cellular events associated with secretion, no simplified sequence can be presented.

Perhaps the best perspective can be attained by considering that stimulus-induced secretory activity mobilizes calcium from more than one pool and thus promotes the formation or activation of one or more cellular messengers. In this manner, any one calcium-mediated step could be bypassed, without necessarily impairing other steps. For example, the elevation of cyclic GMP levels in the exocrine pancreas is predicated upon the mobilization of a specific pool of membranous calcium (Schulz and Stolze, 1980); the subsequent activation of protein kinase(s) promotes the phosphorylation of proteins that may be required in the secretory process. Should the cyclic-nucleotide-mediated steps be bypassed, however, the complexation of calmodulin and calcium derived from an extracellular source may directly activate a calcium-dependent protein kinase capable of triggering phosphorylation reactions.

At first glance, this concept may seem at odds with the longstanding notion of Nature's efficiency and parsimony in developing cellular mechanisms. However, because of the essential role of calcium in diverse cellular functions, a complex, yet coordinated system for handling calcium is warranted. The more diversified the functions of the secretory cell, therefore, the more complex and more varied will be the mechanisms for regulating calcium homeostasis.

Epilogue

Having reviewed such an abundance of information, we require some overview that might be of value in gaining an overall perspective of the role of calcium in the secretory process as it now stands and perhaps, in some general way, in generating ideas for future studies.

The staggering body of established facts and generalizations has led to the elucidation of certain basic features of the secretory process, yet the picture is still incomplete and interpretations of experimental observations often appear highly speculative and sometimes controversial. Nevertheless, in the end, a unity in the mechanism of secretion begins to emerge; for despite the variation on the basic theme from one organ to another, or even within the same cell—depending upon the nature and the duration of the stimulus—there is now no doubt that the modulation of the level of free calcium in the cytoplasm represents a major, if not the principal, control mechanism governing the functional activity of secretory cells. In fact, calcium is unique in having pride of place within the hierarchy of the coordinated process that controls secretion. Cyclic nucleotides and arachidonic acid metabolites, on the other hand, not only depend upon calcium for their synthesis, but also serve only as secondary adjuncts of the secretory process either to regulate calcium homeostasis or to sensitize the secretory apparatus to the actions of calcium (cf. Fig. 32).

While a number of imaginative theories have been promulgated to explain the cellular mechanism of calcium action, no single one has engendered adequate experimental support over the others to be envisioned as the most viable one. Beyond these problems looms the strong possibility of still undiscovered desiderata that will spawn new paradigms. Even so, the rapid growth in this field and the synergistic interaction among those with diverse expertise—while leading to the disappearance of classical biological disciplines—have also culminated in the convergence of ideas and have set the stage for the attainment of profound conceptual advances. We are, therefore, entering a period of scientific renaissance that will probably close many gaps in our knowledge and inevitably superannuate this volume.

So while no final judgment can be made with regard to the cellular mechanism whereby calcium activates the secretory process, I still hope that this volume has proven to be of value to those interested in the unique and obviously pivotal role of calcium in cellular function. A definitive assessment of my efforts to acknowledge those scientists who merit particular recognition by the cogency of their analysis and the accuracy of their predictions must await future studies.

But I must conclude by observing that, despite all of the information and theories that I have dealt with, the words of Isaac Newton are most appropriate to consider at this final juncture—"I seem to have been only like a boy playing on the sea-shore . . . whilst the great ocean of truth lay undiscovered before me."

References

Abe, M., and Sherwood, L. M., 1972, Regulation of parathyroid hormone secretion by adenyl cyclase, *Biochem. Biophys. Res. Commun.* **48**:396.

Abrahamsson, H., Gylfe, E., and Hellman, B., 1981, Influence of external calcium ions on labelled calcium efflux from pancreatic β cells and insulin granules in mice, *J. Physiol. (London)* **311**:541.

Adair, R., and Davidoff, R. A., 1977, Studies of the uptake and release of [³H] β-alanine by frog spinal slices. *J. Neurochem.* **29**:213.

Adelstein, R. S., 1980, Phosphorylation of muscle contractile proteins, *Fed. Proc.* **39**:1544.

Adelstein, R. S., Conti, M. A., and Pato, M. D., 1980, Regulation of myosin light chain kinase by reversible phosphorylation and calcium-calmodulin. *Ann. N.Y. Acad. Sci.* **356**:142.

Albano, J., Bhoola, K. A., Heap, P. F., and Lemon, M. J. C., 1976, Stimulus-secretion coupling: Role of cyclic AMP, cyclic GMP and calcium in mediating enzyme (kallikrein) secretion in the submandibular gland, *J. Physiol. (London)* **258**:631.

Allison, A. C., 1973, The role of microfilaments and microtubules in cell movement, endocytosis and exocytosis, in: *Ciba Fdn. Symp. 14. Locomotion of Tissue Cells,* pp. 109–148, Elsevier, Amsterdam.

Allison, A. C., and Davies, P., 1973, Mechanisms of endocytosis and exocytosis, in: *Transport at the Cellular Level* (M. Sleigh and D. H. Jennings, eds.), pp. 419–446, Cambridge University Press, Cambridge.

Allison, A. C., and Davies, P., 1974, Mechanisms of endocytosis and exocytosis, *Symp. Soc. Exp. Biol.* **28**:419.

Arthur, G., and Sheltawy, A., 1980, The presence of lysophosphatidylcholine in chromaffin granules, *Biochem. J.* **191**:523.

Ashby, J. P., and Speake, R. N., 1975, Insulin and glucagon secretion from isolated islets of Langerhans. The effects of calcium ionophores, *Biochem. J.* **150**:89.

Ashley, C. C., and Campbell, A. K., 1979, Introduction: Why is it necessary to measure intracellular free calcium ions, in: *Detection and Measurement of Free Ca²⁺ in Cells* (C. C. Ashley and A. K. Campbell, eds.), pp. 1–10, Elsevier, Amsterdam.

Atwater, I., Ribalet, B., and Rojas, E., 1978, Cyclic changes in potential and resistance of the β-cell membrane induced by glucose in islets of Langerhans from mouse, *J. Physiol. (London)* **278**:117.

Atwater, I., Dawson, C. M., Ribalet, B., and Rojas, E., 1979, Potassium permeability activated by intracellular calcium ion concentration in the pancreatic β-cell, *J. Physiol. (London)* **288:**575.

Atwood, H. L., 1976, Organization and synaptic physiology of crustacean neuromuscular systems, *Prog. in Neurobiol.* **7:**291.

Aunis, D., Pescheloche, M., and Zwiller, J., 1978, Guanylate cyclase from bovine adrenal medulla: Subcellular distribution and studies on the effect of lysolecithin on enzyme activity, *Neuroscience* **3:**83.

Aunis, D., Hesketh, S. E., and Devilliers, G., 1979, Freeze–fracture study of the chromaffin cell during exocytosis, *Cell Tissue Res.* **197:**433.

Austin, L. A., Heath, H., and Go, V. L. W., 1979, Regulation of calcitonin secretion in normal man by changes of serum calcium within the physiological range, *J. Clin. Invest.* **64:**1721.

Babkin, B. P., 1944, *Secretory Mechanism of the Digestive Glands,* pp. 23–27, Paul B. Hoeber, New York.

Bainton, D. F., and Farquhar, M. G., 1968, Differences in enzyme content of azurophil and specific granules of polymorphonuclear leukocytes, *J. Cell Biol.* **39:**286.

Baker, P. F., 1972, Transport and metabolism of calcium ions in nerve, *Prog. Biophys. Mol. Biol.* **24:**177.

Baker, P. F., 1976, The regulation of intracellular calcium, *Symp. Soc. Exptl. Biol.* **30:**67.

Baker, P. F., and Crawford, A. C., 1972, Mobility and transport of magnesium in squid giant axons, *J. Physiol. (London)* **227:**855.

Baker, P. F., and McNaughton, P. A., 1978, The influence of extracellular calcium binding on the calcium efflux from squid axons, *J. Physiol. (London)* **276:**125.

Baker, P. F., and Rink, T. J., 1975, Catecholamine release from bovine adrenal medulla in response to maintained depolarization, *J. Physiol. (London)* **253:**593.

Baker, P. F., Meves, H., and Ridgway, E. B., 1973, Effects of manganese and other agents on the calcium uptake that follows depolarization of squid axons: Calcium entry in response to maintained depolarization of squid axons, *J. Physiol. (London)* **231:**511.

Balnave, R. J., and Gage, P. W., 1973, The inhibitory effect of manganese on transmitter release at the neuromuscular junction of the toad, *Br. J. Pharmacol.* **47:**339.

Banks, P., and Helle, K., 1965, The release of protein from the stimulated adrenal medulla, *Biochem. J.* **97:**40c.

Banks, P., Biggins, R., Bishop, R., Christian, B., and Currie N., 1969, Sodium ions and the secretion of catecholamines, *J. Physiol. (London)* **200:**797.

Barajas, L., 1979, Anatomy of the juxtaglomerular apparatus, *Am. J. Physiol.* **237:**F333.

Bargmann, W., and Scharrer, E., 1951, The site of origin of the hormones of the posterior pituitary, *Am. Scientist* **39:**255.

Barnes, G. D., Brown, B. L., Gard, T. G., Atkinson, D., and Ekins, R. P., 1978, Effect of TRH and dopamine on cyclic AMP levels in enriched mammotroph and thyrotroph cells, *Mol. Cell. Endocr.* **12:**273.

Barreras, R. F., 1973, Calcium and gastric secretion, *Gastroenterology* **64:**1168.

Bartelt, D. C., and Scheele, G. A., 1980, Calmodulin and calmodulin—Binding proteins in canine pancreas, *Ann. N.Y. Acad. Sci.* **356:**356.

Batzri, S., and Selinger, Z., 1973, Enzyme secretion mediated by the epinephrine β-receptor in parotid slices, *J. Biol. Chem.* **248:**356.

Batzri, S., Selinger, Z., Schramm, M., and Robinovitch, M. R., 1973, Potassium release mediated by the epinephrine α-receptor in rat parotid slices, *J. Biol. Chem.* **248:**361.

Baumbach, L., and Leyssac, P. P., 1977, Studies on the mechanism of renin release from isolated superfused rat glomeruli: Effects of calcium, calcium ionophore and lanthanum, *J. Physiol. (London)* **273:**745.

Bayliss, W. M., and Starling, E. H., 1904, The chemical regulation of the secretory process, *Proc. R. Soc. Ser. B.* **73**:310.

Becker, E. L., 1979, A multifunctional receptor on the neutrophil for synthetic chemotactic oligopeptides, *J. Ret. Endoth. Soc.* **26** (Suppl):701.

Becker, H., Konturek, S. J., Reeder, D. D., and Thompson, J. C., 1973, Effect of calcium and calcitonin on gastrin and gastric secretion in cats, *Am. J. Physiol.* **225**:277.

Bell, N. H., and Queener, S., 1974, Stimulation of calcitonin synthesis and release *in vitro* by calcium and dibutyryl cyclic AMP, *Nature* **248**:343.

Bell, R. L., and Majerus, P. W., 1980, Thrombin-induced hydrolysis of phosphatidylinositol in human platelets, *J. Biol. Chem.* **255**:1790.

Bennett, G. W., and Edwardson, J. A., 1975, Release of corticotrophin releasing factor and other hypophysiotropic substances from isolated nerve-endings (synaptosomes), *J. Endocr.* **65**:33.

Bennett, J. P., Cockcroft, S., and Gomperts, B. D., 1979, Ionomycin stimulates mast cell histamine secretion by forming a lipid-soluble calcium complex, *Nature* **282**:851.

Bennett, J. P., Cockcroft, S., and Gomperts, B. D., 1981, Rat mast cells permeabilized with ATP secrete histamine in response to calcium ions buffered in the micromolar range, *J. Physiol. (London)* **317**:335.

Bennett, M. R., and Florin, T., 1975, An electrophysiological analysis of the effect of Ca ions on neuromuscular transmission in the mouse vas deferens, *Br. J. Pharmacol.* **55**:97.

Bennett, W. F., Belville, J. S., and Lynch, G., 1979, Calcium-dependent serotonin release from blood platelets: A model for neurosecretion. *Neuroscience* **4**:1203.

Bent-Hansen, L., Capito, K., and Hedeskov, C. J., 1979, The effect of calcium on somatostatin inhibition of insulin release and cyclic AMP production in mouse pancreatic islets, *Biochim. Biophys. Acta.* **585**:240.

Benz, L., Eckstein, B., Matthews, E. K., and Williams, J. A., 1972, Control of pancreatic amylase release *in vitro:* Effects of ions, cyclic AMP, and colchicine, *Brit. J. Pharmacol.* **46**:66.

Berggren, P-O., Ostensen, C-G., and Hellman, B., 1979, Evidence for divergent glucose effects on calcium metabolism in pancreatic β- and α$_2$-cells, *Endocrinology* **105**:1463.

Berglindh, T., Dibona, D. R., Ito, S., and Sachs, G., 1980a, Probes of parietal cell function, *Am. J. Physiol.* **238**:G165.

Berglindh, T., Dibona, D. R., Pace, C. S., and Sachs, G., 1980b, ATP dependence of H$^+$ secretion, *J. Cell. Biol.* **85**:392.

Berglindh, T., Sachs, G., and Takeguchi, N., 1980c, Ca^{2+}-dependent secretagogue stimulation in isolated rabbit gastric glands, *Am. J. Physiol.* **239**:G90.

Berl, S., Puszkin, S., and Nicklas, W. J., 1973, Actomyosin-like protein may function in the release of transmitter material at synaptic endings, *Science* **179**:441.

Berlind, A., and Cooke, I. M., 1971, The role of divalent cations in electrically elicited release of a neurohormone from crab pericardial organ, *Gen. Comp. Endocr.* **17**:60.

Berridge, M., 1975, The interaction of cyclic nucleotides and calcium in the control of cellular activity, *Adv. Cyclic Nucleotide. Res.* **6**:1.

Biales, B., Dichter, M., and Tischler, A., 1976, Electrical excitability of cultured adrenal chromaffin cells, *J. Physiol. (London)* **262**:743.

Bigdeli, H., and Snyder, P. J., 1978, Gonadotropin-releasing hormone release from the rat hypothalamus: Dependence on membrane depolarization and calcium influx, *Endocrinology* **103**:281.

Billah, M. M., Lapetina, E. G., and Cuatrecasas, P., 1979, Phosphatidylinositol-specific phospholipase-C of platelets. Association with 1,2 diacylglycerol-kinase and inhibition by cyclic AMP, *Biochem. Biophys. Res. Commun.* **90**:92.

Billah, M. M., Lapetina, E. G., and Cuatrecasas, P., 1980, Phospholipase A₂ and phospholipase C activities of platelets, *J. Biol. Chem.* **255**:10227.

Bills, T. K., Smith, J. B., and Silver, M. J., 1976, Metabolism of [¹⁴C] arachidonic acid by human platelets, *Biochim. Biophys. Acta.* **424**:303.

Birks, R., 1965, The role of sodium ions in the release of acetylcholine, in: *Pharmacology of Cholinergic and Adrenergic Transmission* (G. B. Koelle, W. W. Douglas, and A. Carlsson, eds.), pp. 3–19, Pergamon Press, New York.

Birks, R., and MacIntosh, F. C., 1961, Acetylcholine metabolism of a sympathetic ganglion, *Can. J. Biochem. Physiol.* **39**:787.

Birmingham, M. K., Elliott, F. H., and Valere, P. H. L., 1953, The need for the presence of calcium for the stimulation *in vitro* of rat adrenal glands by adrenocorticotrophic hormone, *Endocrinology* **53**:687.

Bisby, M. A., Fillenz, M., and Smith, A. D., 1973, Evidence for the presence of dopamine-β-hydroxylase in both populations of noradrenaline storage vesicles in sympathetic nerve terminals of the rat vas deferens, *J. Neurochem.* **20**:245.

Black, J. W., 1979, The riddle of gastric histamine, in: *Histamine Receptors* (T. O. Yellin, ed.), pp. 23–33, SP Medical, New York.

Blackmore, P. F., El-Refai, M. F., and Exton, J. H., 1979, α-Adrenergic blockade and inhibition of A23187 mediated Ca²⁺ uptake by the calcium antagonist verapamil in rat liver cells, *Mol. Pharmacol.* **15**:598.

Blaschko, H., Hagen, P., and Welch, A. D., 1955, Observations on the intracellular granules of the adrenal medulla, *J. Physiol. (London)* **129**:27.

Blaschko, H., Comline, R. S., Schneider, F. H., Silver, M., and Smith, A. D., 1967, Secretion of a chromaffin granule protein, chromogranin, from the adrenal gland after splanchnic stimulation, *Nature* **215**:58.

Blaustein, M. P., 1974, The interrelationship between sodium and calcium fluxes across cell membranes, *Rev. Physiol. Biochem. Pharmacol.* **70**:33.

Blaustein, M. P., 1975, Effects of potassium, veratridine and scorpion venom on calcium accumulation and transmitter release by nerve terminals *in vitro*, *J. Physiol. (London)* **247**:617.

Blaustein, M. P., and Ector, A. C., 1975, Barbiturate inhibition of calcium uptake by depolarized nerve terminals *in vitro*, *Mol. Pharmacol.* **11**:369.

Blaustein, M. P., and Goldring, J. M., 1975, Membrane potentials in pinched-off presynaptic nerve terminals monitored with a fluorescent probe: Evidence that synaptosomes have potassium diffusion potentials, *J. Physiol. (London)* **247**:589.

Blaustein, M. P., Johnson, E. M., and Needleman, P., 1972, Calcium-dependent norepinephrine release from presynaptic nerve endings *in vitro*, *Proc. Nat. Acad. Sci.* **69**:2237.

Blaustein, M. P., Ratzlaff, R. W., and Kendrick, N. K., 1978, The regulation of intracellular calcium in presynaptic nerve terminals, *Ann. N.Y. Acad. Sci.* **307**:195.

Blaustein, M. P., Ratzlaff, R. W., and Schweitzer, E. S., 1980, Control of intracellular calcium in presynaptic nerve terminals, *Fed. Proc.* **39**:2790.

Blinks, J. R., 1978, Measurement of [Ca⁺⁺] with photoproteins, *Ann. N.Y. Acad. Sci.* **307**:71.

Blinks, J. R., Prendergast, G., and Allen, D. G., 1976, Photoproteins as biological calcium indicators, *Pharmacol. Rev.* **28**:1.

Bloom, S. R., and Polak, J. M., 1978, Gut hormone overview, in: *Gut Hormones* (S. R. Bloom and M. I. Grossman, eds.), pp. 3–18, Churchill Livingstone, Edinburgh.

Boeynaems, J. M., Waelbroeck, M., and Dumont, J. E., 1979, Cholinergic and α-adrenergic stimulation of prostaglandin release by dog thyroid *in vitro*, *Endocrinology* **105**:988.

Bokoch, G. M., and Reed, P. W., 1980, Stimulation of arachidonic acid metabolism in the polymorphonuclear leukocyte by an N-formylated peptide, *J. Biol. Chem.* **255**:10223.

Bokoch, G. M., and Reed, P. W., 1981, Effect of various lipoxygenase metabolites of arachidonic acid on degranulation of polymorphonuclear leukocytes, *J. Biol. Chem.* **256:**5317.

Borle, A., 1969, Kinetic analysis of calcium movements in Hela cell cultures. II. Calcium efflux. J. Gen. Physiol. **53:**57.

Borle, A. B., 1974, Cyclic AMP stimulation of calcium efflux from kidney, liver and heart mitochondria, *J. Membrane Biol.* **16:**221.

Borle, A. B., 1975, Modulation of mitochondrial control of cytoplasmic calcium activity, in: *Calcium Transport in Contraction and Secretion*, (E. Carafoli, F. Clementi, W. Drabikowski, and A. Margreth, eds.), pp. 77–86, North-Holland, Amsterdam.

Borle, A. B., and Studer, B., 1978, Effects of calcium ionophores on the transport and distribution of calcium in isolated cells and in liver and kidney slices, *J. Membrane Biol.* **38:**51.

Borowitz, J. L., 1972, Effect of lanthanum on catecholamine release from adrenal medulla, *Life Sci.* **1:**959.

Botelho, S. Y., 1964, Tears and the lacrimal gland, *Sci. Am.* **211:**78.

Boucek, M. M., and Snyderman, R., 1976, Calcium influx requirement for human neutrophil chemotaxis: inhibition by lanthanum chloride, *Science* **193:**905.

Bourne, G. A., and Baldwin, D. M., 1980, Extracellular Ca^{+2} independent and dependent components of the biphasic release of LH in response to luteinizing hormone-releasing hormone *in vitro*, *Endocrinology* **107:**780.

Bracho, H., and Orkand, R. K., 1970, Effect of calcium on excitatory neuromuscular transmission in the crayfish, *J. Physiol. (London)* **206:**61.

Brandt, B. L., Hagiwara, S., Kidokoro, Y., and Miyazaki, S., 1976, Action potentials in the rat chromaffin cell and effects of acetylcholine, *J. Physiol. (London)* **263:**417.

Brink, F., 1954, The role of calcium ions in neural processes, *Pharmacol. Rev.* **6:**243.

Brinley, F. S., Jr., 1978, Calcium buffering in squid axons, *Ann. Rev. Biophys. Bioeng.* **7:**363.

Brisson, G. R., Malaisse-Lagae, F., and Malaisse, W. J., 1972, The stimulus-secretion coupling of glucose-induced insulin release. VII. A proposed site of action for adenosine-3'-5'-cyclic monophosphate. *J. Clin. Invest.* **51:**232.

Brockerhoff, H., and Jensen, R. G., 1974, *Lipolytic Enzymes*, pp. 194–243. Academic Press, New York.

Bromberg, B. B., 1981, Autonomic control of lacrimal protein secretion, *Invest. Ophthal. Vis. Science.* **20:**110.

Brown, E. M., 1980, Calcium-regulated phosphodiesterase in bovine parathyroid cells, *Endocrinology* **107:**1998.

Brown, E. M., Hurwitz, S., and Aurbach, G. D., 1976, Preparation of viable isolated bovine parathyroid cells, *Endocrinology* **99:**1582.

Brown, E. M., Gardner, D. G., Windeck, R. A., and Aurbach, G. D., 1978, Relationship of intracellular 3'-5'-adenosine monophosphate accumulation to parathyroid hormone release from dispersed bovine parathyroid cells, *Endocrinology* **103:**2323.

Brown, E. M., Gardner, D. G., and Aurbach, G. D., 1980, Effects of the calcium ionophore A23187 on dispersed bovine parathyroid cells, *Endocrinology* **106:**133.

Burke, B. E., and DeLorenzo, R. J., 1981, Ca^{2+}-and calmodulin-stimulated endogenous phosphorylation of neurotubulin, *Proc. Natl. Acad. Sci.* **78:**991.

Burke, G., 1970, Effects of cations and ouabain on thyroid adenyl cyclase, *Biochim. Biophys. Acta.* **220:**30.

Burke, G., Chang, L-L., and Szabo, M., 1973, Thyrotropin and cyclic nucleotide. Effects on prostaglandin levels in isolated thyroid cells, *Science* **180:**872.

Burn, J. H., and Gibbons, W. R., 1964, The part played by calcium in determining the response to stimulation of sympathetic postganglionic nerves, *Br. J. Pharmcol.* **22:**540.

Butcher, F. R., 1975, The role of calcium and cyclic nucleotides in α-amylase release from slices of rat parotid: Studies with the divalent cation ionophore A23187, *Metabolism* **24**:409.

Butcher, F. R., 1978, Regulation of exocytosis, in: *Biochemical Actions of Hormones,* Vol. V (G. Litwack, ed.), pp. 53–99, Academic Press, New York.

Butcher, F. R., and Putney, J. W., 1980, Regulation of parotid gland function by cyclic nucleotides and calcium, *Adv. Cyclic Nucleotide Res.* **13**:215.

Butcher, F. R., Rudich, L., Emler, C., and Nemerovski, M., 1976, Adrenergic regulation of cyclic nucleotide levels, amylase release and potassium efflux in rat parotid gland, *Mol. Pharmacol.* **12**:862.

Bygrave, F. L., 1977, Mitochondrial calcium transport, in: *Current Topics in Bioenergetics,* Vol. 6 (D.R. Sanadi, ed.), pp. 260–318, Academic Press, London.

Bygrave, F. L., 1978, Mitochondria and the control of intracellular calcium, *Biol. Rev.* **53**:43.

Capito, K., and Hedeskov, C. J., 1977, Effects of glucose, glucose metabolites and calcium ions on adenylate cyclase activity in homogenates of mouse pancreatic islets, *Biochem. J.* **162**:569.

Carafoli, E., and Crompton, M., 1978a, The regulation of intracellular calcium by mitochondria, *Ann. N.Y. Acad. Sci.* **307**:269.

Carafoli, E., and Crompton, M., 1978b, The regulation of intracellular calcium, in: *Current Topics in Membranes and Transport,* Vol. 10 (F. Bronner and A. Kleinzeller, eds.), pp. 151–216, Academic Press, New York.

Carbone, E., 1979, Aequorin and fluorescent chelating probes to detect free calcium influx and membrane-associated calcium in excitable cells, in: *Detection and Measurement of Free Ca^{2+} in Cells* (C.C. Ashley and A.K. Campbell, eds.), pp. 355–371, Elsevier, New York.

Carchman, R. A., Shen, J. C., Bilgin, S., and Rubin, R. P., 1980, Diverse effects of Ca^{2+} on the prostacyclin and corticotropin modulation of cyclic AMP and steroid production in normal cat and mouse tumor cells of the adrenal cortex, *Biochem. Pharmacol.* **29**:2213.

Care, A. D., Cooper, C. W., Duncan, T., and Orimo, H., 1968, A study of thyrocalcitonin secretion by direct measurement of *in vitro* secretion rates in pigs, *Endocrinology* **83**:161.

Care, A. D., Bell, N. H., and Bates, R. F. L., 1971, The effects of hypermagnesemia on calcitonin secretion *in vivo, J. Endocrinol.* **51**:381.

Carpentier, J. L., Malaisse-Lagae, F., Muller, W. A., and Orci, L., 1977, Glucagon release from rat pancreatic islets. A combined morphological and functional approach. *J. Clin. Invest.* **60**:1174.

Carsten, M. E., and Miller, J. D., 1977, Effects of prostaglandins and oxytocin on calcium release from a uterine microsomal fraction. Hypothesis for ionophoretic action of prostaglandins, *J. Biol. Chem.* **252**:1576.

Case, R. M., 1978, Synthesis, intracellular transport and discharge of exportable proteins in the pancreatic acinar cell and other cells, *Biol. Rev.* **53**:211.

Case, R. M., and Clausen, T., 1973, The relationship between calcium exchange and enzyme secretion in the isolated rat pancreas, *J. Physiol.* **235**:75.

Castellucci, V. F., Kandel, E. R., Schwartz, J. H., Wilson, F. D., Nairn, A. C., and Greengard, P., 1980, Intracellular injection of the catalytic subunit of cyclic AMP-dependent protein kinase stimulates facilitation of transmitter release underlying behavioral sensitization in *Aplysia, Proc. Natl. Acad. Sci.* **77**:7492.

Ceccarelli, B., and Hurlbut, W. R., 1980a, Ca^{2+}-dependent recycling of synaptic vesicles at the frog neuromuscular junction, *J. Cell. Biol.* **87**:297.

Ceccarelli, B., and Hurlbut, W. P., 1980b, Vesicle hypothesis of the release of quanta of acetylcholine, *Physiol. Rev.* **60**:396.

Ceccarelli, B., Hurlbut, W. P., and Mauro, A., 1972, Depletion of vesicles from frog neuromuscular junctions by prolonged tetanic stimulation, *J. Cell. Biol.* **54**:30.

Cervetto, L., and Piccolino, M., 1974, Synaptic transmission between photoreceptors and horizontal cells in the turtle retina, *Science* **183**:417.

Chakravarty, N., and Echetebu, Z., 1978, Plasma membrane adenosine triphosphatases in rat pertioneal mast cells and macrophages—the relation of the mast cell enzyme to histamine release, *Biochem. Pharm.* **27**:1561.

Chambaut-Guerin, A. M., Muller, P., and Rossignol, B., 1978, Microtubules and protein secretion in rat lacrimal glands, *J. Biol. Chem.* **253**:3870.

Champion, S., and Jacquemin, C., 1978, The role of cyclic GMP in the action of carbamylcholine on the cyclic AMP level in the dog thyroid, *FEBS Lett.* **92**:156.

Chance, B., 1965, The energy linked reaction of calcium with mitochondria, *J. Biol. Chem.* **240**:2729.

Chandler, D. E., 1978, Control of pancreatic enzyme secretion: A critique of the role of calcium, *Life Sci.* **23**:323.

Chandler, D. E., and Williams, J. A., 1974, Pancreatic acinar cells: Effects of lanthamum ions on amylase release and calcium ion fluxes, *J. Physiol. (London)* **243**:831.

Chandler, D. E., and Williams, J. A., 1977, Intracellular uptake and α-amylase and lactate dehydrogenase releasing actions of the divalent cation ionophore A23187 in dissociated pancreatic acinar cells, *J. Membrane Biol.* **32**:201.

Chandler, D. E., and Williams, J. A., 1978, Intracellular divalent cation release in pancreatic acinar cells during stimulus-secretion coupling. II. Subcellular localization of the fluorescent probe chlorotetracycline, *J. Cell. Biol.* **76**:386.

Chap, H. J., Zwaal, R. F. A., and Van Deenen, L. L. M., 1977, Action of highly purified phospholipases on blood platelets. Evidence for an asymmetric distribution of phospholipids in the surface membrane, *Biochim. Biophys. Acta.* **467**:146.

Chapman, D. B., and Way, E. L., 1980, Metal ion interactions with opiates, *Ann. Rev. Pharm. Tox.* **20**:553.

Charles, M. A., Lawecki, J., Pictet, R., and Grodsky, G. M., 1975, Insulin secretion. Interrelationships of glucose, cyclic adenosine 3'5' monophosphate and calcium, *J. Biol. Chem.* **250**:6134.

Chen, D. S., and Poisner, A. M., 1976, Direct stimulation renin release by calcium, *Proc. Soc. Exp. Biol. Med.* **152**:565.

Chenoweth, D. E., and Hugli, T. E., 1978, Demonstration of specific C5a receptor on intact human polymorphonuclear leukocytes, *Proc. Natl. Acad. Sci.* **75**:3943.

Cheung, W. Y., 1979, Calmodulin plays a pivotal role in cellular regulation, *Science* **207**:19.

Chew, G. S., Hersey, S. J., Sachs, G., and Berglindh, T., 1980, Histamine responsiveness of isolated gastric glands, *Am. J. Physiol.* **238**:G312.

Christiansen, J., Rehfeld, J. F., and Kirkegaard, P., 1979, Interaction of calcium, magnesium and gastrin on gastric acid secretion, *Gastroenterology* **76**:57.

Christophe, J., Frandsen, E. K., Conlon, T. P., Krishna, G., and Gardner, J. D., 1976, Action of cholecystokinin, cholinergic agents and A23187 on accumulation of guanosine 3'5'-monophosphate in dispersed guinea pig pancreatic acinar cells, *J. Biol. Chem.* **251**:4640.

Chubb, I. W., DePotter, W. P., and DeSchaepdryver, A. F., 1972, Tyramine does not release noradrenaline from splenic nerve by exocytosis, *Naunyn-Schmiedeberg's Arch. Pharmacol.* **274**:281.

Churchill, P. C., 1979, Possible mechanism of the inhibitory effect of ouabain on renin secretion from rat renal cortical slices, *J. Physiol. (London)* **294**:123.

Churchill, P. C., 1980, Effect of D-600 on inhibition of *in vitro* renin release in the rat by high extracellular potassium and angiotensin II. *J. Physiol. (London)* **304**:449.

Churchill, M. C., and Churchill, P. C., 1980, Separate and combined effects of ouabain and extracellular potassium on renin secretion from rat renal cortical slices, *J. Physiol. (London)* **300**:105.

Churchill, P. C., and Lyons, H. J., 1976, Effect of intrarenal arterial infusion of magnesium on renin release in dogs, *Proc. Soc. Exp. Biol. Med.* **152**:6.

Churchill, P. C., McDonald, F. D., and Churchill, M. C., 1979, Phenytoin stimulates renin secretion from rat kidney slices, *J. Pharmacol. Exptl. Ther.* **211**:615.

Clark, M. R., Marsh, J. M., and LeMaire, W. J., 1978, Mechanism of luteinizing hormone regulation of prostaglandin synthesis in rat granulosa cells, *J. Biol. Chem.* **253**:7757.

Clark, R. M., and Collins, G. G. S., 1976, The release of endogenous amino acids from the rat visual cortex, *J. Physiol. (London)* **262**:383.

Clarke, M., and Spudich, J. A., 1977, Nonmuscle contractile proteins: The role of actin and myosin in cell motility and shape determination, *Ann. Rev. Biochem.* **46**:797.

Clemente, F., and Meldolesi, J., 1975, Calcium and pancreatic secretion. I. Subcellular distribution of calcium and magnesium in the exocrine pancreas of the pig, *J. Cell. Biol.* **65**:88.

Cochrane, D. E., and Douglas, W. W., 1974, Calcium-induced extrusion of secretory granules (exocytosis) in mast cells exposed to 48/80 or the ionophore A23187 and X-537A, *Proc. Natl. Acad. Sci.* **71**:408.

Cochrane, D. E., and Douglas, W. W., 1976, Histamine release by exocytosis from rat mast cells on reduction of extracellular sodium: A secretory response inhibited by calcium strontium, barium, or magnesium, *J. Physiol. (London)* **257**:433.

Cockcroft, S., and Gomperts, B. D., 1979, Activation and inhibition of calcium-dependent histamine secretion by ATP ions applied to rat mast cells, *J. Physiol. (London)* **296**:229.

Cockcroft, S., Bennett, J. P., and Gomperts, B. D., 1980, F-Met-Leu-Phe-Induced phosphatidylinositol turnover in rabbit neutrophils is dependent on extracellular calcium, *FEBS. Lett.* **110**:115.

Coleman, R., 1973, Membrane-bound enzymes and membrane ultrastructure, *Biochim. Biophys. Acta.* **300**:1.

Collier, B., 1969, The preferential release of newly synthesized transmitter by a sympathetic ganglion, *J. Physiol. (London)* **205**:341.

Collins, F., and Epel, D., 1977, The role of calcium ions in the acrosome reaction of sea urchin sperm, *Exp. Cell. Res.* **106**:211.

Collins, G. G. S., 1974, The spontaneous and electrically evoked release of [³H]-GABA from the isolated hemisected frog spinal cord, *Brain Res.* **66**:121.

Collins, G. G. S., 1977, The release of endogenous amino acids from rat visual cortex by calcium ions in the presence of the calcium ionophores X537A and A23187, *J. Neurochem.* **28**:461.

Conigrave, A. D., Treiman, M., Saermark, T., and Thorn, N., 1981, Stimulation by calmodulin of Ca^{2+} uptake and (Ca^{2+}-Mg^{2+}) ATPase activity in membrane fractions from ox neurohypophyses, *Cell Calcium* **2**:125.

Conn, P. M., and Rogers, D. C., 1980, Gonadotropin release from pituitary cultures following activation of endogenous ion channels, *Endocrinology* **107**:2133.

Conn, P. M., Marian, J., McMillan, M., Stern, J., Rogers, D., Hamby, U., Penna, A., and Grant, E., 1981, Gonadotropin releasing hormone action in the pituitary: a three step mechanism, *Endocrine Rev.* **2**:174.

Cooper, C. W., 1975, Ability of several cations to promote secretion of thyrocalcitonin in the pig, *Proc. Soc. Exp. Biol.* **148**:449.

Cooper, P. H., and Stanworth, D. R., 1976, Characterization of calcium-ion activated adenosine triphosphatase in the plasma membrane of rat mast cells, *Biochem. J.* **156**:691.

Copp, D. H., 1969, Endocrine control of calcium homeostasis, *J. Endocrinol.* **43**:137.

Corbin, D. E., and Sun, G. Y., 1978, Characteryation of the enzymic transfer of arachidonyl groups to 1-acyl-phosphoglycerides in mouse synaptosome fraction, *J. Neurochem.* **30**:77.

Corrodi, H., and Jonsson, G., 1967, The formaldehyde fluorescence method for the histochemical demonstration of biogenic amines, *J. Histochem. Cytochem.* **15**:65.

Couturier, E., and Malaisse, W. J., 1980, Insulinotropic effects of hypoglycaemic and hyperglycaemic sulphonamides: The ionopheretic hypothesis, *Diabetologia* **19**:335.

Cramer, W., 1918, Further observations on the thyroid-adrenal apparatus. A histochemical method for the demonstration of adrenalin granules in the suprarenal gland, *J. Physiol. (London)* **52**:VIII.

Crawford, N., Chahal, H., and Jackson, P., 1980, The isolation and characterisation of guinea-pig polymorphonuclear leucocyte actin and myosin, *Biochim. Biophys. Acta.* **626**:218.

Cunningham, J., and Neal, M. J., 1981, On the mechanism by which veratridine causes a calcium-independent release of γ-aminobutyric acid from brain slices, *Brit. J. Pharmacol.* **73**:655.

Curry, D. L., Bennett, L. L., and Grodsky, G. M., 1968a, Dynamics of insulin secretion by the perfused rat pancreas, *Endocrinology* **83**:572.

Curry, D. L., Bennett, L. L., and Grodsky, G. M., 1968b, Requirement for calcium ion in insulin secretion by the perfused rat pancreas, *Am. J. Physiol.* **214**:174.

Cushing, H., 1901, Concerning the poisonous effects of pure sodium chloride solution upon the nerve–muscle preparation, *Am. J. Physiol.* **6**:77.

Dahlquist, R., 1974, Relationship of uptake of sodium and ^{45}Calcium to ATP-induced histamine release from rat mast cells, *Acta. Pharm. Tox.* **35**:11.

Dahlström, A., 1973, Aminergic transmission: Introduction and short review, *Brain Res.* **62**:441.

Davis, B., and Lazarus, N. R., 1976, An *in vitro* system for studying insulin release caused by secretory granules-plasma membrane interaction: Definition of the system, *J. Physiol. (London)* **256**:709.

Davis, J. O., and Freeman, R. H., 1976, Mechanisms regulating renin release, *Physiol. Rev.* **56**:1.

DeBelleroche, J. S., and Bradford, H. F., 1972, The stimulus-induced release of acetylcholine from synaptosome beds and its calcium dependence, *J. Neurochem.* **19**:1817.

Dedman, J. R., Brinkley, B. R., and Means, A. R., 1979, Regulation of microfilaments and microtubules by calcium and cyclic AMP, *Adv. Cyclic Nucleotide Res.* **11**:131.

Deftos, L. J., Roos, B. A., Knecht, G. L., Lee, J. C., Paulinac, D., Bone, H. G., and Parthemore, J. G., 1978, Calcitonin secretion, in: *Endocrinology of Calcium Metabolism* (B. H. Copp and R. V. Talmage, eds.), pp. 134–142, Excerpta Medica, Amsterdam.

Dekker, A., and Field, J. B., 1970, Correlation of effects of thyrotropin, prostaglandins and ions on glucose oxidation, cyclic AMP, and colloid proplet formation in dog thyroid slices, *Metabolism* **19**:453.

DeLorenzo, R. J., Freedman, S. D., Yohe, W. B., and Maurer, S. C., 1979, Stimulation of Ca^{2+}-dependent neurotransmitter release and presynaptic nerve terminal protein phosphorylation by calmodulin and a calmodulin-like protein isolated from synaptic vesicles, *Proc. Natl. Acad. Sci.* **76**:1838.

Denef, J. F., Björkman, U., and Ekholm, R., 1980, Structural and functional characteristics of isolated thyroid follicles, *J. Ultrast. Res.* **71**:185.

DePotter, W. P., and Chubb, I. W., 1977, Biochemical observations on the formation of small noradrenergic vesicles in the splenic nerve of the dog, *Neuroscience* **2**:167.

Der, O. M., and Sun, G. V., 1981, Degradation of arachidonoyl-labeled phosphatidyl-inositols by brain synaptosomes, *J. Neurochem.* **36**:355.

DeRubertis, F. R., and Craven, P. A., 1976, Calcium-independent modulation of cyclic GMP and activation of guanylate cyclase by nitrosamines, *Science* **193**:897.

Dibona, D. R., Berglindh, T., and Sachs, G., 1979, Cellular site of gastric secretion, *Proc. Natl. Acad. Sci.* **76**:6689.

Dicker, S. E., 1966, Release of vasopressin and oxytocin from isolated pituitary glands of adult and new-born, *J. Physiol. (London)* **185**:429.

DiPolo, R., 1978, Ca pump driven-by ATP in squid axons, *Nature* **274**:390.

Dixon, W. R., Garcia, A. G., and Kirpekar, S. M., 1975, Release of catecholamines and dopamine-β-hydroxylase from the perfused adrenal gland of the cat, *J. Physiol. (London)* **244**:805.

Dormer, R. L., and Ashcroft, S. J. H.: 1974, Studies on the role of calcium ions in the stimulation by adrenaline of amylase release from rat parotid, *Biochem. J.* **144**:543.

Dormer, R. L., Poulsen, J. H., Licko, V., and Williams, J. A., 1981, Calcium fluxes in isolated pancreatic acini: Effects of secretagogues, *Am. J. Physiol.* **240**:G38.

Dorsling, J. E., and Ripps, H., 1973, Effect of magnesium on horizontal cell activity in the skate retina, *Nature* **242**:101.

Douglas, W. W., 1968, Stimulus–secretion coupling: The concept and clues from chromaffin and other cells, *Br. J. Pharmacol.* **34**:451.

Douglas, W. W., 1974a, Involvement of calcium in exocytosis and the exocytosis-vesiculation sequence, *Biochem. Soc. Symp.* **39**:1.

Douglas, W. W., 1974b, Mechanisms of release of neurohypophysial hormones: stimulus-secretion coupling, in: *Handbook of Physiology*, Section 7, *Endocrinology*, Vol. IV, *The Pituitary Gland and its Neuroendocrine Control*, Part 1 (R. O. Greep and E. B. Astwood, Sect. eds.), pp. 191–224, Amer. Physiol. Soc., Washington, D.C.

Douglas, W. W., 1974c, Exocytosis and the exocytosis—vesiculation sequence: With special reference to neurohypophysis, chromaffin and mast cells, calcium and calcium ionophores, in: *Secretory Mechanisms of Exocrine Glands* (N. A. Thorn and O. H. Petersen eds.), pp. 116–136, Munksgaard, Copenhagen.

Douglas, W. W., 1975, Secretomotor control of adrenal medullary secretion: Synaptic, membrane, and ionic events in stimulus–secretion coupling, in: *Handbook of Physiology*, Section 7, Volume VI (H. Blaschko, G. Sayers and A. D. Smith, eds.), pp. 367–388, Washington, D.C., Amer. Physiol. Soc.

Douglas, W. W., 1978, Stimulus–secretion coupling: Variations on the theme of calcium-activated exocytosis involving cellular and extracellular sources of calcium, *Ciba. Foundation Symp.* **54**, pp. 61–90, Elsevier, Amsterdam.

Douglas, W. W., and Kagayama, M., 1977, Calcium and stimulus–secretion coupling in the mast cell: Stimulant and inhibitory effects of calcium-rich media on exocytosis, *J. Physiol. (London)* **270**:691.

Douglas, W. W., and Kanno, T., 1967, The effect of amethocaine on acetycholine-induced depolarization and catecholamine secretion in the adrenal chromaffin cell, *Br. J. Pharmacol.* **30**:612.

Douglas, W. W., and Poisner, A. M., 1962, On the mode of acetylcholine in evoking adrenal medullary secretion: increased uptake of calcium during the secretory response, *J. Physiol. (London)* **162**:385.

Douglas, W. W., and Poisner, A. M., 1963, The influence of calcium on the secretory response of the submaxillary gland to acetylcholine or to noradrenaline, *J. Physiol. (London)* **165**:528.

Douglas, W. W., and Poisner, A. M., 1964a, Calcium movement in the neurohypophysis of the rat and its relation to the release of vasopressin, *J. Physiol. (London)* **172**:19.

Doublas, W. W., and Poisner, A. M., 1964b, Stimulus–secretion coupling in a neurosecretory organ: The role of calcium in the release of vasopressin from the neurohypophysis, *J. Physiol. (London)* **172**:1.

Douglas, W. W., and Rubin, R. P., 1964, The stimulant action of barium on the adrenal medulla, *Nature* **203**:305.

Dreifuss, J. J., 1975, A review on neurosecretory granules: their contents and mechanisms of release, *Ann. N.Y. Acad. Sci.* **248**:184.

Dreifuss, J. J., Kalnins, I., Kelly, J. S., and Ruf, K. B., 1971, Action potentials and release of neurohypophysial hormones *in vitro*, *J. Physiol. (London)* **215**:805.

Dretcher, K. L., Standaert, F. G., Skirboll, L. R., and Morgenroth, V. H., 1976, Evidence for a prejunctional role of cyclic nucleotides in neuromuscular transmission, *Nature* **264**:79.

Drouva, S. V., Epelbaum, J., Hery, M., Tapia-Arancibia, L., Laplante, E., and Kordon, C., 1981, Ionic channels involved in the LHRH and SRIF release from rat mediobasal hypothalamus, *Neuroendocrinology* **32**:155.

Dumont, J. E., 1971, The action of thyrotropin on thyroid metabolism, *Vit. Horm.* **29**:287.

Dumont, J. E., Williams, C., Van Sande, J., and Neve, P., 1971, Regulation of the release of thyroid hormones: Role of cyclic AMP, *Ann. N.Y. Acad. Sci.* **185**:291.

Dumont, J. E., Boeynaems, J. M., Van Sande, J., Erneux, C., Decosker, C., Van Cauter, E., and Mockel, J., 1977, Cyclic AMP, calcium and cyclic GMP in the regulation of thyroid function, in: *Hormones and Cell Regulation* (J. E. Dumont and J. Nunez, eds.), pp. 171–194, Elsevier North Holland, Amsterdam.

Duncan, C. J., and Statham, H. E., 1977, Interacting effects of temperature and extracellular calcium on the spontaneous release of transmitter at the frog neuromuscular junction, *J. Physiol. (London)* **268**:329.

Eaton, D. C., and Brodwick, M. S., 1980, Effects of barium on the potassium conductance of squid axon, *J. Gen. Physiol.* **75**:727.

Ekholm, R., and Ericson, L. E., 1968, The ultrastructure of the parafollicular cells of the thyroid gland in the rat, *J. Ultrast. Res.* **23**:378.

Elsbach, P., Patriarca, P., Pettis, P., Stossel, T. P., Mason, R. J., and Vaughan, M., 1972, The appearance of lecithin-^{32}P, synthesized from lysolecithin-^{32}P, in phagosomes of polymorphonuclear leukocytes, *J. Clin. Invest.* **51**:1910.

Emson, P. C., Fahrenkrug, J., Schaffalitzky-DeMuckadell, O. B., Jessell, T. M., and Iversen, L. L., 1978, Vasoactive intestinal polypeptides (VIP); vesicular localization and potassium-evoked release from rat hypothalamus. *Brain Res.* **143**:174.

Endo, M., 1976, Calcium release from the sarcoplasmic reticulum, *Physiol. Rev.* **57**:71.

Ennis, M., Truneh, A., White, J. R., and Pearce, F. L., 1980, Calcium pools involved in histamine release from rat mast cells, *Int. Archs. Allergy Appl. Immun.* **62**:467.

Ericson, L. E., 1981, Exocytosis and endocytosis in the thyroid follicle cell, *Mol. Cell. Endocrinol.* **22**:1.

Erulkar, S. D., Dambach, G. E., and Mender, D., 1974, The effect of magnesium at motorneurons of the isolated spinal cord of the frog, *Brain Res.* **66**:413.

Esterhuizen, A. C., and Howell, S. L., 1970, Ultrastructure of the A-cells of cat islets of Langerhans following sympathetic stimulation of glucagon secretion, *J. Cell. Biol.* **46**:593.

Eto, S., Wood, J. M., Hutchins, M., and Fleischer, N., 1974, Pituitary $^{45}Ca^{2+}$ uptake and release of ACTH, GH, and TSH: effect of verapamil, *Am. J. Physiol.* **226**:1315.

Ettienne, E. M., and Fray, J. C. S., 1979, Influence of potassium, sodium, calcium, perfusion pressure and isoprenaline on renin release induced by high concentrations of magnesium, *J. Physiol. (London)* **292**:373.

Euler, U. S. von, and Hillarp, N. A., 1956, Evidence for the presence of noradrenaline in submicroscopic structures of adrenergic axons, *Nature* **177**:44.

Eyzaguirre, C., and Fidone, S. J., 1980, Transduction mechanisms in carotid body: glomus cells, putative neurotransmitters and nerve endings, *Am. J. Physiol.* **239**:C135.

Fabiato, A., and Fabiato, F., 1978, Calcium-induced release of calcium from the sarcoplasmic reticulum of skinned cells from adult human, dog, cat, rabbit, rat, and frog hearts and from fetal and new-born rat ventricles, *Ann. N.Y. Acad. Sci.* **307**:491.

Fain, J. N., and Berridge, M. J., 1979, Relationship between phosphatidylinositol synthesis and recovery of 5-hydroxytryptamine-responsive Ca^{2+} flux in blowfly salivary glands, *Biochem. J.* **180**:655.

Fakunding, J. L., and Catt, K. J., 1980, Dependence of aldosterone stimulation in adrenal glomerulosa cells on calcium uptake: Effects of lanthanum and verapamil, *Endocrinology* **107**:1345.

Fakunding, J. L., Chow, R., and Catt, K. J., 1979, The role of calcium in the simulation of aldosterone production by adrenocorticotropin, angiotensin II, and potassium in isolated glomerulosa cells, *Endocrinology* **105**:327.

Farese, R. V., and Prudente, W. J., 1978, On the role of calcium in adrenocorticotropin-induced changes in mitochondrial pregnenolone synthesis, *Endocrinology* **103**:1264.

Farquhar, M. G., 1971, Processing of secretory products by cells of the anterior pituitary gland, *Mem. Soc. Endocrinol.* **19**:79.

Farquhar, M. G., Skutelsky, E. H., and Hopkins, C. R., 1975, Structure and function of the anterior pituitary and dispersed pituitary cells *in vitro* studies, in: *Ultrastructure in Biological Systems* Vol. 7 (A. Tixier-Vidal and M. G. Farquhar, eds.), pp. 45–61, Academic Press, New York.

Fatt, P., and Ginsborg, B. L., 1958, The ionic requirements for the production of action potentials in crustacean muscle fibers, *J. Physiol. (London)* **142**:516.

Feinman, R. D., and Detwiler, T. C., 1974, Platelet secretion induced by divalent cation ionophores, *Nature* **249**:172.

Feinstein, M. B., 1964, Reaction of local anesthetics with phospholipids, a possible chemical basis for anesthesia, *J. Gen. Physiol.* **48**:357.

Feinstein, M. B., 1966, Inhibition of contraction and calcium exchangeability in rat uterus by local anesthetics, *J. Pharmacol. Exptl. Therap.* **152**:516.

Feinstein, M. B., 1978a, The role of calcium in blood platelet function, in: *Calcium in Drug Action* (G. B. Weiss, ed.), pp. 197–239, Plenum Press, New York.

Feinstein, M. B., 1978b, Role of Ca^{2+} ions in the regulation of platelet function, *Rec. Prog. in Cell. Biol.* **45**:1.

Feinstein, M. B., 1980, Release of intracellular membrane-bound calcium precedes the onset of stimulus–induced exocytosis in platelets, *Biochim. Biophys. Res. Comm.* **93**:593.

Feinstein, M. B., Fiekers, J., and Fraser, C., 1976, An analysis of the mechanism of local anesthetic inhibition of platelet aggregation and secretion, *J. Pharmacol. Exptl. Therap.* **197**:215.

Feinstein, M. B., Becker, E. L., and Fraser, C., 1977, Thrombin, collagen and A23187 stimulated endogenous platelet arachidonate metabolism: Differential inhibition by PGE_1, local anesthetics and a serine-protease inhibitor, *Prostaglandins* **14**:1075.

Feinstein, M. B., Rodan, G. A., and Cutler, L. S., 1981, Cyclic AMP and calcium in platelet function, in: *Platelets in Biology and Pathology* (J. L. Gordon, ed.), pp. 437–472, Elsevier, Amsterdam.

Feng, T. P., 1936, Studies on the neuromuscular junction. II. The universal antagonism between calcium and eurarizing agencies, *Chin. J. Physiol.* **10**:513.

Ferrendelli, J. A., Rubin, E. H., and Kinscherf, D. A., 1976, Influence of divalent cations on regulation of cyclic GMP and cyclic AMP levels in brain tissue, *J. Neurochem.* **26**:741.

Field, J. B., Dekker, A., Zor, U., and Kaneko, T., 1971, *In vitro* effects of prostaglandins on thyroid gland metabolism, *Ann. N.Y. Acad. Sci.* **180**:278.

Field, M., 1979, Intracellular mediators of secretion in the small intestine, in: *Mechanisms of Intestinal Secretion* (H. J. Binder, ed.), pp. 83–91, Alan R. Liss, Inc., New York.

Fishman, M. C., 1976, Membrane potential of juxtaglomerular cells, *Nature* **260**:542.

Flatt, P. R., Boquist, L., and Hellman, B., 1980, Calcium and pancreatic β-cell function, The mechanism of insulin secretion studied with the aid of lanthanum, *Biochem. J.* **190**:361.

Fleckenstein, A., 1977, Specific pharmacology of calcium in myocardium, cardiac pacemakers and vascular smooth muscle, *Ann. Rev. Pharmacol. Toxicol.* **17**:149.

Fleckman, A., Erlichman, J., Schubart, U. K., and Fleischer, N., 1981, Effect of trifluoperazine, D600 and phenytoin on depolarization- and thyrotropin-releasing hormone-induced thyrotropin release from rat pituitary tissue, *Endocrinology* **108**:2072.

Fleischer, N., Schubart, U. K., Fleckman, A., and Erlichman, J., 1981, Calcium antagonists and the secretion of insulin and thyrotropin, in: *New Perspectives on Calcium Antagonists* (G. B. Weiss, ed.), pp. 177–190, Waverly Press, Baltimore.

Flemström, G., and Garner, A., 1980, Stimulation of gastric acid and bicarbonate secretions by calcium in guinea pig stomach and amphibian isolated mucosa, *Acta. Physiol. Scand.* **110**:419.

Flower, R. J., 1974, Drugs which inhibit prostaglandin biosynthesis, *Pharmacol. Rev.* **26**:33.

Foreman, J. C., 1977, Spontaneous histamine secretion from mast cells in the presence of strontium, *J. Physiol. (London)* **271**:215.

Foreman, J. C., 1981, The pharmacological control of immediate hypersensitivity, *Ann. Rev. Pharmacol.* **21**:63.

Foreman, J. C., and Garland, L. G., 1974, Densensitization in the process of histamine secretion induced by antigen and dextran, *J. Physiol. (London)* **239**:381.

Foreman, J. C., and Mongar, J. L., 1972, The role of alkaline earth ions in anaphylactic histamine secretion, *J. Physiol. (London)* **224**:753.

Foreman, J. C., and Mongar, J. L., 1973, The action of lanthanum and manganese on anaphylactic histamine secretion, *Br. J. Pharmacol.* **48**:527.

Foreman, J. C., Mongar, J. L., and Gomperts, B. D., 1973, Calcium ionophores and movement of calcium ions following the physiological stimulus to a secretory process, *Nature* **245**:249.

Foreman, J. C., Garland, L. G., and Mongar, J. L., 1976, The role of calcium in secretory processes: model studies in mast cells, *Symp. Soc. Exp. Biol.* **30**:193.

Foreman, J. C., Hallett, M. B., and Mongar, J. L., 1977a, The relationship between histamine secretion and ^{45}calcium uptake by mast cells, *J. Physiol. (London)* **271**:193.

Foreman, J. C., Hallett, M. B., and Mongar, J. L., 1977b, Movement of strontium ions into mast cells and its relationship to the secretory process, *J. Physiol. (London)* **271**:233.

Foreman, J. C., Sobotka, A. K., and Lichtenstein, L., 1980, Modulation of the rate of histamine release from basophils by cyclic AMP, *Europ. J. Pharmacol.* **63**:341.

Forte, S. G., and Nauss, A. H., 1963, Effects of calcium removal on bullfrog gastric mucosa, *Am. J. Physiol.* **205**:631.

Franco-Saenz, R., Suzuki, S., and Tan. S. Y., 1980a, Prostaglandins and renin production: A review, *Prostaglandins* **20**:1131.

Franco-Saenz, R., Suzuki, S., Tan, S. Y., and Mulrow, P. S., 1980b, Prostaglandin stimulation of renin release: Independence of β-adrenergic receptor activity and possible mechanism of action, *Endocrinology* **106**:1400.

Frankenhaeuser, B., and Hodgkin, A. L., 1957, The action of calcium on the electrical properties of squid axons, *J. Physiol. (London)* **137**:218.

Fray, J. C. S., 1980, Stimulus–secretion coupling of renin: Role of hemodynamic and other factors, *Circ. Res.* **47**:485.

Freer, R. J., 1975, Calcium and angiotensin tachyphylaxis in rat uterine muscle, *Am. J. Physiol.* **228**:1423.

Fried, R. C., and Blaustein, M. P., 1978, Retrieval and recycling of synaptic vesicle membrane in pinched-off nerve terminals (synaptosomes), *J. Cell. Biol.* **78**:685.

Friedman, Y., Levasseur, S., and Burke, G., 1976, Demonstration of a thyrotropin responsive prostaglandin E_2-9-ketoreductase in bovine thyroid, *Biochim. Biophys. Acta.* **431**:615.

Fujita, T., and Kobayashi, S., 1977, Structure and function of gut endocrine cells, *Int. Rev. Cytology* (Suppl.) **6**:187.

Fujita, T., and Kobayashi, S., 1979, Current views on the paraneuron concept, *Trends in Neurosci.* **2**:27.

Furchgott, R. F., Kirpekar, S. M., Rieker, M., and Schwab, A., 1963, Actions and interactions of norepinephrine, tyramine and cocaine on aortic strips of rabbit and left atria of guinea pig and cat, *J. Pharmacol. Exptl. Therap.* **142**:39.

Fynn, M., Onomakpome, N., and Peart, W. S., 1977, The effects of ionophores (A23187 and RO2-2985) on renin secretion and renal vasoconstriction, *Proc. R. Soc. Lond. Ser. B.* **199**:199.

Gagel, R. F., Zeytinoglu, F. N., Voelkel, E. F., and Tashjian, A. H., 1980, Establishment of a calcitonin-producing rat medullary thyroid carcinoma cell line. II. Secretory studies of the tumor and cells in culture, *Endocrinology* **107**:516.

Gallardo, E., and Ramirez, V. D., 1977, A method for the superfusion of rat hypothalami: Secretion of luteinizing hormone-releasing hormone, *Proc. Soc. Exp. Biol.* **155**:79.

Garcia, A. G., Kirpekar, S. M., and Prat, J. C., 1975, A calcium ionophore stimulating the secretion of catecholamines from the cat adrenal, *J. Physiol. (London)* **244**:253.

Gard, T. G., Bernstein, B., and Larsen, P. R., 1981, Studies on the mechanism of 3′,5′,3′-triiodothyronine-induced suppression of secretogogue-induced thyrotropin release *in vitro*, *Endocrinology* **108**:2046.

Gardner, D. G., Brown, E. M., Windeck, R., and Aurbach, G. D., 1978, Prostaglandin E_2 stimulation of adenosine 3′,5′-monophosphate accumulation and parathyroid hormone release in dispersed bovine parathyroid cells, *Endocrinology* **103**:577.

Gardner, J. D., 1979, Regulation of pancreatic exocrine function *in vitro:* Initial steps in the actions of secretagogues, *Ann. Rev. Physiol.* **41**:55.

Gardner, J. D., Walker, M. D., and Rottman, A. J., 1980, Effect of A23187 on amylase release from dispersed acini prepared from guinea pig pancreas, *Am. J. Physiol.* **238**:G458.

Garrett, J. R., 1975, Changing attitudes on salivary secretion—A short history on spit, *Proc. R. Soc. Med.* **68**:553.

Gautvik, K. M., Iversen, J. G., and Sand, O., 1980, On the role of extracellular Ca^{2+} for prolactin release and adenosine 3′,5′-monophosphate formation induced by thyroliberin in cultured rat pituitary cells, *Life. Sci.* **26**:995.

Gemmell, R. T., Stacy, B. D., and Thorburn, G. D., 1974, Ultrastructural study of secretory granules in the corpus luteum of the sheep during the estrous cyclic, *Biol. Reprod.* **11**:447.

Gemmell, R. T., Laychock, S. G., and Rubin, R. P., 1977, Ultrastructural and biochemical evidence for a steroid-containing secretory organelle in the perfused cat adrenal gland, *J. Cell. Biol.* **72**:209.

Gerich, J. E., Charles, M. A., and Grodsky, G. M., 1976, Regulation of pancreatic insulin and glucagon secretion, *Ann. Rev. Physiol.* **38**:353.

Gerich, J. E., and Lorenzi, M., 1978, The role of the autonomic nervous system and somatostatin in the control of insulin and glucagon secretion, in: *Frontiers in Neuroendocrinology* (W. F. Ganong and L. Martini, eds.), Vol. 5. pp. 265–288, Raven Press, New York.

Gerich, J. E., Frankel, B. J., Franska, R., West, L., Forsham, P. H., and Grodsky, G. M., 1974, Calcium dependency of glucagon secretion from the *in vitro* perfused rat pancreas, *Endocrinology* **94**:1381.

Gerrard, J. M., White, J. G., and Peterson, D. A., 1978, The platelet dense tubular system: Its relationship to prostaglandin synthesis and calcium flux, *Thrombos. Haemostas.* **40**:224.

Gerrard, J. M., Peterson, D. A., and White, J. G., 1981, Calcium mobilization, in: *Platelets in Biology and Pathology* (J. L. Gordon, ed.), pp. 407–436, Elsevier, Amsterdam.

Gershengorn, M. C., 1980, Thyrotropin releasing hormone stimulation of prolactin release, *J. Biol. Chem.* **255**:1801.

Gershengorn, M. C., Hoffstein, S. T., Rebecchi, M. J., Geras, E., and Rubin, B. G., 1981, Thyrotropin-releasing hormone stimulation of prolactin release from clonal rat pituitary cells, *J. Clin. Invest.* **67**:1769.

Geschwind, I. I., 1969, Mechanism of action of releasing factors, in: *Frontiers in Neuroendocrinology* (W. F. Ganong and L. Martini, eds.) pp. 389–431, Oxford University Press.

Geschwind, I., 1971, Mechanisms of release of anterior pituitary hormones: Studies *in vitro*, *Mem. Soc. Endocrinology* **19**:221.

Getz, D., Gibson, J. F., Sheppard, R. N., Micklem, K. J., and Pasternak, C. A., 1979, Manganese as a calcium probe: Electron paramagnetic resonance and nuclear magnetic resonance spectroscopy of intact cells, *J. Membrane Biol.* **50**:311.

Giachetti, A., Said, S. I., Reynolds, R. C., and Koniges, F. C., 1977, Vasoactive intestinal polypeptide in brain: Localization in and release from isolated nerve terminals, *Proc. Nat. Acad. Sci.* **74**:3424.

Gill, D. L., Grollman, E. F., and Kohn, L. D., 1981, Calcium transport mechanisms in membrane vesicles from guinea pig brain synaptosomes, *J. Biol. Chem.* **256**:184.

Gilula, N. B., and Epstein, M. L., 1976, Cell-to-cell communication, gap junctions and calcium, *Symp. Soc. Exp. Biol.* **30**:257.

Ginsborg, B. L., and House, C. R., 1980, Stimulus response coupling in gland cells, *Ann. Rev. Biophys. Bioeng.* **9**:55.

Ginsborg, B. L., House, C. R., and Mitchell, M. R., 1980, A calcium-readmission response recorded from *Nauphoeta* salivary gland acinar cells, *J. Physiol. (London)* **304**:437.

Glick, D. M., and Mockel, J., 1980, The disposition of calcium within parathyroid tissue, *Horm. Met. Res.* **12**:475.

Goldberg, N. D., and Haddox, M. K., 1977, Cyclic GMP metabolism and involvement in biological regulation, *Ann. Rev. Biochem.* **46**:823.

Goldberg, N. D., O'Dea, R. F., and Haddox, M. K., 1973, Cyclic GMP, *Adv. Cyclic Nucleotide Res.* **3**:155.

Goldberg, N. D., Haddox, M. K., Nicol, S. E., Glass, D. B., Sanford, G. H., Kuehl, F. A. Jr., and Estensen, R., 1975, Biologic regulation through opposing influences of cyclic GMP and cyclic AMP. The Yin-Yang hypothesis, *Adv. Cyclic Nucleotide Res.* **5**:307.

Goldberg, N. D., Graff, G., Haddox, M. K., Stephenson, J. H., Glass, D. B., and Moser, M. E., 1978, Redox modulation of splenic cell soluble guanylate cyclase activity: Activation by hydrophilic and hydrophobic oxidants represented by ascorbic and dehydroascorbic acids, fatty acid hydroperoxides and prostaglandin endoperoxides, *Adv. Cyclic Nucleotide Res.* **9**:101.

Goldstein, I. M., Hoffstein, S. T., and Weissmann, G., 1975, Mechanisms of lysosomal enzyme release from human polymorphonuclear leukocytes: Effects of phorbol myristate acetate, *J. Cell. Biol.* **66**:647.

Goldstein, I. M., Lind, S., Hoffstein, S., and Weissmann, G., 1977, Influence of local anesthetics upon human polymorphonuclear leukocyte function *in vitro*, *J. Exp. Med.* **146**:483.

Gomperts, B., 1976, Calcium and cell activation, in: *Receptors and Recognition*, Series A, Vol. 2 (P. Cuatrecasas and M. F. Greaves, eds.), pp. 44–102, Chapman and Hall, London.

Gorman, R. R., 1979, Modulation of human platelet function by prostacyclin and thromboxane A_2, *Fed. Proc.* **38**:83.

Goth, A., Adams, H. R., and Knoohuizen, M., 1971, Phosphatidylserine: selective enhancer of histamine release, *Science* **173**:1034.

Göthert, M., 1977, Effects of presynaptic modulators on Ca^{2+}-induced noradrenaline release from cardiac sympathetic fibers, *Naunyn-Schmiedeberg's Arch. Pharmacol.* **300**:267.

Grahame-Smith, D. G., Butcher, R. W., Ney, R. L., and Sutherland, E. W., 1967, Adenosine 3′,5′-monophosphate as the intracellular mediator of the action of adrenocorticotropic hormone on the adrenal cortex, *J. Biol. Chem* **242**:5535.

Grette, K., 1962, Studies on the mechanism of thrombin-catalyzed hemostatic reactions in blood platelets, *Acta. Physiol. Scand.* **56** (Suppl. 195) :5.

Grill, V., and Cerasi, E., 1974, Stimulation by D-glucose of cyclic adenosine 3′-5′-monophosphate accumulation and insulin release in isolated pancreatic islets of the rat, *J. Biol. Chem.* **249**:4196.

Grodsky, G. M., 1972, A threshold distribution hypothesis for packet storage of insulin. II. Effect of calcium, *Diabetes* **21**(Suppl):584.

Grodsky, G. M., Epstein, G. H., Fanska, R., and Karam, J. H., 1977, Pancreatic action of the sulfonylureas, *Fed. Proc.* **36**:2714.

Gröschel-Stewart, U., 1980, Immunochemistry of cytoplasmic contractile proteins, *Int. Rev. Cytol.* **65**:193.

Grynszpan-Winograd, O., 1975, Ultrastructure of the chromaffin cell, in: *Handbook of Physiology*, Section 7, Vol. VI, (H. Blaschko, G. Sayers, and A. D. Smith, eds.), pp. 295–308, American Physiological Society, Washington, D.C.

Guerrero-Munoz, F., DeLourdes Guerrero, M., and Way, E. L., 1979, Effect of β-endorphin on calcium uptake in the brain, *Science* **206**:89.

Guillemin, R., 1978, Biochemical and physiological correlates of hypothalamic peptides. The New endocrinology of the neuron, in: *The Hypothalamus* (S. Reichlin, R. J. Baldessarini, and J. B. Martin, eds.), pp. 155–194, Raven Press, New York.

Gwatkin, R. B. L., 1977, *Fertilization Mechanisms in Man and Mammals*, Plenum Press, New York.

Habener, J. F., and Potts, J. T., 1976, Relative effectiveness of magnesium and calcium on the secretion and biosynthesis of parathyroid hormone *in vitro*, *Endocrinology* **98**:197.

Habener, J. F., and Potts, J. T., 1978, Biosynthesis of parathyroid hormone, *New Eng. J. Med.* **299**:580.

Habener, J. F., and Potts, J. T., 1979, Subcellular distributions of parathyroid hormone, hormonal precursors and parathyroid secretory protein, *Endocrinology* **104**:265.

Habener, J. F., Kemper, B. W., Rich, A., and Potts, J. T., 1977, Biosynthesis of parathyroid hormone, *Rec. Prog. Horm. Res.* **33**:249.

Habener, J. F., Amherdt, M., Ravazzola, M., and Orci, L., 1979, Parathyroid hormone biosynthesis, *J. Cell Biol.* **80**:715.

Hackett, J. T., 1976, Selective antagonism of frog cerebellar synaptic transmission by manganese and cobalt ions, *Brain Res.* **114**:47.

Hagan, H. H., and Ormond, J. K., 1912, Relation of calcium to the cardio-inhibitory function of the vagus, *Am. J. Physiol.* **30**:105.

Hagiwara, S. and Byerly, L., 1981, Calcium channel, *Ann. Rev. Neurosci.* **4**:69.

Hahn, H. J., Gylfe, E., and Hellman, B., 1980, Calcium and pancreatic β-cell function. Evidence for cyclic AMP induced translocation of intracellular calcium, *Biochim. Biophys. Acta.* **630**:425.

Haksar, A., Maudsley, D. V., Peron, F. G., and Bedigian, E., 1976, Lanthanum: Inhibition of ACTH-stimulated cyclic AMP and corticosterone synthesis in isolated rat adrenocortical cells, *J. Cell. Biol.* **68**:142.

Hales, C. N., and Milner, R. D. G., 1968a, Cations and secretion of insulin from rabbit pancreas *in vitro*, *J. Physiol (London)* **199**:177.

Hales, C. N., and Milner, R. D. G., 1968b, The role of sodium and potassium in insulin secretion from rabbit pancreas, *J. Physiol. (London)* **194**:725.

Halkerston, I. D. K., 1975, Cyclic AMP and adrenocortical function, *Adv. Cyclic Nucleotide Res.* **6**:100.

Hallett, M., and Carbone, E., 1972, Studies of calcium influx into squid giant axons with aequorin, *J. Cell Physiol.* **80**:219.

Hammerschlag, R., 1980, The role of calcium in the initiation of fast axonal transport, *Fed. Proc.* **39**:2809.

Hanson, J. P., Repke, D. I., Katz, A. M., and Aledort, L. M., 1973, Calcium ion control of platelet thrombosthenin ATPase activity, *Biochim. Biophys. Acta.* **314**:382.

Harada, E., and Rubin, R. P., 1978, Stimulation of renin secretion and calcium efflux from the isolated perfused cat kidney by noradrenaline after prolonged calcium deprivation, *J. Physiol. (London)* **274**:367.

Harada, E., Lester, G. E., and Rubin, R. P., 1979, Stimulation of renin secretion from the intact kidney and from isolated glomeruli by the calcium ionophore A23187, *Biochim. Biophys. Acta.* **583**:20.

Harfield, B., and Tenenhouse, A., 1973, Effect of EGTA in protein release and cyclic AMP accumulation in rat parotid gland, *Can. J. Physiol. Pharmacol.* **51**:997.

Hargis, G. K., Williams, G. A., Reynolds, W. A., Chertow, B. S., Kukreja, S. C., Bowser, E. N., and Henderson, W. J., 1978, Effect of somatostatin on parathyroid hormone and calcitonin secretion, *Endocrinology* **102**:745.

Harris, G. W., 1955, *Neural Control of the Pituitary Gland*, Edward Arnold, London.

Harris, R. A., and Hood, W. F., 1980, Inhibition of synaptosomal calcium uptake by ethanol, *J. Pharm. Exp. Therap.* **213**:562.

Hartter, D. E., and Raimirez, V. D., 1980, The effects of ions, metabolic inhibitors, and colchicine on luteinizing hormone releasing hormone release from superfused rat hypothami, *Endocrinology* **107**:375.

Harvey, A. M., and MacIntosh, F. C., 1940, Calcium and synaptic transmission in a sympathetic ganglion, *J. Physiol. (London)* **97**:408.

Haslam, R. J., Lynham, J. A., and Fox, J. E. B., 1979, Effects of collagen, ionophore A23187, and prostaglandin E_1 on the phosphorylation of specific proteins in blood platelets, *Biochem. J.* **178**:397.

Hatano, S., and Oosawa, F., 1978, Cell motility and the organization of action and myosin in non-muscle cells, in *Cell Motility: Molecules and Organization* (S. Hatano, H. Lshikawa, and H. Sato, eds.) pp. 675–683, University Park Press, Baltimore.

Häusler, R., Burger, A, and Niedermaier, W., 1981, Evidence for an inherent, ATP-stimulated uptake of calcium into chromaffin granules, *Naunyn-Schmiedeberg's Arch. Pharmacol,* **315**:255.

Hawthorne, J. N., and Pickard, M. R., 1979, Phospholipids in synaptic function, *J. Neurochem.* **32**:5.

Hayasaki-Kimura, N., and Takahashi, K., 1979, Studies on action of somatostatin on growth hormone release in relation to calcium and cAMP, *Proc. Soc. Exp. Biol. Med.* **161**:312.

Haycock, J. W., Levy, W. B., and Cotman, C. W., 1977, Pentobarbital depression of stimulus-secretion coupling in brain-selective inhibition of depolarization-induced calcium dependent release, *Biochem. Pharm.* **26**:159.

Haye, B., and Jacquemin, C., 1977, Incorporation of [^{14}C] arachidonate in pig thyroid lipids and prostaglandins, *Biochim. Biophys. Acta.* **487**:231.

Hedeskov, C. J., 1980, Mechanism of glucose-induced insulin secretion, *Physiol. Rev.* **60**:442.

Hedge, G. A., 1977, Roles for the prostaglandins in the regulation of anterior pituitary secretion, *Life Sci.* **20**:17.

Hedqvist, P., 1977, Basic mechanisms of prostaglandin action on autonomic neurotransmission, *Ann. Rev. Pharmacol. Toxicol.* **17**:259.

Heilbrunn, L. V., 1956, *Dynamic Aspects of Living Protoplasm*, Academic Press, New York.

Heisler, S., and Grondin, G., 1973, Effect of lanthanum on ^{45}Ca flux and secretion of protein from rat exocrine pancreas, *Life Sci.* **13**:783.

Heisler, S., and Lambert, M., 1978, Dissociation of cyclic GMP synthesis from cholinergic-stimulated secretion of protein from rat exocrine pancreas, *Can. J. Physiol. and Pharmacol.* **56**:395.

Hellman, B., 1978, Calcium and pancreatic β-cell function. 3. Validity of the La^{3+}-wash technique for discriminating betweeen superficial and intracellular ^{45}Ca, *Biochim. Biophys. Acta.* **540**:534.

Hellman, B., 1981, Tolbutamide stimulation of ^{45}Ca fluxes in microdissected pancreatic islets rich in β-cells, *Mol. Pharmacol.* **20**: 83.

Hellman, B., Sehlin, J., and Täljedal, I-B, 1971, Calcium uptake by pancreatic β-cells as measured with the aid of ^{45}Ca and mannitol-^{3}H, *Am. J. Physiol.* **221**:1795.

Hellman, B., Idahl, L-Å, Lernmark, A., Sehlin, J., and Täljedal, I-B, 1974a, The pancreatic β-cell recognition of insulin secretagogues. Effects of calcium and sodium on glucose metabolism and insulin release, *Biochem. J.* **138**:33.

Hellman, B., Idahl, L-Å, Lernmark, A., and Täljedal, I-B, 1974b, The pancreatic β-cell recognition of insulin secretagogues: Does cycles AMP mediate the effect of glucose?, *Proc. Natl. Acad. Sci.* **71**:3405.

Hellman, B., Sehlin, J., and Täljedal, I-B., 1978, Effects of Na^+, K^+ and Mg^{2+} on $^{45}Ca^{2+}$ uptake by pancreatic islets, *Pflugers Arch.* **378**:93.

Hellman, B., Anderson, T., Berggren, P. O., Flatt, P., Gylfe, E., and Kohnert, K. D., 1979, The role of calcium in insulin secretion, in *Hormones and Cell Regulation*, Vol. 3 (J. Dumont and J. Nunez, eds.), pp. 69–96, Elsevier, Amsterdam.

Henderson, G., Hughes, J., and Kosterlitz, H. W., 1978, In vitro release of leu- and met-enkephalin from the corpus striatum, Nature 271:677.

Henkart, M., 1980, Identification and function of intracellular calcium stores in axons and cell bodies of neurons, Fed. Proc. 39:2783.

Henquin, J. C., 1980, Tolbutamide stimulation and inhibition of insulin release: Studies of the underlying ionic mechanisms in isolated rat islets, Diabetologia 18:151.

Henquin, J. C., and Lambert, A. E., 1974, Cationic environment and dynamics of insulin secretion. III. Effect of the absence of potassium, Diabetologia 10:789.

Henson, P. M., Ginsberg, M. H., and Morrison, D. C., 1978, Mechanisms of mediator release by inflammatory cells, in: Cell Surface Reviews, Vol. 5 (G. Poste and G. L. Nicolson, eds.), pp. 411–508, North-Holland, Amsterdam.

Herchuelz, A., and Malaisse, W. J., 1980, Regulation of calcium fluxes in pancreatic islets: Two calcium movements' dissociated response to glucose. Am. J. Physiol. 238:E87.

Herchuelz, A., Couturier, E., and Malaisse, W. J., 1980, Regulation of calcium fluxes in pancreatic islets; glucose-induced calcium–calcium exchange. Am. J. Physiol. 238:E96.

Herman, L., Sato, T., and Hales, C. N., 1973, The electron microscopic localization of cations to pancreatic islets of Langerhans and their possible role in insulin secretion, J. Ultrast. Res. 42:298.

Hermansen, K., 1980, The role of sodium in somatostatin secretion: Evidence for the involvement of Na channels in the release mechanism, Endocrinology 106:1843.

Hermansen, K., and Iversen, J., 1978, Dual action of Mn^{++} upon the secretion of insulin and glucagon from the isolated, perfused canine pancreas. Diabetologia 15:475.

Hermansen, K., Christensen, S. E., and Orskov, H., 1979, Characterization of somatostatin release from the pancreas, Diabetologia 16:261.

Hermansen, K., Christensen, S. E., and Orskov, H., 1980, The significance of the Na^+/K^+ pump for somatostatin release, Horm. Metab. Res. 12:23.

Hersey, S. J., 1974, Interactions between oxidative metabolism and acid secretion in gastric mucosa, Biochim. Biophys. Acta. 344:157.

Hertelendy, F., Todd, H., and Narconis, R. J., 1978, Studies on growth hormone secretion. IX. Prostaglandins do not act like ionophores. Prostaglandins 15:575.

Herzog, V., Sies, H., and Miller, F., 1976, Exocytosis in secretory cells of rat lacrimal gland, J. Cell Biol. 70:692.

Heuser, J., and Miledi, R., 1971, Effect of lanthanum ions on function and structure of frog neuromuscular junctions, Proc. R. Soc. Lond. Ser. B. 179:247.

Heuser, J. E., Reese, T. S., Dennis, M. J., Jan, Y., Jan, L., and Evans, L., 1979, Synaptic vesicle exocytosis captured by quick-freezing and correlated with quantal transmitter release, J. Cell Biol. 81:275.

Higuchi, T., Kaneko, A., Abel, J. H., and Niswender, G. D., 1976, Relationship between membrane potential and progesterone release in ovine corpora lutea, Endocrinology 99:1023.

Hille, B., 1970, Ionic channels in nerve membranes, Prog. Biophys. Mol. Biol. 21:1.

Hille, B., 1976, Gating in sodium channels of nerve, Ann. Rev. Physiol. 38:139.

Hille, B., 1977, Local anesthetics: Hydrophilic and hydrophobic pathways for the drug-receptor reaction, J. Gen. Physiol. 69:497.

Hirata, F., and Axelrod, J., 1980, Phospholipid methylation and biological signal transmission, Science 209:1082.

Hirsch, P. F., and Munson, P. L., 1969, Thyrocalcitonin, Physiol. Rev. 49:548.

Hitchcock, S. E., 1977, Regulation of motility in non-muscle cells, J. Cell. Biol. 74:1.

Höber, T., 1945, Physical Chemistry of Cells and Tissues, Blakiston, Philadelphia.

Hodgkin, A. L., and Huxley, A. F., 1952, Currents carried by sodium and potassium ions through the membrane of the giant axon of *Loligo, J. Physiol. (London)* 116:449.

Hodgkin, A. L., and Keynes, R. D., 1957, Movement of labelled calcium in squid giant axons, *J. Physiol. (London)* 138:253.

Hoffstein, S., Korchak, H. M., and Smolen, J. E., 1981, Rapid membrane and cytoskeletal changes in human neutrophils during stimulus induced secretion, *Fed. Proc.* 40:209.

Hogberg, B., and Uvnäs, B., 1957, The mechanism of the disruption of mast cells produced by compound 48/80, *Acta. Physiol. Scand.* 41:345.

Hokin, L. E., 1955, Isolation of the zymogen granules of dog pancreas and a study of their properties, *Biochim. Biophys. Acta* 18:379.

Hokin, L. E., 1966, Effects of calcium omission on acetylcholine-stimulated amylase secretion and phospholipid synthesis in pigeon pancreas slices, *Biochim. Biophys. Acta* 115:219.

Hokin, L. E., 1968, Dynamic aspects of phospholipids during protein secretion, *Int. Rev. Cytol.* 23:187.

Holmsen, H., 1975, Biochemistry of the platelet release reaction, in *Biochemistry and Pharmacology of Platelets* (Ciba Foundation Symp. 35), pp. 175–205, Elsevier, Amsterdam.

Holmsen, H., and Day, H. J., 1970, The selectivity of the thrombin-induced platelet release reaction: Subcellular localization of released and retained constituents, *J. Lab. Clin. Med.* 75:840.

Holmsen, H., Day, H. J., and Stormorken, H., 1969, The blood platelet release reaction, *Scand. J. Haematol.* (Suppl.)8:1.

Hopkins, C. R., and Walker, A. M., 1978, Calcium as a second messenger in the stimulation of luteinizing hormone secretion, *Mol. Cell. Endocr.* 12:189.

Hostetler, K. Y., and Hall, L. B., 1980, Phospholipase C activity of rat tissues, *Biochem. Biophys. Res. Commun.* 96:388.

Hotz, J., Goebell, H., and Ziegler, R., 1975, The role of calcium in gastric acid and pepsin secretion, in: *Stimulus–Secretion Coupling in the Gastrointestinal Tract* (R. M. Case and H. Goebell, eds.), pp. 159–170. University Park Press, Baltimore.

Houslay, M. D., and Palmer, R. W., 1979, Lysophosphatidylcholine can modulate the activity of the glucagon-stimulated adenylate cyclase from rat liver plasma membranes, *Biochem. J.* 178:217.

Howell, S. L., Montague, W., and Tyhurst, M., 1975, Calcium distribution in islets of Langerhams. A study of calcium concentrations and of calcium accumulation in β-cell organelles, *J. Cell. Sci.* 19:395.

Howell, W. H., 1905, Inhibition of the heart in relation to the inorganic salts of the blood, *Am. J. Physiol.* 15:280.

Hsu, L. S., and Becker, E. L., 1975, Volume decrease of glycerinated polymorphonuclear leukocytes induced by ATP and Ca^{2+}, *Exp. Cell. Res.* 91:469.

Hubel, K. A., and Callanan, D., 1980, Effects of Ca^{2+} on ileal transport and electrically induced secretion, *Am. J. Physiol.* 239:G18.

Hukovic, S., and Muscholl, E., 1962, Die Noradrenalin-Abgabe aus dem isolierten Kaninchenherzen bei sympathischer Nervenreizung und ihre pharmakologische Beeinflussung, *Naunyn-Schmiedeberg's Arch. Pharmakol.* 244:81.

Hutter, O. F., and Kostial, K., 1955, The relationship of sodium ions to the release of acetylcholine, *J. Physiol. (London)* 129:159.

Huxley, H. E., 1976, The relevance of studies on muscle to problems of cell motility, in *Cell Motility*, Book A, (R. Goldman, T. Pollard, and J. Rosenbaum, eds.), Vol. 3, pp. 115–126, Cold Spring Harbor Laboratory.

Huxley, H. E., 1978, Behavior of contractile proteins in muscle and other systems, in *Cell Motility: Molecules and Organization* (S. Hatano, H. Ishikawa, and H. Sato, eds.), pp. 3–12, University Park Press, Baltimore.

Ignarro, L. J., 1978, Interference with stimulus–secretion coupling by glucocorticoids, *Adv. Cyclic Nucleotide Res.* **9**:677.

Ignarro, L. J., and George, W. J., 1974, Hormonal control of lysosomal enzyme release from human neutrophils: Elevation of cyclic nucleotide by autonomic neurohormones, *Proc. Natl. Acad. Sci.* **71**:2027.

Impraim, C. C., Micklem, K. J., and Pasternak, C. A., 1979, Calcium, cells and virus-alterations caused by paramyxoviruses, *Biochem. Pharmacol.* **28**:1963.

Irvine, R. F., and Dawson, R. M. C., 1978, The distribution of calcium-dependent phosphatidylinositol-specific phosphodiesterase in rat brain, *J. Neurochem.* **31**:1427.

Irvine, R. F., Letcher, A. J., and Dawson, R. M. C., 1980, Thyrotropin-stimulated phosphatidylinositol-specific phospholipase A_2 in pig thyroid, a re-examination, *FEBS Lett.* **119**:287.

Ishida, A., 1968, Stimulus–secretion coupling on the oxytocin release from the isolated posterior pituitary lobe, *Jap. J. Physiol.* **18**:471.

Ishizaka, T., Foreman, J. C., Sterk, A. R., and Ishizaka, K., 1979, Induction of calcium flux across the rat mast cell membrane by bridging IgE receptors, *Proc. Natl. Acad. Sci.* **76**:5858.

Israel, M., and Dunant, Y., 1979, Mechanism of acetylcholine release, *Prog. Brain. Res.* **49**:125.

Iversen, J., and Hermansen, K., 1977, Calcium, glucose and glucagon release, *Diabetologia* **13**:297.

Iversen, L. L., Iversen, S. D., Bloom, F., Douglas, C., Brown, M., and Vale, W., 1978a, Calcium dependent release of somatostatin and neurotensin from rat brain *in vitro*, *Nature* **273**:161.

Iversen, L. L., Iversen, S. D., Bloom, F. E., Vargo, T., and Guillemin, R., 1978b, Release of enkephalin from rat globus pallidus *in vitro*, *Nature* **271**:679.

Iwatsuki, N., and Petersen, O. H., 1978a, Membrane potential resistance and intercellular communication in the lacrimal gland: Effects of acetylcholine and adrenaline, *J. Physiol. (London)* **275**:507.

Iwatsuki, N., and Petersen, O. H., 1978b, Electrical coupling and uncoupling of exocrine acinar cells, *J. Cell Biol.* **79**:533.

Jaanus, S. D., and Rubin, R. P., 1971, The effect of ACTH on calcium distribution in the perfused cat adrenal gland, *J. Physiol. (London)* **213**:581.

Jaanus, S. D., and Rubin, R. P., 1974, Analysis of the role of cyclic adenosine 3' 5' monophosphate in catecholamine release, *J. Physiol. (London)* **237**:465.

Jacobson, A., Schwartz, M., and Rehm, W. S., 1965, Effects of removal of calcium from bathing media on frog stomach, *Am. J. Physiol.* **209**:134.

Jacobus, W. E., Tiozzo, R., Lugli, G., Lehninger, A. L., and Carafoli, E., 1975, Aspects of energy-linked calcium accumulation by rat heart mitochondria, *J. Biol. Chem.* **250**:7863.

Jackowski, S., and Sha'afi, R. I., 1979, Response of adenosine cyclic 3', 5'-monophosphate level in rabbit neutrophils to the chemotactic peptide formyl-methionyl-leucyl-phenylalanine, *Mol. Pharmacol.* **16**:473.

Jackowski, S., Petro, K., and Sha'afi, R. I., 1979, A Ca^{2+}-stimulated ATPase activity in rabbit neutrophil membranes, *Biochim. Biophys. Acta* **558**:348.

Jacobs, T. P., Henry, D. P., Johnson, D. G., and Williams, R. H., 1978, Epinephrine and dopamine β-hydroxylase secretion from bovine adrenal, *Am. J. Physiol.* **234**:E600.

Jamieson, J. D., and Palade, G. E., 1971, Synthesis, intracellular transport and discharge of secretory proteins in stimulated pancreatic exocrine cells, *J. Cell. Biol.* **50**:135.

Janszen, F. H. A., Cooke, B. A., Van Driel, M. J. A., and Van der Molen, H. J., 1976, The effect of calcium ions on testosterone production in Leydig cells from rat testis, *Biochem. J.* **160**:433.

Jesse, R. L., and Franson, R. C., 1979, Modulation of purified phospholipase A_2 from human platelets by calcium and indomethacin, *Biochim. Biophys. Acta* **575**:467.

Jirakulsomchok, D., and Schneyer, C. A., 1979, Effects on rat parotid amylase and Ca of α- and β-adrenergic sympathetic stimulation, *Amer. J. Physiol.* **236**:E371.

Johansen, T., 1979, Adenosine triphosphate levels during anaphylactic release in rat mast cells: *in vitro* effects of glycolytic and respiratory inhibitors, *Eur. J. Pharmacol.* **58**:107.

Johansen, T., and Chakravarty, N., 1972, Dependence of histamine release from rat mast cells on adenosine triphosphate, *Naunyn-Schmiedeberg's Arch. Pharmacol.* **275**:457.

Johnson, D. G., Thoa, N. B., Weinshilboum, R., Axelrod, J., and Kopin, I. J., 1975, Enhanced release of dopamine-beta hydroxylase from sympathetic nerves by calcium and phenoxybenzamine and its reversal by prostaglandins, *Proc. Natl. Acad. Sci.* **68**:2227.

Johnston, A. G., and Miller, M. D., 1979, Inhibition of histamine release and ionophore-induced calcium flux in rat mast cells by lidocaine and chlorpromazine, *Agents and Actions* **9**:239.

Jones, G. S., VanDyke, K., and Castranova, V., 1981, Transmembrane potential changes associated with superoxide release from human granulocytes, *J. Cell. Physiol.* **106**:75.

Josephson, I., and Sperelakis, N., 1976, Local anesthetic blockade of Ca^{2+}-mediated action potentials in cardiac muscle, *Europ. J. Pharmacol.* **40**:201.

Kagayama, M., and Douglas, W. W., 1974, Electron microscope evidence of calcium-induced exocytosis in mast cells treated wirh 48/80 or the ionophores A-23187 and X-537A, *J. Cell. Biol.* **62**:519.

Kanagasuntheram, P., and Randle, P. J., 1976, Calcium metabolism and amylase release in rat parotid cells, *Biochem. J.* **160**:547.

Kanatsuka, A., Makino, H., Matsushima, Y., Kasanuki, J., Osegawa, M., and Kumagai, A., 1981, Effect of calcium on the secretion of somatostatin and insulin from pancreatic islets, *Endocrinology* **108**:2254.

Kanno, T., Cochrane, D. E., and Douglas, W. W., 1973, Exocytosis (secretory granule extrusion) induced by injection of calcium into mast cells, *Can. J. Physiol. Pharmacol.* **51**:1001.

Kanno, T., Saito, A., and Sato, Y., 1977, Stimulus–secretion coupling in pancreatic acinar cells: Influences of external sodium and calcium, on responses to cholecystokinin-pancreozymin and ionophore A23187, *J. Physiol. (London)* **270**:9.

Kanno, T., Saito, A., Yonezawa, H., Sato, H., Yanaihara, C., and Yanaihara, N., 1979, Calcium dependent release of the gut hormones, kitten VIP and rat CCK-PZ, in: *Gut Peptides* (A. Miyoshi, ed.), pp. 59–65, Elsevier-North Holland, Amsterdam.

Kaplan, K. L., 1981, Platelet granule proteins: Localization and secretion, in: *Platelets in Biology and Pathology* (J. L. Gordon, ed.), pp. 77–90, Elsevier, Amsterdam.

Karl, R. C., Zawalich, W. S., Ferrendelli, J. A., and Matschinsky, F. M., 1975, The role of Ca^{2+} and cyclic adenosine 3′,5′-monophosphate in insulin release induced *in vitro* by the divalent cation ionophore A23187, *J. Biol. Chem.* **250**:4575.

Kasbekar, D. K., 1974, Calcium-secretagogue interaction in the stimulation of gastric acid secretion, *Proc. Soc. Exp. Biol.* **145**:234.

Kasbekar, D. K., and Chugani, H., 1974, Role of calcium ion in *in vitro* gastric acid secretion, in: *Gastric Hydrogen Ion Secretion* (D. K. Kasbekar, G. Sachs, W. S. Rehm eds.), pp. 187–211, Marcel Dekker, New York.

Käser-Glanzmann, R., Jakabova, M., George, J. N., and Lüscher, E. F., 1978, Further characterization of calcium-accumulating vesicles from human blood platelets, *Biochim. Biophys. Acta* **512**:1.

Kater, S., Rued, J. R., and Murphy, A. P., 1978, Propagation of action potentials through electrotonic junctions in the salivary glands of the pulmonate mollusc, *Heliosoma Trivolvis*, *J. Exp. Biol.* **72**:77.

Katz, B., 1958, Microphysiology of the neuromuscular junction, *Bull. Johns Hopkins Hosp.* **102**:275.

Katz, B., 1969, *The Release of Neural Transmitter Substances*, Charles C. Thomas, Springfield, Illinois.

Katz, B., and Miledi, R., 1965, The effect of calcium on acetylcholine release from motor nerve terminals, *Proc. Roy. Soc. B.* **161**:496.

Katz, B., and Miledi, R., 1967a, The timing of calcium action during neuromuscular transmission, *J. Physiol. (London)* **189**:535.

Katz, B., and Miledi, R., 1967b, A study of synaptic transmission in the absence of nerve impulses, *J. Physiol. (London)* **192**:407.

Katz, B., and Miledi, R., 1969, Tetrodotoxin-resistant electrical activity in presynaptic terminals, *J. Physiol. (London)* **203**:459.

Kauffman, R. F., Taylor, R. W., and Pfeiffer, D. R., 1980, Cation transport and specificity of ionomycin, *J. Biol. Chem.* **255**:2735.

Kazimierczak, W., and Diamant, B., 1978, Mechanisms of histamine release in anaphylactic and anaphylactoid reactions, *Prog. Allergy* **24**:295.

Keeton, T. K., and Campbell, W. B., 1981, The pharmacologic alteration of renin release, *Pharmacol. Rev.* **31**:81.

Kelly, K. L., and Laychock, S. G., 1981, Prostaglandin synthesis and metabolism in isolated pancreatic islets of the rat, *Prostaglandins* **21**:759.

Kemper, B., Habener, J. F., Rich, A., and Potts, J. T., 1974, Parathyroid secretion: Discovery of a major calcium dependent protein, *Science* **184**:167.

Keryer, G. and Rossignol, B., 1976, Effect of carbachol on ^{45}Ca uptake and protein secretion in rat lacrimal gland, *Am. J. Physiol.* **230**:99.

Kidokoro, Y., and Ritchie, A. K., 1980, Chromaffin cell action potentials and their possible role in adrenaline secretion from rat adrenal medulla, *J. Physiol. (London)* **307**:199.

Kilpatrick, D. L., Slepetis, R. J., Corcoran, J. J., and Kirshner, N., 1982, Calcium uptake and catecholamine secretion by cultured bovine adrenal medulla cells, *J. Neurochem.* **38**:427.

Kirpekar, S. M., 1975, Factors influencing transmission at adrenergic synapses, *Prog. in Neurobiol.* **4**:163.

Kirpekar, S. M., and Prat, J. C., 1978, Effect of tetraethylammonium on noradrenaline release from cat spleen treated with tetrodotoxin, *Nature* **276**:623.

Kirpekar, S. M., and J. C. Prat, 1979, Release of catecholamines from perfused cat adrenal gland by veratridine, *Proc. Natl. Acad. Sci.* **76**:2081.

Klausner, R. D., Kleinfeld, A. M., Hoover, R. L., and Karnovsky, M. J., 1980, Lipid domains in membranes, *J. Biol. Chem.* **255**:1286.

Klee, C. B., Crouch, T. H., and Richman, P. G., 1980, Calmodulin, *Ann. Rev. Biochem.* **49**:489.

Kobayashi, S., 1979, Cellular background in gut hormone secretion, in: *Gut Peptides: Secretion, Function and Chemical Aspects* (A. Miyoshi, ed.), pp. 53–58, Elsevier, North Holland, Amsterdam.

Koelz, H. R., Kondo, S., Blum, A. L., and Schulz, I., 1977, Calcium uptake induced by cholinergic and α adrenergic stimulation in isolated cells of rat salivary glands, *Pflugers Arch.* **370**:37.

Kohnert, K. D., Hahn, H. J., Gylfe, E., Borg, H., and Hellman, B., 1979, Calcium and pancreatic β-cell function. 6. Glucose and intracellular ^{45}Ca distribution, *Mol. Cell. Endocr.* **16**:205.

Kopp, L. L., Aurell, M., Nilsson, I-M, and Ablad, B., 1980, The role of beta-1-adrenoceptors in the renin release response to graded renal sympathetic nerve stimulation, *Pflugers Arch.* **387**:107.

Korchak, H. M., and Weissmann, G., 1980, Stimulus–response coupling in the human neutrophil, Transmembrane potential and the role of extracellular Na^+, *Biochim. Biophys. Acta* **601**:180.

Kraicer, J., 1975, Mechanisms involved in the release of adenohypophysial hormones, in: *Ultrastructure in Biological Systems*, Vol. 7, (A. Tixier-Vidal and M. G. Farquhar, eds.), pp. 21–43, Academic Press, New York.

Kraicer, J., and Spence, J. W., 1981, Release of growth hormone from purified somatotrophs: Use of high K and the ionophore A23187 to elucidate interrelations among Ca^{++}, adenosine $3',5'$-monophosphate and somatostatin, *Endocrinology* **108**:651.

Kramer, R. E., Gallant, S., and Brownie, A. C., 1979, The role of cytochrome P-450 in the action of sodium depletion on aldosterone biosynthesis in rats, *J. Biol. Chem.* **254**:3953.

Kretsinger, R. H., 1975, Hypothesis: Calcium-modulated proteins contain EF hands, in: *Calcium Transport in Contraction and Secretion* (E. Carafoli, F. Clementi, W. Drabikowski, and A. Margreth, eds.), pp. 469–478, North Holland, Amsterdam.

Kuehl, F. A., Jr., Oien, H. G., and Ham, E. A., 1974, Prostaglandins and prostaglandin synthetase inhibitors: Actions on cell function, in: *Prostaglandin Synthetase Inhibitors* (H. J. Robinson and J. R. Vane, eds.), pp. 53–65, Raven Press, New York.

Kuffler, S., 1944, The effect of calcium on the neuromuscular junction, *J. Neurophysiol.* **7**:17.

Kumakura, K., Guidotti, A., and Costa, E., 1979, Primary cultures of chromaffin cells: Molecular mechanisms for the induction of tyrosine hydroxylase mediated by 8-Br-cyclic AMP, *Mol. Pharmacol.* **16**:865.

Kuo, J. F., Andersson, R. G. G., Wise, B. C., Mackerlova, L., Salomonsson, I., Brackett, N. L., Katoh, N., Shoji, M., and Wrenn, R. W., 1980, Calcium-dependent protein kinase: Widespread occurrence in various tissues and phyla of the animal kingdom and comparison of the effects of phospholipid, calmodulin and trifluoperazine, *Proc. Natl. Acad. Sci.* **77**:7039.

Kurebe, M., 1978, Interaction of dibucaine and calcium ion on a calcium pump reconstituted from defined components of intestinal brush border, *Mol. Pharmacol.* **14**:138.

Labrie, F., Borgeat, P., Godbout, M., Barden, N., Beaulieu, M., Lagace, L., Massicotte, J., and Veilleux, R., 1978, Mechanism of action of hypothalamic hormones and interactions with sex steroids in the anterior pituitary gland, in: *Synthesis and Release of Adenohypophyseal Hormones* (M. Jutisz and K. W. McKerns, eds.), pp. 67–103, Plenum Press, New York.

Labrie, F., Borgeat, P., Drouin, J., Beaulieu, M., Lagace, L., Ferland, L., and Raymond, V., 1979, Mechanism of action of hypothalamic hormones in the adenohypophysis, *Ann. Rev. Physiol.* **41**:555.

Lacy, P. E., and Malaisse, W. J., 1973, Microtubules and beta cell secretion, *Rec. Prog. Horm. Res.* **29**:199.

Lagercrantz, H., 1976, On the composition and function of large dense cored vesicles in sympathetic nerves, *Neuroscience* **1**:81.

Lambert, A. E., 1976, The regulation of insulin secretion, *Rev. Physiol. Biochem. Pharmacol.* **75**:97.

Lands, W. E. M., and Crawford, C. G., 1976, Enzymes of membrane phospholipid metabolism in animals, in: *The Enzymes of Biological Membranes*, Vol. 2 (A. Martonosi, ed.), pp. 3–85, Plenum Press, New York.

Langer, G. A., and Frank, J. S., 1972, Lanthanum in heart cell culture: Effect on calcium exchange correlated with its localization, *J. Cell. Biol.* **54**:441.

Langer, G. A., Serena, S. D., and Nudd, L. M., 1975, Localization of contractile-dependent Ca: comparison of Mn and verapamil in cardiac and skeletal muscle, *Am. J. Physiol.* **229**:1003.

Langer, S. Z., 1980, Presynaptic regulation of the release of catecholamines, *Pharmacol. Rev.* **32**:337.

Lastowecka, A., and Trifaro, J. M., 1974, The effect of sodium and calcium ions on the release of catecholamines from the adrenal medulla: sodium deprivation induces release by exocytosis in the absence of extracellular calcium, *J. Physiol. (London)* **236**:681.

Laugier, R., and Petersen, O. H., 1980, Pancreatic acinar cells: electrophysiological evidence for stimulant evoked increase in membrane calcium permeability in the mouse, *J. Physiol. (London)* **303**:61.

Laychock, S. G., 1981, Evidence for guanosine 3′,5′-monophosphate as a putative mediator of insulin secretion from isolated rat islets, *Endocrinology* **108**:1197.

Laychock, S. G., and Hardman, J. G., 1978, Effects of sodium nitroprusside and ascorbic acid on rat adrenocortical cell cGMP levels and steroidogenesis, *J. Cyclic Nucleotide. Res.* **5**:335.

Laychock, S. G., and Rubin, R. P., 1976, Indomethacin-induced alterations in corticosteroid and prostaglandin release by isolated adrenocortical cells of the cat, *Br. J. Pharmacol.* **57**:273.

Laychock, S. G., Warner, W., and Rubin, R. P., 1977, Further studies on the mechanisms controlling prostaglandin biosynthesis in the cat adrenal cortex: the role of calcium and cyclic AMP, *Endocrinology* **100**:74.

Laychock, S. G., Landon, E. J., and Hardman, J. G., 1978, The effect of ACTH and nucleotides on Ca^{2+} uptake in adrenal cortical, microsomal vesicles. *Endocrinology* **103**:2198.

Lebowitz, E. A., and Cooke, R., 1978, Contractile properties of actomyosin from human blood platelets, *J. Biol. Chem.* **253**:5443.

Leclercq-Meyer, V., Herchuelz, A., Valverde, I., Couturier, F., Marchand, J., and Malaisse, W. J., 1980, Mode of action of clonidine upon islet function, *Diabetes* **29**:193.

Lee, A. G., 1975, Functional properties of biological membranes: a physical-chemical approach, *Prog. Biophys. Mol. Biol.* **29**:5.

Lehman, W., 1976, Calcium regulation in invertebrate muscles, in: *Cell Motility (Book A)* (R. Goldman, T. Pollard and J. Rosenbaum, eds.), Vol. 3, pp. 151–152, Cold Spring Harbor Laboratory.

Lehninger, A. L., Reynafarje, B., Vercesi, A., and Tew, W. P., 1978, Transport and accumulation of calcium in mitochondria, *Ann. N.Y. Acad. Sci.* **307**:160.

Leier, D. J., and Jungmann, R. A., 1973, Adrenocorticotropic hormone and dibutyryl adenosine cyclic monophosphate-mediated Ca^{2+} uptake by rat adrenal glands, *Biochim. Biophys. Acta* **329**:196.

Leitner, J. W., Sussman, K. E., Vatter, A. E., and Schneider, F. H., 1975, Adenine nucleotides in the secretory granule fraction of rat islets, *Endocrinology* **96**:662.

Lester, G. E., and Rubin, R. P., 1977, The role of calcium in renin secretion from the isolated perfused cat kidney, *J. Physiol. (London)* **269**:93.

Levy, W. B., Haycock, J. W., and Cotman, C. W., 1974, Effects of polyvalent cations on stimulus-coupled secretion of [^{14}C]-γ-aminobutyric acid from isolated brain synaptosomes, *Mol. Pharmacol.* **10**:438.

Lew, V. L., and Ferreira, H. G., 1978, Calcium transport and the properties of a calcium-activated potassium channel in red cell membranes, *Current Topics in Membranes and Transport* **10**:218.

Lichtenstein, L. M., 1971, The role of cyclic AMP in inhibiting the IgE-mediated release of histamine, *Ann. N.Y. Acad. Sci.* **185**:403.

Lichtenstein, L. M., 1975, The mechanism of basophil histamine release induced by antigen and by the calcium ionophore A23187, *J. Immunol.* **114**:1692.

Lichtenstein, L. M., and Osler, A. G., 1964, Studies on the mechanisms of hypersensitivity phenomena. IX histamine release from human leukocytes by ragweed pollen antigen, *J. Exp. Med.* **120**:507.

Lincoln, T. M., and Corbin, J. D., 1978, On the role of the cAMP and cGMP dependent protein kinases in cell function, *J. Cyclic Nucleotide Res.* **4**:3.

Lincoln, T. M., and Keely, S. L., 1980, Effects of acetylcholine and nitroprusside on cGMP-dependent protein kinase in the perfused rat heart, *J. Cyclic Nucleotide Res.* **6**:83.

Linehan, W. M., Cooper, C. W., Bolman, R. M., and Wells, S. A., 1979, Inhibition of *in vivo* secretion of calcitonin in the pig by somatostatin, *Endocrinology* **104**:1602.

Ling, W. Y., Williams, M. T., and Marsh, J. M., 1980, Correlation of cyclic AMP and cyclic GMP accumulation and steroidogenesis during stimulation of bovine luteal cells with luteinizing hormone, *J. Endocr.* **86**:45.

Locke, F. S., 1894, Notiz über den Einfluss physiologischer Kochsalzlösung auf die Erregbarkeit von Muskel und Nerv, *Zentbl. Physiol.* **8**:166.

Loeb, J., 1900, On the different effect of ions upon neurogenic rhythmical contractions and upon embryonic and muscular tissue, *Am. J. Physiol.* **3**:383.

Loewenstein, W. R., and Rose. B., 1978, Calcium in (junctional) intercellular communication and a thought on its behavior in intracellular communication, *Ann. N.Y. Acad. Sci.* **307**:285.

Loewi, O., 1945, Chemical transmission of nerve impulses, *Am. Scientist* **33**:159.

Logan, A. G., and Chatzilias, A., 1980, The role of calcium in the control of renin release from the isolated rat kidney, *Can. J. Physiol. Pharmacol.* **58**:60.

Loubatieres, A., 1977, Effects of sulfonylureas on the pancreas, in: *The Diabetic Pancreas* (B. W. Volk and K. F. Wellmann, eds.), pp. 489–515, Plenum Press, New York.

Loubatieres-Mariani, M-M. Chapal, J., Lignon, F., and Valette, G., 1979, Structural specificity of nucleotides for insulin secretory action from the isolated perfused rat pancreas, *Europ. J. Pharmacol.* **59**:277.

Low, P. S., Lloyd, D. H., Stein, T. M., and Rogers, J. A., 1979, Calcium displacement by local anesthetics, *J. Biol. Chem.* **254**:4119.

Lowe, D. A., Richardson, B. P., Taylor, P., and Donatsch, P., 1976, Increasing intracellular sodium triggers calcium release from bound pools, *Nature* **260**:337.

Lucke, B., and McCutcheon, M., 1932, The living cell as an osmotic system and its permeability to water, *Physiol. Rev.* **12**:68.

Lucy, J. A., 1978, Mechanisms of chemically induced cell fusion, in: *Membrane Fusion* (G. Poste and G. L. Nicolson, eds.), pp. 268–304, North Holland Publishing Company, Amsterdam.

Lundquist, I., Fanska, R., and Grodsky, G. M., 1976, Interaction of calcium and glucose on glucagon secretion, *Endocrinology* **99**:1304.

Lüscher, E. F., Probst, E., and Bettex-Galland, M., 1972, Thrombosthenin: Structure and function, *Ann. N.Y. Acad. Sci.* **201**:122.

Lyons, R. M., and Shaw, J. O., 1980, Interaction of Ca^{2+} and protein phosphorylation in the rabbit platelet release reaction, *J. Clin. Invest.* **65**:242.

MacGregor, R. R., and Cohn, D. V., 1978, The intracellular pathway for parathormone biosynthesis and secretion, *Clin. Orth. Rel. Res.* **137**:244.

MacGregor, R. R., Chu, L. L. H., Hamilton, J. W., and Cohn, D. V., 1973, Studies on the subcellular localization of proparathyroid hormone and parathyroid hormone in bovine parathyroid gland: Separation of newly synthesized from mature forms, *Endocrinology* **93**:1387.

MacGregor, R. R., Hamilton, J. W., and Cohn, D. V., 1975, The by-pass of tissue hormone stores during the secretion of newly synthesized parathyroid hormone, *Endocrinology* **97**:178.

Mackie, C., Warren, R. L., and Simpson, E. R., 1978, Investigations into the role of calcium ions in the control of steroid production by isolated adrenal zona glomerulosa cells of the rat, *J. Endocr.* **77**:119.

Mahaffee, D. D., and Ontjes, D. A., 1980, The role of calcium in the control of adrenal adenylate cyclase, *J. Biol. Chem.* **255**:1565.

Main, I. H. M. and Pearce, J. B., 1978, Effects of calcium on acid secretion from the rat isolated gastric mucosa during stimulation with histamine, pentagastrin, methacholine and dibutyryl cyclic adenosine-3',5'-monophosphate, *Br. J. Pharmacol.* **64**:359.

Malaisse-Lagae, F., and Malaisse, W. J. 1971, Stimulus–secretion coupling of glucose-induced insulin release. III. Uptake of ^{45}calcium by isolated islets of Langerhans, *Endocrinology* **88**:72.

Malaisse, W. J., Herchuelz, A., Levy, J., Somers, G., Devis, G., Ravazzola, M. Malaisse-Lagae, F., and Orci, L., 1975, Insulin release and movements of calcium in pancreatic islets, in: *Calcium Transport in Contraction and Secretion* (E. Carafoli, F. Clementi, W. Drabikowski and A. Margreth, eds.), pp. 211–226, North Holland, Amsterdam.

Malaisse, W., Herchuelz, A., Devis, G., Somers, G., Boschew, C., Hutton, J. C., Kawazu, S., Sener, A., Atwater, I. J., Duncan, G., Ribalet, B., and Rojas, E., 1978, Regulation of calcium fluxes and their regulatory roles in pancreatic islets, *Ann. N.Y. Acad. Sci.* **307**:562.

Malamed, S., 1975, Ultrastructure of the mammalian adrenal cortex in relation to secretory function, in: *Handbook of Physiology*, Vol. VI, Section 7, (H. Blaschko, G., Sayers, and A. D. Smith, eds.), pp. 25–39, American Physiological Society, Washington, D.C.

Mandarino, L., Itoh, M., Blanchard, W., Patton, G., and Gerich, J., 1980, Stimulation of insulin release in the absence of extracellular calcium by isobutylmethylxanthine and its inhibition by somatostatin, *Endocrinology* **106**:430.

Marchbanks, R. M., 1976, Turnover and release of acetylcholine, in: *Synapses* (G. A. Cottrell and P. N. R. Usherwood, eds.), pp. 81–101, Academic Press, New York.

Marshall, P. J., Dixon, J. F., and Hokin, L. E., 1980, Evidence for a role in stimulus-secretion coupling of prostaglandins derived from release of arachidonoyl residues as a result of phosphatidylinositol breakdown, *Proc. Natl. Acad. Sci.* **77**:3292.

Marshall, P. J., Boatman, D. E., and Hokin, L. E., 1981, Direct demonstration of the formation of prostaglandin E_2 due to phosphatidylinositol breakdown associated with stimulation of enzyme secretion in the pancreas, *J. Biol. Chem.* **256**:844.

Martin, R. B., and Richardson, F. S., 1979, Lanthanides as probes for calcium in biological systems, *Quart Rev. Biophys.* **12**:181.

Martin, T. W., and Lagunoff, D., 1979, Interaction of lysophospholipids and mast cells, *Nature* **279**:250.

Massini, P., 1977, The role of calcium in the stimulation of platelets, in: *Platelets and Thrombosis* (Mills, D. C. B., and Pareti, F. I., eds.), pp. 33–43, Academic Press, New York.

Massini, P., and Lüscher, E. F., 1976, On the significance of the influx of calcium ions into stimulated human blood platelets, *Biochim. Biophys. Acta* **436**:652.

Massini, P., and Näf, U., 1980, Ca^{2+}-ionophores and the activation of human blood platelets, *Biochim. Biophys. Acta* **598**:575.

Massini, P., Käser-Glanzmann, R., and Lüscher, E. F., 1978, Movement of calcium ions and their role in the activation of platelets, *Thrombos. Haemostas.* **40**:212.

Matsuzaki, S., and Dumont, J. E., 1972, Effect of calcium ion on horse parathyroid gland adenyl cyclase, *Biochim. Biophys. Acta* **284**:227.

Matthews, E. K., 1977, Insulin secretion, in: *First European Symposium on Hormones and Cell Regulation* (Dumont, J., and Nunez, E. J., eds), pp. 57–76, Elsevier, Amsterdam.

Matthews, E. K., and Saffran, M., 1968, Effect of ACTH on the electrical properties of adrenocortical cells, *Nature* **219**:1369.

Matthews, E. K., and Saffran, M., 1973, Ionic dependence of adrenal steroidogenesis and ACTH-induced changes in the membrane potential of adrenocortical cells, *J. Physiol. (London)* **234**:43.

Matthews, E. K., Petersen, O. H., and Williams, J. A., 1973, Pancreatic acinar cells: acetylcholine-induced membrane depolarization calcium efflux and amylase release, *J. Physiol. (London)* **234**:689.

McAfee, D. A., Henon, B. K., Horn, J. P., and Yarowsky, P., 1981, Calcium currents modulated by adrenergic receptors in sympathetic neurons, *Fed. Proc.* **40**:2246.

McDonald, T. F., Pelzer, D., and Trautwein, W., 1980, On the mechanism of slow calcium channel block in heart, *Pflugers Arch.* **385**:175.

McGee, R., and Schneider, J. E., 1979, Inhibition of high affinity synaptosomal uptake systems by verapamil, *Mol. Pharmacol.* **16**:877.

McGraw, C. F., Somlyo, A. V., and Blaustein, M. P., 1980, Probing for calcium at presynaptic nerve terminals, *Fed. Proc.* **39**:2796.

McLaughlin, S., and Eisenberg, M., 1975, Antibiotics and membrane biology, *Ann. Rev. Biophys. Bioeng.* **4**:335.

McShan, W. H., and Meyer, R. K., 1952, Gonadotrophic activity of granule fractions obtained from anterior pituitary glands of castrate rats, *Endocrinology* **50**:294.

Means, A. R., and Dedman, J. R., 1980, Calmodulin—an intracellular Ca receptor, *Nature* **285**:73.

Meda, P., Halban, P., Perrelet, A., Renold, A. E., and Orci, L, 1980, Gap junction development is correlated with insulin content in the pancreatic β-cell, *Science* **209**:1026.

Meech, R. W., 1978, Calcium-dependent potassium activation in nervous tissues, *Ann. Rev. Biophys. Bioeng.* **7**:1.

Meissner, H. P., Preissler, M., and Henquin, J. C., 1979, Possible ionic mechanisms of the electrical activity induced by glucose and tolbutamide in pancreatic β-cells, in: *Proceedings of the 10th Congress of the International Diabetes Foundation* (W. K. Waldhausl et al., eds.), pp. 166–171, Excerpta Medica, Amsterdam.

Mela, L., 1977, Mechanisms and physiological significance of calcium transport across the mammalian mitochondrial membrane, in: *Current Topics in Membranes and Transport*, Vol. 9 (F. Bronner and A. Kleinzeller, eds.), pp. 332–354, Academic Press, New York.

Mela, L., and Chance, B., 1968, Spectrophotometric measurements of the kinetics of Ca^{2+} and Mn^{2+} accumulation in mitochondria, *Biochemistry* **7**:4059.

Melander, A., 1977, Aminergic regulation of thyroid activity: importance of the sympathetic innervation and the mast cells of the thyroid gland, *Acta Med. Scand.* **201**:257.

Meldolesi, J., Borgese, N., DeCamilli, P., and Ceccarelli, B., 1978, Cytoplasmic membranes and the secretory process, in: *Membrane Fusion* (G. Poste and G. L. Nicolson, eds.), pp. 510–627, North Holland Publishing Company, Amsterdam.

Metz, S. A., Robertson, R. P., and Fujimoto, W. Y., 1981, Inhibition of prostaglandin E synthesis augments glucose-induced insulin secretion in cultured pancreas, *Diabetes* **30**:551.

Meves, H., and Vogel, W., 1973, Calcium inward currents in internally perfused giant axons, *J. Physiol. (London)* **235**:225.

Meyer, D. I., and Burger, M. M., 1979, The chromaffin granule surface: The presence of actin and the nature of its interaction with the membrane, *FEBS Lett.* **101**:129.

Michaelson, D. M., and Avissar, S., 1979, Ca^{2+}-dependent protein phosphorylation of purely cholinergic Torpedo synaptosomes, *J. Biol. Chem.* **254**:12542.

Michaelson, D. M., and Ophir, I., 1980, Sidedness of (calcium, magnesium) adenosine triphosphatase of purified *Torpedo* synaptic vesicles, *J. Neurochem.* **34**:1483.

Michaelson, D. M., Ophir, I., and Angel, I., 1980, ATP-stimulated Ca^{2+} transport in cholinergic Torpedo synaptic vesicles, *J. Neurochem* **35**:116.

Michell, R. H., 1975, Inositol phospholipids and cell surface receptors function, *Biochim. Biophys. Acta* **415**:81.

Mikkelsen, R. B., 1976, Lanthanides as calcium probes in biomembranes, in: *Biological Membranes*, Vol. 3, (D. Chapman and D. F. H. Wallach, eds.), pp. 153–190, Academic Press, New York.

Miledi, R., 1971, Lanthanum ions abolish the "calcium response" of nerve terminals, *Nature* **229**:410.

Miledi, R., 1973, Transmitter release induced by injection of calcium ions into nerve terminals, *Proc. R. Soc. Ser. B.* **183**:421.

Miledi, R., and Slater, C. R., 1966, The action of calcium on neuronal synapses in the squid, *J. Physiol. (London)* **184**:473.

Miller, B. E., and Nelson, D. L., 1977, Calcium fluxes in isolated acinar cells from rat parotid, *J. Biol. Chem.* **252**:3629.

Milutinovic, S., Argent, B. E., Schulz, I., and Sachs, G., 1977, Studies on isolated sub-cellular components of cat pancreas, *J. Membrane Biol.* **36**:281.

Mines, G. R., 1911, On the replacement by calcium in certain neuro-muscular mechanisms by allied substances, *J. Physiol. (London)* **42**:251.

Mongar, J. L., and Schild, H. O., 1958, The effect of calcium and pH on the anaphylactic reaction, *J. Physiol. (London)* **140**:72.

Moore, J. W., Blaustein, M. P., Anderson, N. C., and Narahashi, T., 1967, Basis of tetrodotoxin's selectivity in blockage of squid axons, *J. Gen. Physiol.* **50**:1401.

Moore, L, Chen, T., Knapp, H. R., and Landon, E. J., 1975, Energy-dependent calcium sequestration activity in rat liver microsomes, *J. Biol. Chem.* **250**:4562.

Moriarty, C. M., 1977, Role of calcium in the regulation of adenohypophysial hormone release, *Life Sci.* **23**:185.

Morimoto, S., Tanaka, K., and Kitano, R., 1975, Distribution of renin in subcellular fractions from the rabbit kidney, *Jap. J. Pharm.* **25**:295.

Morris, B. J., and Johnston, C. I., 1976, Isolation of renin granules from rat kidney cortex and evidence for an inactive form of renin (prorenin) in granules and plasma, *Endocrinology* **98**:1466.

Morris, J. F., Nordmann, J. J., and Dyball, R. E. J., 1978, Structure–function correlation in mammalian neurosecretion, *Int. Rev. Exp. Path.* **18**:1.

Morrissey, J. J., and Cohn, D. V., 1979a, Regulation of secretion of parathormone and secretory protein-1 from separate intracellular pools by calcium, dibutyryl cyclic AMP and (1)-isoproterenol, *J. Cell. Biol.* **82**:93.

Morrissey, J. J., and Cohn, D. V., 1979b, Secretion and degradation of parathormone as a function of intracellular maturation of hormone pools, *J. Cell. Biol.* **83**:521.

Morrissey, J. J., Shofstall, R. E., Hamilton, J. W., and Cohn, D. V., 1980, Synthesis, intracellular distribution, and secretion of multiple forms of parathyroid secretory protein-I, *Proc. Natl. Acad. Sci.* **77**:6406.

Mueller, E., and Van Breemen, C., 1979, Role of intracellular Ca^{2+} sequestration in β-adrenergic relaxation of a smooth muscle, *Nature* **281**:682.

Mulder, A. H., and Snyder, S. H., 1974, Potassium-induced release of amino acids from cerebral cortex and spinal cord slices of the rat, *Brain Res.* **76**:297.

Mullins, L. J., 1978, The mechanisms of Ca transport in squid axons, in: *Membrane Transport Processes*, Vol. 2 (D. C. Tosteson, Y. A. Ovchinnikov, and R. Latorre, eds.), pp. 371–381, Raven Press, New York.

Mullins, L. J., 1979, The generation of electric currents in cardiac fibers by Na/Ca exchange, *Amer. J. Physiol.* **236**:C103.

Munger, B. L., and Roth, S. I., 1963, The cytology of the normal parathyroid glands of man and Virginia deer, *J. Cell Biol.* **16**:379.

Munson, P. L., 1976, Physiology and pharmacology of throcalcitonin, in: *Handbook of Physiology*, Section 7, Vol. VII (G. D. Aurbach, ed.), pp. 443–464, American Physiological Society, Washington, D.C.

Murad, F., Arnold, W. P., Mittal, C. K., and Braughler, J. M., 1979, Properties and regulation of guanylate cyclase and some proposed functions for cyclic GMP, *Adv. Cyclic Nucleotide. Res.* **11**:175.

Mustard, J. F., and Packham, M. A., 1970, Factors influencing platelet function: adhesion, release, and aggregation, *Pharmacol. Rev.* **22**:97.

Naccache, P. H., Showell, H. J., Becker, E. L, and Sha'afi, R. I. 1977a, Transport of sodium, potassium and calcium across rabbit polymorphonuclear leukocyte membranes, *J. Cell. Biol.* **73**:428.

Naccache, P. H., Showell, H. J., Becker, E. L., and Sha'afi, R. I., 1977b, Changes in ionic movements across rabbit polymorphonuclear leukocyte membranes during lysosomal enzyme release, *J. Cell. Biol.* **75**:635.

Naccache, P. H., Showell, H. J., Becker E. L., and Sha'afi, R. I., 1979, Involvement of membrane calcium in the response of rabbit neutrophils of chemotactic factors as evidenced by the fluorescence of chlorotetracycline, *J. Cell. Biol.* **83**:179.

Naccache, P. H., Sha'afi, R. I., Borgeat, P., and Goetzl, E. J., 1981, Mono- and dihydroxyeicosatetraenoic acids alter calcium homeostasis in rabbit neutrophils, *J. Clin. Invest.* **67**:1584.

Nachsen, D. A., and Blaustein, M. P., 1979, The effects of some organic "calcium antagonists" on calcium influx in presynaptic nerve terminals, *Mol. Pharmacol.* **16**:579.

Nakano, H., Fawcett, C. P., and McCann, S. M., 1976, Enzymatic dissociation and short-term culture of isolated rat anterior pituitary cells for studies on the control of hormone secretion, *Endocrinology* **98**:278.

Nakazato, Y., and Onoda, Y., 1980, Barium and strontium can substitute for calcium in noradrenaline output induced by excess potassium in the guinea pig, *J. Physiol. (London)* **305**:59.

Naor, Z., and Catt, K. J., 1980, Independent actions of gonadotropin releasing hormone upon cyclic AMP production and luteinizing hormone release, *J. Biol. Chem.* **255**:342.

Naor, Z., and Catt, K. J., 1981, Mechanism of action of gonadotropin releasing hormone. Involvement of phospholipid turnover in luteinizing hormone release, *J. Biol. Chem.* **256**:2226.

Naor, Z., Fawcett, C. P., and McCann, S. M., 1978, Involvement of cGMP in LHRH-stimulated gonadotropin release, *Am. J. Physiol.* **235**:E586.

Narahashi, T., Frazier, D. T., and Takeno, K., 1976, Effects of calcium on the local anesthetic suppression of ionic conductances in squid axon membranes, *J. Pharmacol. Exptl. Therap.* **197**:426.

Natke, Jr., E., and Kabela, E., 1979, Electrical responses in cat adrenal cortex: possible relation to aldostrone secretion, *Am. J. Physiol.* **237**:E158.

Nemeth, E. F., and Douglas, W. W., 1978, Effects of microfilament-active drugs phalloidin and the cytochalasins A and B, on exocytosis in mast cells evoked by 48/80 or A23187, *Naunyn Schmiedeberg's Arch. Pharmacol.* **302**:153.

Nestler, E. J., Bean, K. G., and Greengard, P., 1978, Nicotinic cholinergic stimulation increases cyclic GMP levels in vertebrate skeletal muscle, *Nature* **275**:451.

Newton, C., Pangborn, W., Nir, S., and Papahadjopoulos, D., 1978, Specificity of Ca^{2+} and Mg^{2+} binding to phosphatidylserine vesicles and resultant phase changes of bilayer membrane structure, *Biochim. Biophys. Acta* **506**:281.

Nicoll, R. A., and Alger, B. E., 1979, Presynaptic inhibition: Transmitter and ionic mechanisms, *Int. Rev. Neurobiol.* **21**:217.

Nicolson, G. L., and Poste, G., 1976, Cell shape changes and transmembrane receptor uncoupling induced by tertiary amine local anesthetics, *J. Supramol. Str.* **5**:65.

Nieuw-Amerongen, A. V., Roukema, P. A., and Vreugdenhil, A. P., 1980, Cyclic AMP in the sublingual glands of the mouse, *J. Physiol. (London)* **303**:83.

Nishimura, S., Sorimachi, M., and Yamagami, K., 1981, Exocytotic secretion of catecholamines from the cat adrenal medulla by sodium deprivation: Involvement of calcium influx mechanism, *Br. J. Pharmacol.* **72**:305.

Nishizuka, Y., Takai, Y., Hashimoto, E., Kishimoto, A., Kuroda, Y., Sakai, K., and Yamamura, H., 1979, Regulatory and functional compartment of three multifunctional protein kinase systems, *Mol. Cell. Biochem.* **23**:153.

Nordmann, J. J., 1976, Evidence for calcium inactivation during hormone release in the rat neurohypophysis, *J. Exp. Biol.* **65**:669.

Nordmann, J. J., and Currell, G. A., 1975, The mechanism of calcium ionophore-induced secretion from the rat neurohypophysis, *Nature* **253**:646.

Nordmann, J. J., and Dyball, R. E. J., 1978, Effects of veratridine on Ca fluxes and the release of oxytocin and vasopressin from the isolated rat neurohypophyses, *J. Gen. Physiol.* **72**:297.

Northover, B. J., 1977, Effect of indomethacin and related drugs on the calcium ion dependent secretion of lysosomal and other enzymes by neutrophil polymorphonuclear leucocytes *in vitro*, *Br. J. Pharmacol.* **59**:253.

Nussdorfer, G. C., Mazzocchi, G., and Meneghelli, V., 1978, Cytophysiology of the adrenal zona fasciculata, *Int. Rev. Cytol.* **55**:291.

Oberg, S. G., and Robinovitch, M. R., 1980, ATP-induced lysis of rat parotid secretory granules: Possible role of ATP in exocytotic release, *J. Supramol. Str.* **13**:295.

O'Doherty, J., Youmans, S. J., Armstrong, W. M., and Stark, R. J., 1980, Calcium regulation during stimulus-secretion coupling: Continuous measurement of intracellular calcium activities, *Science* **209**:510.

Oka, T., and Sawa, A., 1979, Calcium requirements for electrically-induced release of an endogenous opiate receptor ligand from the guinea pig ileum, *Br. J. Pharm.* **65**:3.

Olsen, R. W., Lamar, E. E., and Bayless, J. D., 1977, Calcium-induced release of γ-aminobutyric acid from synaptosomes. Effects of tranquilizer drugs, *J. Neurochem.* **28**:299.

Ontjes, D. A., 1980, The pharmacologic control of adrenal steroidogenesis, *Life Sci.* **26**:2023.

Opmeer, F. A., and van Ree, J. M., 1979, Competitive antagonism of morphine action in *vitro* by calcium, *Eur. J. Pharmacol.* **53**:395.

Orci, L., and Perrelet, A., 1977, Morphology of membrane systems in pancreatic islets, in: *The Diabetic Pancreas* (B. W. Volk and K. F. Wellmann, eds.), pp. 171–210, Plenum Press, New York.

Orci, L., and Perrelet, A., 1978, Ultrastructural aspects of exocytotic membrane fusion, *Cell. Surface Rev.* **5**:629.

Ornberg, R. L., and Reese, T. S., 1980, A freeze-substitution method for localizing divalent cations: examples from secretory systems, *Fed. Proc.* **39**:2802.

Osborn, J. L., Noordewier, B., Hook, J. B., and Bailie, M. D., 1978, Mechanism of prostaglandin E_2 stimulation of renin secretion, *Proc. Soc. Exp. Biol.* **159**:249.

Osborne, M. P., 1975, The ultrastructure of nerve muscle synapses, in: *Insect Muscle* (P. N. R. Usherwood, ed.), pp. 151–205, Academic Press, New York.

Otsuka, M., Iversen, L. L., Hall, Z. W., and Kravitz, E. A., 1966, Release of gamma-aminobutyric acid from inhibitory nerves of lobster, *Proc. Natl. Acad. Sci.* **56**:1110.

Ozawa, S., and Kimura, N., 1979, Membrane potential changes caused by thyrotropin-releasing hormone in the clonal GH_3 cell and their relationship to secretion of pituitary hormone, *Proc. Natl. Acad. Sci.* **76**:6017.

Ozawa, S., and Miyazaki, S., 1979, Electrical excitability in the rat clonal pituitary cell and its relation to hormone secretion, *Jap. J. Physiol.* **29**:411.

Pace, C. S., 1979, Activation of Na channels in islet cells: Metabolic and secretory effects, *Am. J. Physiol.* **237**:E130.

Palade, G. E., 1959, Functional changes in the structure of cell components, in: *Subcellular Particles* (Soc. of Gen. Physiologists Symp.) (T. Hayashi, ed.), pp. 64–80, Ronald Press, New York.

Papahadjopoulos, D., 1972, Studies on the mechanism of action of local anesthetics with phospholipid model membranes, *Biochim. Biophys. Acta* **265**:169.

Papahadjopoulos, D., 1978, Calcium-induced phase changes and fusion in natural and model membranes, in: *Membrane Fusion* (G. Poste and G. L. Nicolson, eds.), pp. 766–790, North Holland Publishing Company, Amsterdam.

Papahadjopoulos, D., Jacobson, K., Poste, G., and Shepherd, G., 1975, Effects of local anesthetics on membrane properties. 1. Changes in the fluidity of phospholipid bilayers, *Biochim. Biophys. Acta* **394**:504.

Papahadjopoulos, D., Portis, A., and Pangborn, W., 1978, Calcium-induced lipid phase transitions and membrane fusion, *Ann. N.Y. Acad. Sci.* **308**:50.

Park, C. S., and Malvin, R. L., 1978, Calcium in the control of renin release, *Am. J. Physiol.* **235**:F22.

Parod, R. S., and Putney, Jr., J. W., 1978, The role of calcium in the receptor mediated control of potassium permeability in the rat lacrimal gland, *J. Physiol. (London)* **281**:371.

Parod, R. S., and Putney, Jr. J. W., 1979, Stimulation of ^{45}Ca efflux from rat lacrimal gland slices by carbachol and epinephrine, *Life Sci.* **25**:2211.

Parod, R. J., and Putney, Jr. J. W., 1980, Stimulus–permeability coupling in rat lacrimal gland, *Am. J. Physiol.* **239**:G106.

Parod, R. J., Leslie, B. A., and Putney, Jr. J. W., 1980a, Muscarinic and α-adrenergic stimulation of Na and Ca uptake by dispersed lacrimal cells, *Am. J. Physiol.* **239**:G99.

Parod, R. S., Dambach, G. E., and Putney, Jr. J. W., 1980b, Membrane potential changes in lacrimal gland acinar cells elicited by carbachol and epinephrine, *J. Pharmacol. Exp. Ther.* **213**:473.

Parsons, J. A., 1970, Effects of cations on prolactin and growth hormone secretion by rat adenohypophyses *in vitro*, *J. Physiol. (London)* **210**:973.

Paton, W. D. M., Vizi, E. S., and Aboo Zar, M., 1971, The mechanism of acetylcholine release from parasympathetic nerves, *J. Physiol. (London)* **215**:819.

Patt, H. M., and Luckhardt, A. B., 1942, Relationship of a low blood calcium to parathyroid secretion, *Endocrinology* **31**:384.

Patzelt, C., Brown, D., and Jeanrenaud, B., 1977, Inhibitory effect of colchicine on amylase secretion by rat parotid glands. Possible localization in the Golgi area, *J. Cell. Biol.* **73**:578.

Payne, A. N., and Garland, L. G., 1978, Interaction between barium, strontium and calcium in histamine release by compound 48/80, *Europ. J. Pharmacol.* **52**:329.

Pearce, R. B., Cronshaw, J., and Holmes, W. N., 1977, The fine structure of the interrenal cells of the duck (Anas Platyrhynchos) with evidence for the possible exocytotic release of steroids, *Cell. Tissue Res.* **183**:203.

Pearse, A. G. E., 1976, Morphology and cytochemistry of thyroid and ultimobranchial C cells, in: *Handbook of Physiology*, Section 7, Vol. VII (G. D. Aurbach, ed.), pp. 411–421, American Physiological Society, Washington, D.C.

Pearse, A. G. E., and Takor, T. T., 1976, Neuroendocrine embryology and the APUD concept, *Clin. Endocrinology* **5**: (Suppl.):2295.

Pearson, G. T., Davison, J. S., Collins, R. C., and Petersen, O. H., 1981a, Control of enzyme secretion by non-cholinergic, non-adrenergic nerves in guinea pig pancreas, *Nature* **290**:259.

Pearson, G. T., Singh, J., Daoud, M. S., Davison, J. S., and Petersen, O. H., 1981b, Control of pancreatic cyclic nucleotides levels and amylase secretion by non-cholinergic, non-adrenergic nerves, *J. Biol. Chem.* **256**:11025.

Pecot-Dechavassine, M., 1976, Action of vinblastine on the spontaneous release of acetylcholine at the frog neuromuscular junction, *J. Physiol. (London)* **261**:31.

Pento, J. T., Glick, S. M., Kagan, A., and Gorfein, P. C., 1974, The relative influence of calcium, strontium and magnesium on calcitonin secretion in the pig, *Endocrinology* **94**:1176.

Peracchia, C., and Peracchia, L. L., 1980, Gap junction dynamics: Reversible effects of divalent cations, *J. Cell. Biol.* **87**:708.

Perchellet, J-P, and Sharma, R. K., 1979, Mediatory role of calcium and guanosine 3'5'-monophosphate in adrenocorticotropin induced steroidogenesis by adrenal cells, *Science* **203**:1259.

Perlman, R. L., Cossi, A. F., and Role, L. W., 1980, Mechanisms of ionophore-induced catecholamine secretion, *J. Pharmacol. Exptl. Therap.* **213**:241.

Pershadsingh, H. A., McDaniel, M. L., Landt, M., Bry, C. G., Lacy, P. E., and McDonald, J. M., 1980, Ca^{2+}-activated ATPase and ATP-dependent calmodulin-stimulated Ca transport in islet cell plasma membrane, *Nature* **288**:492.

Peterson, C., 1974, Role of energy metabolism in histamine release. *Acta Physiol. Scand.* (Suppl.)**413**:1.

Petersen, O. H., 1971, The effect of dinitrophenol on secretory potentials, secretion and potassium accumulation in the perfused cat submandibular gland, *Acta Physiol. Scand.* **80**:117.

Petersen, O. H., 1976, Electrophysiology of mammalian gland cells, *Physiol. Rev.* **56**:535.

Petersen, O. H., 1980, Exploring the exocrine pancreas by microelectrodes: Lessons from the first decade of pancreatic electrophysiology, in: *Biology of Normal and Cancerous Exocrine Pancreatic Cells* (A. Ribet, L. Pradayrol, and C. Susini, eds.), pp. 65–87, Elsevier, North Holland.

Petersen, O. H., and Iwatsuki, N., 1978, The role of calcium in pancreatic cell stimulus-secretion coupling: An electrophysiological approach. *Ann. N.Y. Acad. Sci.* **307**:599.

Petersen, O. H., and Iwatsuki, N., 1979, Hormonal control of cell to cell coupling in the exocrine pancreas, in: *Hormone Receptors in Digestion and Nutrition* (G. Rosselin, P. Fromageot, and S. Bonfils, eds.), pp. 191–202, North Holland, Biomedical Press, Amsterdam.

Petersen, O. H., and Ueda, N., 1976, Pancreatic acinar cells: the role of calcium in stimulus–secretion coupling, *J. Physiol. (London)* **254**:583.

Petroski, R. J., Naccache, P. H., Becker, E. L., and Sha'afi, R. I., 1979, Effect of chemotactic factors on calcium levels of rabbit neutrophils, *Am. J. Physiol.* **237**:C43.

Pfeiffer, D. R., Taylor, R. W., and Lardy, H. A., 1978, Ionophore A23187: Cation binding and transport properties, *Ann. N.Y. Acad. Sci.* **307**:402.

Piccolino, M., 1976, Strontium–calcium substitution in synaptic transmission in turtle retina, *Nature* **261**:504.

Pickett, W. C., Jesse, R. L., and Cohen, P., 1977, Initiation of phospholipase A_2 activity in human platelets by the calcium ionophore A23187, *Biochim. Biophys. Acta* **486**:209.

Pinto, J. E. B., and J. M. Trifaro, 1976, The different effects of D-600 (methoxyverapamil) on the release of adrenal catecholamines induced by acetylcholine, high potassium, or sodium deprivation, *Br. J. Pharmacol.* **57**:127.

Plattner, H., 1978, Fusion of cellular membranes, in: *Transport of Macromolecules in Cellular Systems* (S. C. Silverstein, ed.), pp. 465–488, Abahon, Verlagsgesellschaft, Berlin.

Plattner, H., Reichel, K., Matt, H., Beisson, J., Lefort-Tran, M., and Pouphile, M., 1980, Genetic dissection of the final exocytosis steps in *Paramecium Tetraurelia* cells: Cytochemical determination of Ca^{2+}-ATPase activity over preformed exocytosis sites, *J. Cell. Sci.* **46**:17.

Podesta, E. J., Dufau, M. L., Solano, A. R., and Catt, K. J., 1978, Hormonal activation of protein kinase in isolated Leydig cells, *J. Biol. Chem.* **253**:8994.

Podesta, E. J., Milani, A., Steffan, H., and Neher, R., 1980, Steroidogenic action of calcium ions in isolated adrenocortical cells, *Biochem. J.* **186**:391.

Poisner, A. M., 1973, Stimulus–secretion coupling in the adrenal medulla and posterior pituitary gland, in: *Frontiers in Neuroendocrinology* (W. F. Ganong and L. Martini, eds.), pp. 33–59, Oxford University Press, New York.

Pollard, H. B., Pazoles, C. J., Creutz, C. E., and Zinder, O., 1979, The chromaffin granule and possible mechanisms of exocytosis, *Int. Rev. Cytol.* **58**:159.

Pollard, H. B., Creutz, C. E., and Pazoles, C. J., 1981, Mechanisms of calcium action and release of vesicle-bound hormones during exocytosis, *Rec. Prog. Horm. Res.* **37**:299.

Ponnappa, B. C., and Williams, J. A., 1980, Effects of ionophore A23187 on calcium flux and amylase release in isolated mouse pancreatic acini, *Cell. Calcium* **1**:267.

Ponnappa, B. C., Dormer, R. L., and Williams, J. A., 1981, Characterization of an ATP-dependent Ca^{2+} uptake system in mouse pancreatic microsomes, *Am. J. Physiol.* **240**:G122.

Porter, K. R., 1976, Introduction: Motility in cells, in: *Cell Motility* (Book A) (R. Goldman, T. Pollard and J. Rosenbaum, eds.), pp. 1–28, Cold Spring Harbor Laboratory, Cold Spring Harbor.

Poste, G., and Allison, A. C., 1973, Membrane fusion, *Biochim. Biophys. Acta* **300**:421.

Prendergast, F. G., Allen, D. G., and Blinks, J. R., 1977, Properties of the calcium sensitive bioluminescent protein aequorin, in: *Calcium-Binding Proteins and Calcium Function* (R. H. Wasserman *et al.* eds.), pp. 469–490, North Holland, Amsterdam.

Pressman, B. C., 1976, Biological applications of ionophores, *Ann. Rev. Biochem.* **45**:501.

Prompt, C. A., and Quinton, P. M., 1978, Functions of calcium in sweat secretion, *Nature* **272**:171.

Prusch, R. D., and Hannafin, J. A., 1979, Sucrose uptake by pinocytosis in *Amoeba Proteus* and the influence of external calcium, *J. Gen. Physiol.* **74**:523.

Putney, Jr. J. W., 1978, Stimulus–permeability coupling: Role of calcium in the receptor regulation of membrane permeability, *Pharmacol. Rev.* **30**:209.

Putney, Jr. J. W., 1981, Recent hypotheses regarding the phosphatidylinositol effect, *Life Sci.* **29**:1183.

Putney, Jr. J. W., Parod, R. S., and Marier, S. H., 1977a, Control by calcium of protein discharge and membrane permeability to potassium in the rat lacrimal gland, *Life Sci.* **20**:1905.

Putney, Jr. J. W., Weiss, S. J., Leslie, B. A., and Marier, S. H., 1977b, Is calcium the final mediator of exocytosis in the rat parotid gland? *J. Pharmacol. Exp. Therap.* **203**:144.

Putney, Jr. J. W., VandeWalle, C. M., and Leslie, B. A., 1978a, Receptor control of calcium influx in parotid acinar cells, *Mol. Pharmacol.* **14**:1046.

Putney, Jr. J. W., VandeWalle, C. M., Leslie, B. A., 1978b, Stimulus–secretion coupling in the rat lacrimal gland, *Am. J. Physiol.* **235**:C188.

Quissell, D. O., 1980, Secretory response of dispersed rat submandibular glands. I. Potassium release, *Am. J. Physiol.* **230**:C90.

Quissell, D. O., and Barzen, K. A., 1980, Secretory response of dispersed rat submandibular glands. II. Mucin secretion. *Am. J. Physiol.* **238**:C99.

Rahamimoff, R., 1976, The role of calcium in neurotransmitter release at the neuromuscular junction, in: *Motor Innervation of Muscle* (S. Thesleff, ed.), pp. 117–149, Academic Press, New York.

Rahamimoff, R., Erulkar, S. D., Lev-Tov, A., and Meiri, H., 1978, Intracellular and extracellular calcium ions in transmitter release at the neuromuscular synapse, *Ann. N.Y. Acad. Sci.* **307**:583.

Ramp, W. K., Cooper, C. W., Ross, A. J., and Wells, S. A., 1979, Effects of calcium and cyclic nucleotides on rat calcitonin and parathyroid hormone secretion, *Mol. Cell. Endocr.* **14**:205.

Rasmussen, H., 1966, Mitochondrial ion transport: Mechanism and physiological significance, *Fed. Proc.* **25**:903.

Rasmussen, H., and Nagata, N., 1970, Hormones, cell calcium, and cyclic AMP, in: *Calcium and Cellular Function* (A. W. Cuthbert, ed.), pp. 198–213, MacMillan, New York.

Ravazzola, M., Malaisse-Lagae, M., Amherdt, M., Ravazzola, M., Malaisse, W. J., and Orci, L., 1976, Patterns of calcium localization in pancreatic endocrine cells, *J. Cell. Sci.* **27**:107.

Redburn, D. A., Stramler, J., and Potter, L. T., 1979, Inhibition by reserpine of calcium-dependent release of [³H]norepinephrine from synaptosomes depolarized with potassium or veratridine, *Biochem. Pharmacol.* **28**:2091.

Reed, P. W., and Lardy, H. A., 1972, A23187: A divalent cation ionophore. *J. Biol. Chem.* **247**:6970.

Reichlin, S., 1978, Introduction, in: *The Hypothalamus* (S. Reichlin, R. J. Baldessarini and J. B. Martin, eds.), pp. 1–14, Raven Press, New York.

Renckens, B. A. M., Van Emst-DeVries, S. E., DePont, J. J. H. H. M., and Bonting, S. L., 1980. Rat pancreas adenylate cyclase. VII. Effect of extracellular calcium on pancreozymin-induced cyclic AMP formation, *Biochim. Biophys. Acta* **630**:511.

Repke, D. I., Spivak, J. C., and Katz, A. M., 1976, Reconstitution of an active calcium pump in sarcoplasmic reticulum, *J. Biol. Chem.* **251**:3169.

Reuter, H., 1973, Divalent cations as charge carriers in excitable membranes, *Prog. Biophys. Mol. Biol.* **26**:1.

Ribalet, B., and Beigelman, P., 1979, Cyclic variation of K^+ conductance in pancreatic β-cells: Ca^{2+} and voltage dependence, *Am. J. Physiol.* **237**:C137.

Ribalet, B., and Beigelman, P. M., 1980, Calcium action potentials and potassium permeability activation in pancreatic β-cells, *Am. J. Physiol.* **239**:C124.

Richards, C. D., and Sercombe, R., 1970, Calcium, magnesium and the electrical activity of guinea pig olfactory cortex *in vitro*, *J. Physiol. (London)* **211**:571.

Richardson, U. I., and Schonbrunn, A., 1981, Inhibition of adrenocorticotropin secretion by somatostatin in pituitary cells in culture, *Endocrinology* **108**:281.

Rigler, G. L., Peake, G. T., and Ratner, A., 1978, Effect of luteinizing hormone releasing hormone on accumulation of pituitary cyclic AMP and GMP *in vitro*, *J. Endocr.* **76**:367.

Ringer, S., 1883, A further contribution regarding the influence of the different constituents of the blood on the contraction of the heart, *J. Physiol. (London)* **4**:29.

Rink, J. T., 1977, The influence of sodium on calcium movements and catecholamine release in thin slices of bovine adrenal medulla, *J. Physiol. (London)* **266**:297.

Ritchie, A. K., 1979, Catecholamine secretion in a rat pheochromocytoma cell line: Two pathways for calcium entry, *J. Physiol. (London)* **286**:541.

Rittenhouse-Simmons, S., 1979, Production of diglyceride from phosphatidylinositol in activated human platelets, *J. Clin. Invest.* **63**:580.

Robberecht, P., Conlon, T. P., and Gardner, J. D., 1976, Interaction of porcine vasoactive intestinal peptide with dispersed pancreatic acinar cells from the guinea pig, *J. Biol. Chem.* **251**:4635.

Robblee, L. S., and Shepro, D., 1976, The effect of external calcium and lanthanum on platelet calcium content and on the release reaction, *Biochim. Biophys. Acta.* **436**:448.

Roberts, M. L., and Petersen, O. H., 1978, Membrane potential and resistance changes induced in salivary gland acinar cells by microiontophoretic application of acetylcholine and adrenergic agonists, *J. Membrane Biol.* **39**:297.

Roberts, M. L., Iwatsuki, N., and Petersen, O. H., 1978, Parotid acinar cells: Ionic dependence of acetylcholine-evoked membrane potential changes, *Pflugers Arch.* **376**:159.

Robertson, R. P., 1979, Prostaglandins as modulators of pancreatic islet function, *Diabetes* **28**:943.

Robertson, R. P., and Chen, M., 1977, A role for prostaglandin E in defective insulin secretion and carbohydrate intolerance in diabetes mellitus, *J. Clin. Invest.* **60**:747.

Robison, G. A., Butcher, R. W., and Sutherland, E. W., 1971, *Cyclic AMP*, Academic Press, New York.

Rodan, G. A., and Feinstein, M. B., 1976, Interrelationship between Ca^{2+} and adenylate and guanylate cyclases in the control of platelet secretion and aggregation, *Proc. Natl. Acad. Sci.* **73**:1829.

Rodesch, F., Bogaert, Ch., and Dumont, J. E., 1976, Compartmentalization and movement of calcium in the thyroid, *Mol. Cell. Endocrinol.* **5**:303.

Roos, B. A., and Deftos, L. J., 1976, Calcitonin secretion *in vitro*. II. Regulatory effects of enteric mammalian polypeptide hormones on trout C-cell cultures, *Endocrinology* **98**:1284.

Roos, B. A., Bundy, L. L., Miller, E. A., and Deftos, L. J., 1975, Calcitonin secretion by monolayer cultures of human C-cells derived from medullary thyroid carcinoma, *Endocrinology* **97**:39.

Rose, B., and Loewenstein, W. R., 1975, Permeability of cell junction depends on local cytoplasmic calcium activity, *Nature* **254**:250.

Ross, E. M., and Gilman, A. E., 1980, Biochemical properties of hormone-sensitive adenylate cyclase, *Ann. Rev. Biochem.* **49**:533.

Rossignol, B., Herman, G., Chambaut, A. M., and Keryer, G., 1974, The calcium ionophore A23187 as a probe for studying the role of Ca^{2+} ions in the mediation of carbachol effects on rat salivary glands, *FEBS. Lett.* **43**:241.

Rothman, J. E., and Lenard, J., 1977, Membrane asymmetry, *Science* **195**:743.

Rothman, S. S., 1975, Enzyme secretion in the absence of zymogen granules, *Am. J. Physiol.* **228**:1828.

Rousset, B., Poncet, C., Dumont, J. E., and Mornex, R., 1980, Intracellular and extracellular sites of iodination in dispersed hog thyroid cells, *Biochem. J.* **192**:801.

Rubin, R. P., 1969, The metabolic requirements for catecholamine release from the adrenal medulla, *J. Physiol. (London)* **202**:197.

Rubin, R. P., 1970, The role of energy metabolism in calcium-evoked secretion from the adrenal medulla, *J. Physiol. (London)* **206**:181.

Rubin, R. P., 1974a, *Calcium and the Secretory Process,* Plenum Press, New York.

Rubin, R. P., 1974b, The role of calcium in drug action, in *Drug Interactions* (P. L. Morselli, S. Garattini, and S. N. Cohen, eds.), pp. 163–172, Raven Press, New York.

Rubin, R. P., 1981, Actions of calcium antagonists on secretory cells, in: *New Perspectives on Calcium Antagonists* (G. B. Weiss, ed.), pp. 147–158, Waverly Press, Baltimore.

Rubin, R. P., and Laychock, S. G., 1978a, Prostaglandins and calcium-membrane interactions in secretory glands, *Ann. N.Y. Acad. Sci.* **307**:377.

Rubin, R. P., and Laychock, S. G., 1978b, Prostaglandins, calcium and cyclic nucleotides in stimulus-response coupling, in: *Calcium in Drug Action* (G. B. Weiss, ed.), pp. 135–155, Plenum Press, New York.

Rubin, R. P., Feinstein, M. B., Jaanus, S. D., and Paimre, M., 1967, Inhibition of catecholamine secretion and calcium exchange in perfused cat adrenal glands by tetracaine and magnesium, *J. Pharmacol. Exptl. Therap.* **155**:643.

Rubin, R. P., Sheid, B., McCauley, R., and Laychock, S. G., 1974, ACTH-induced protein release from the perfused cat adrenal gland: Evidence for exocytosis? *Endocrinology* **95**:370.

Rubin, R. P., Sink, L. E., and Freer, R. J., 1981a, On the relationship between the formylmethionyl-leucyl-phenylalanine stimulation of arachidonyl phosphatidylinositol turnover and lysosomal enzyme secretion by rabbit neutrophils, *Mol. Pharmacol.* **19**:31.

Rubin, R. P., Sink, L. E., and Freer, R. J., 1981b, Activation of arachidonyl phosphatidylinositol turnover in rabbit neutrophils by the calcium ionophore A23187, *Biochem. J.* **194**:497.

Ruskell, G. L., 1975, Nerve terminals and epithelial cell variety in the human lacrimal gland, *Cell. Tiss. Res.* **158**:121.

Russell, J. T., and Thorn, N. A., 1975, Adenosine triphosphate dependent calcium uptake by subcellular fractions from bovine neurohypophyses, *Acta Physiol. Scand.* **93**:364.

Said, S. I., 1978, VIP: overview, in: *Gut Hormones* (S. R. Bloom and M. I. Grossman, eds.), pp. 465–469, Churchill Livingstone, Edinburgh.

Samuels, L. T., and Nelson, D. H., 1975, Biosynthesis of corticosteroids, in: *Handbook of Physiology,* Section 7, Vol. VI (H. Blaschko, G. Sayers, and A. D. Smith, eds.), pp. 55–68, American Physiological Society, Washington, D.C.

Samuelsson, B., Borgeat, P., Hammarstrom, S., and Murphy, R. C., 1980, Leukotrienes: A new group of biologically active compounds, *Adv. Prostaglandin Thromboxane Res.* **6**:1.

Sandermann, H., 1978, Regulation of membrane enzymes of lipids, *Biochim. Biophys. Acta* **515**:209.

Sawyer, H. R., Abel, J. H., McClellan, M. C., Schmitz, M., and Niswender, G. D., 1979, Secretory granules and progesterone secretion by ovine corpora lutea *in vitro*, *Endocrinology* **104**:476.

Sayers, G., Beall, R. J., and Seelig, S., 1972, Isolated adrenal cells: Adrenocorticotropic hormone, calcium, steroidogenesis and cyclic adenosine monophosphate, *Science* **175**:1131.

Scarpa, A., 1979, Measurement of calcium ion concentrations with metallochromic indicators, in: *Detection and Measurement of Free Ca²⁺ in Cells* (C. C. Ashley and A. K. Campbell, eds.), pp. 85–115, Elsevier, New York.

Scarpa, A., Brinley, F. J., Tiffert, T., and Dubyak, G. R., 1978, Metallochromic indicators of ionized calcium, *Ann. N.Y. Acad. Sci.* **307:**86.

Schaeffer, J. M., Axelrod, J., and Brownstein, M. J., 1977, Regional differences in dopamine mediated release of TRH-like material from synaptosomes, *Brain Res.* **138:**571.

Schanne, F. A. X., Kane, A. B., Young, E. E., and Farber, J. L., 1979, Calcium dependence of toxic cell death: A final common pathway, *Science* **206:**700.

Scharrer, E., and Scharrer, B., 1945, Neurosecretion, *Physiol. Rev.* **10:**171.

Scharrer, E., and Scharrer, B., 1954, Hormones produced by neurosecretory cells, *Rec. Prog. Horm. Res.* **10:**183.

Schatzmann, H. J., and Burgin, H., 1978, Calcium in human red blood cells, *Ann. N.Y. Acad. Sci.* **307:**125.

Scheele, G., and Haymovits, A., 1979, Cholinergic and peptide-stimulated discharge of secretory protein in guinea pig pancreatic lobules, *J. Biol. Chem.* **254:**10346.

Scheid, C. R., Honeyman, T. W., and Fay, F. S., 1979, Mechanism of β-adrenergic relaxation of smooth muscle, *Nature* **277:**32.

Schimmer, B., 1980, Cyclic nucleotides in hormonal regulation of adrenocortical function, *Adv. Cyclic Nucleotide Res.* **13:**181.

Schneyer, L. H., Young, J. A., and Schneyer, C. A., 1972, Salivary secretion of electrolytes, *Physiol. Rev.* **52:**720.

Schoepflin, G. S., Pickett, W., Austen, K. F., and Goetzl, E. J., 1977, Elevation of the cyclic GMP concentration in human platelets by sodium ascorbate and 5-hydroxytryptamine, *J. Cyclic Nucleotide Res.* **3:**355.

Schofield, J. G., and Bicknell, R. J., 1978, Effects of somatostatin and verapamil on growth hormone release and ⁴⁵Ca fluxes, *Mol. Cell. Endocr.* **9:**255.

Schramm, M., and Danon, D., 1961, The mechanism of enzyme secretion by the cell. I. Storage of amylase in the zymogen granules of the rat-parotis gland, *Biochim. Biophys. Acta* **50:**102.

Schramm, M., and Selinger, Z., 1974, The function of α- and β-adrenergic receptors and a cholinergic receptor in the secretory cell of rat parotid gland, in: *Advances in Cytopharmacology*, Vol. 2, *Cytopharmacology of Secretion* (B. Ceccarelli, J. Meldolesi, and F. Clementi, eds.), pp. 29–32, Raven Press, New York.

Schramm, M., and Selinger, Z., 1975, The functions of cyclic AMP and calcium as alternative second messengers in parotid gland and pancreas, *J. Cyclic Nucleotide Res.* **1:**181.

Schrey, M. P., and Rubin, R. P., 1979, Characterization of a calcium mediated activation of arachidonic acid turnover in adrenal phospholipids by corticotropin, *J. Biol. Chem.* **254:**11.

Schrey, M. P., Brown, B. L., and Ekins, R. P., 1978, Studies on the role of calcium and cyclic nucleotides in the control of TSH secretion, *Mol. Cell. Endocr.* **11:**249.

Schubart, U. K., Erlichman, J., and Fleischer, N., 1980a, The role of calmodulin in the regulation of protein phosphorylation and insulin release in hamster insulinoma cells, *J. Biol. Chem.* **255:**4120.

Schubart, U. K., Fleischer, N., and Erlichman, J., 1980b, Ca²⁺-dependent protein phosphorylation and insulin release in intact hamster insulinoma cells, *J. Biol. Chem.* **255:**11063.

Schulman, H., and Greengard, P., 1978, Stimulation of brain membrane protein phosphorylation by calcium and an endogenous heat-stable protein, *Nature* **271:**478.

Schulster, D., and Schwyzer, R., 1980, ACTH receptors, in: *Cellular Receptors for Hormones and Neurotransmitters* (D. Schulster and A. Levitzki, eds.), pp. 197–217, John Wiley and Sons, New York.

Schulz, I., 1980, Messenger role of calcium in function of pancreatic acinar cells, *Am. J. Physiol.* **239**:G335.

Schulz, I., and Stolze, H. H., 1980, The exocrine pancreas: The role of secre-tagogues, cyclic nucleotides and calcium in enzyme secretion, *Ann. Rev. Physiol.* **42**:127.

Scott, J. H., Creutz, C. E., and Pollard, H. B., 1982, Synexin binds in a calcium dependent fashion to chromaffin cell plasma membrane, *J. Cell. Biol.* (in press).

Seeman, P., 1972, The membrane actions of anesthetics and tranquilizers, *Pharmacol. Rev.* **24**:583.

Sehlin, J., and Taljedal, I. B., 1974, Transport of rubidium and sodium in pancreatic islets, *J. Physiol. (London)* **242**:505.

Sehlin, J., and Taljedal, I. B., 1979, ^{45}Ca $^{2+}$ uptake up dispersed pancreatic islet cells: Effects of d-glucose and the calcium probe, chlorotetracycline, *Pflügers. Arch.* **381**:281.

Seiler, D., and Hasselbach, W., 1971, Essential fatty acid deficiency and the activity of the sarcoplasmic calcium pump, *Eur. J. Biochem.* **21**:385.

Selinger, Z., and Naim, E., 1970, The effect of calcium on amylase secretion by rat parotid slices, *Biochim. Biophys. Acta.* **203**:335.

Selinger, Z., Eimerl, S., and Schramm, M., 1974, A calcium ionophore stimulating the action of epinephrine on the α-adrenergic receptor, *Proc. Natl. Acad. Sci.* **71**:128.

Sener, A., Kawazu, S., and Malaisse, W. J., 1980, The stimulus–secretion coupling of glucose-induced insulin release. Metabolism of glucose in K$^+$ deprived islets, *Biochem. J.* **186**:183.

Serhan, C. N., Korchak, H. M., and Weissmann, G., 1980, PGBx, a prostaglandin derivative, mimics the action of the calcium ionophore A23187 on human neutrophils, *J. Immunol.* **125**:2020.

Sha'afi, R. I., and Naccache, P. H., 1981, Ionic events in neutrophil chemotaxis and secretion, in: *Advances in Inflammation Research*, Vol. 3 (G. Weissmann, ed.), pp. 115–148, Raven Press, New York.

Sha'afi, R. I., Naccache, P. H., Alobaidi, T., Molski, T. F. P., and Volpi, M., 1981, Effect of arachidonic acid and the chemotactic factor F-Met-Leu-Phe on cation transport in rabbit neutrophils, *J. Cell. Physiol.* **106**:215.

Shanes, A. M., 1958, Electrochemical aspects in excitable cells, *Pharmacol. Rev.* **10**:59.

Shapiro, S., Kaneko, Y., Baum, S. G., and Fleischer, N., 1977, The role of calcium in insulin release from hamster insulinoma cells, *Endocrinology* **101**:485.

Sharp, G. W. G., Weidenkeller, D. E., Kaelin, D., Siegel, E. G., and Wollheim, C. B., 1980, Stimulation of adenylate cyclase by Ca^{2+} and calmodulin in rat islets of Langerhans, *Diabetes* **29**:74.

Shaw, F. D., and Morris, J. F., 1980, Calcium localization in the rat neurohypophysis, *Nature* **287**:56.

Sheppard, M. S., Kraicer, J., and Milligan, J. V., 1980, Mechanisms governing the release of growth hormone from acutely dispersed purified somatotrophs, in: *Synthesis and Release of Adenohypophyseal Hormones* (M. Jutisz and K. W. McKerns, eds.), pp. 495–522, Plenum Press, New York.

Sherwood, L. M., Potts, J. T., Care, A. D., Mayer, G. P., and Aurbach, G. D., 1966, Evaluation by radioimmunoassay of factors controlling the secretion of parathyroid hormone, *Nature* **209**:52.

Sherwood, L. M., Mayer, G. P., Ramberg, C. F., Kronfeld, D. S., Aurbach, G. D., and
Potts, J. T., 1968, Regulation of parathyroid hormone secretion: proportional control
by calcium, lack of effect of phosphate, *Endocrinology* **83**:1043.

Shier, W. T., Baldwin, J. H., Hamilton, M. N., Hamilton, R. T., and Thanass, N. M.,
1976, Regulation of guanylate and adenylate cyclase activities by lysolecithin, *Proc.
Nat. Acad. Sci.* **73**:1586.

Shima, S., Kawashima, Y., and Hirai, M., 1978, Studies on cyclic nucleotides in the adrenal
gland. VIII. Effects of angiotensin on adenosine 3',5'-monophosphate and steroidogenesis in the adrenal cortex, *Endocrinology* **103**:1361.

Shimizu, H., Creveling, C. R., and Daly, J. W., 1970, Cyclic adenosine 3',5'-monopho-
sphate formation in brain slices: Stimulation by batrachotoxin, ouabain, veratridine
and potassium ions, *Mol. Pharmacol.* **6**:184.

Showell, H. J., Naccache, P. H., Sha'afi, R. I., and Becker, E. L., 1977, The effects of
extracellular K^+, Na^+, Ca^{++} on lysosomal enzyme secretion from polymorphonuclear
leukocytes, *J. Immunol.* **119**:804.

Siegel, E. G., Wollheim, C. B., and Sharp, G. W. G., 1980, Glucose-induced first phase
insulin release in the absence of extracellular Ca^{2+} in rat islets, *FEBS. Lett.* **109**:213.

Sihusdziarra, V., Ipp, E., Harris, V., Dobbs, R. E., Raskin, P., Orci, L., and Unger, R. H.,
1978, Studies of the physiology and pathophysiology of the pancreatic D cell, *Metabolism* **27** (Suppl. 1):1227.

Silinsky, E. M., 1981, On the calcium receptor that mediates depolarization-secretion
coupling at cholinergic motor nerve terminals, *Brit. J. Pharmacol.* **73**:413.

Simon, W., Ammann, D., Oehme, M., and Morf, W. E., 1978, Calcium-selective electrodes,
Ann. N.Y. Acad. Sci. **307**:52.

Simons, T. J. B., 1975, Resealed ghosts used to study the effect of intracellular calcium
ions on the potassium permeability of human red cell membranes, *J. Physiol. (London)*
246:52P.

Simpson, E. R., Waters, J., and Williams-Smith, D., 1975, Effect of calcium on pregnenolone formation and cytochrome P-450 in rat adrenal mitochondria, *J. Steroid. Biochem.*
6:395.

Singh, M., 1980, Stimulus–secretion coupling in rat pancreas: Role of sodium, calcium and
cyclic nucleotides studied by X-537A and BrX-537A, *J. Physiol. (London)* **302**:1.

Siraganian, R. P., and Hook, W. A., 1977, Mechanism of histamine release by formyl-
methionine-containing peptides, *J. Immunol.* **119**:2078.

Skirboll, L. R., Baizer, L., and Dretchen, K. L., 1977, Evidence for a cyclic nucleotide-
mediated calcium flux in motor nerve terminals, *Nature* **268**:352.

Smith, A. D., 1972, Subcellular localization of noradrenaline in sympathetic neurons,
Pharmacol. Rev. **24**:435.

Smith, R. E., and Farquhar, M. G., 1966, Lysosome function in the regulation of the
secretory processes in cells of the anterior pituitary gland, *J. Cell. Biol.* **31**:319.

Smith, R. J., and S. S. Iden, 1979, Phorbol myristate acetate-induced release of granule
enzymes from human neutrophils: Inhibition of the calcium antagonist 8-(N,N-dieth-
ylamino)-octyl, 3,4,5 trimethylbenzoate hydrochloride, *Biochem. Biophys. Res. Commun.*
91:263.

Smith, R. J., and Ignarro, L. J., 1975, Regulation of lysosomal enzyme secretion from
human neutrophils: Roles of guanosine 3',5'-monophosphate and calcium in stimulus-
secretion coupling, *Proc. Natl. Acad. Sci.* **72**:108.

Smith, R. J., Wierenga, W., and Iden, S. S., 1980, Characteristics of N-formyl-methionyl-
leucyl-phenylalanine as an inducer of lysosomal enzyme release from human neutrophils, *Inflammation* **4**:73.

Smolen, J. E., and Weissmann, G., 1981, The secretion of lysosomal enzymes from human neutrophils: The first events in stimulus–secretion coupling, in: *Lysosomes and Lysosomal Storage Disease* (Callahan, J. W., and Lowden, J. A., eds.), pp. 31–62, New York, Raven Press.

Sneddon, J. M., 1972, Divalent cations and the blood platelet release reaction, *Nature* **236**:103.

Snyder, G., Naor, Z., Fawcett, C. P., and McCann, S. M., 1980, Gonadotropin release and cyclic nucleotides: evidence for luteinizing hormone-releasing hormone-induced elevation of guanosine 3',5'-monophosphate levels in gonadotrophs, *Endocrinology* **107**:1627.

Snyder, S. H., 1980, Brain peptides as neurotransmitters, *Science* **209**:976.

Snyderman, R., and Pike, M. C., 1980, N-formylmethionyl peptide receptors on equine leukocytes initiate secretion but not chemotaxis, *Science* **209**:493.

Soll, A. H., and Walsh, J. H., 1979, Regulation of gastric acid secretion, *Ann. Rev. Physiol.* **41**:35.

Soll, A. H., and Whittle, B. J. R., 1981, Prostacyclin analogues inhibit canine parietal cell activity and cyclic AMP formation, *Prostaglandins* **21**:353.

Soll, A. H., Rodrigo, R., and Ferrari, J. C., 1981, Effects of chemical transmitters on function of isolated canine parietal cells, *Fed. Proc.* **40**:2519.

Somlyo, A. P., and Somlyo, A. V., 1968, Vascular smooth muscle, *Pharmacol. Rev.* **20**:197.

Spaulding, S. W., 1979, Nitrogenous compounds reversibly inhibit the adenosine 3',5'-monophosphate response to thyrotropin: This effect is dissociated from altered guanine 3',5'-monophosphate levels, *Endocrinology* **105**:697.

Spearman, T. N., and Pritchard, E. T., 1979, Role of cyclic GMP in submandibular gland secretion, *Biochim. Biophys. Acta* **588**:55.

Spence, J. W., Sheppard, M. S., and Kraicer, J., 1980, Release of growth hormone from purified somatotrophs: interrelation between Ca^{++} and adenosine 3',5'-monophosphate, *Endocrinology* **106**:764.

Starke, K., 1977, Regulation of noradrenaline release by presynaptic receptor systems, *Rev. Physiol. Biochem. Pharmacol.* **77**:1.

Strong, C. G., and Bohr, D. F., 1967, Effect of prostaglandins, E_1, E_2, A_1, and $F_2\alpha$ on isolated vascular smooth muscle, *Amer. J. Physiol.* **213**:725.

Stroobant, P., Dame, J. B., and Scarborough, G. A., 1980, The *Neurospora* plasma membrane Ca^{2+} pump, *Fed. Proc.* **39**:2437.

Sugden, M. C., and Ashcroft, S. J. H., 1978, Effects of phosphoenolpyruvate, other glycolytic intermediates and methylxanthines on calcium uptake by a mitochondrial fraction from rat pancreatic islets, *Diabetologia* **15**:173.

Sugden, M. C., Christie, M. R., and Ashcroft, S. J. H., 1979, Presence and possible role of calcium-dependent regulator (calmodulin) in rat islets of Langerhans, *FEBS. Lett.* **105**:95.

Sugiyama, K., 1971, Calcium-dependent histamine release with degranulation from isolated rat mast cells by adenosine 5'-triphosphate, *Jap. J. Pharmacol.* **21**:209.

Sulakhe, P. V., and St. Louis, P. J., 1976, Membrane phosphorylation and calcium transport in cardiac and skeletal muscle membranes, *Gen. Pharmacol.* **7**:313.

Sun, G. Y., 1979, On the membrane phospholipids and their acyl group profiles of adrenal glands, *Lipids* **14**:918.

Suszkiw, J. B., Zimmermann, H., and Whittaker, V. P., 1980, Vesicular storage and release of acetylcholine in *Torpedo* electroplaque synapses, *J. Neurochem.* **30**:1269.

Sydbom, A., and Uvnäs, B., 1976, Potentiation of anaphylactic histamine release from isolated rat pleural mast cells by rat serum phospholipids, *Acta Physiol. Scand.* **97**:222.

Szerb, J. C., 1979, Relationship between Ca^{2+}-dependent and independent release of [^{3}H] GABA evoked by high K^+, veratridine or electrical stimulation from rat cortical slices, *J. Neurochem.* **32**:1565.

Tada, M., Ohmori, F., Kinoshita, N., and Abe, H., 1978, Cyclic AMP regulation of active calcium transport across membranes of sarcoplasmic reticulum: role of the 22,000 dalton protein phospholamban, *Adv. Cyclic Nucleotide. Res.* **9**:355.

Takai, Y., Kishimoto, A., Iwasa, Y., Kawahara, Y., Mori, T., and Nishizuka, Y., 1979, Calcium-dependent activation of a multifunctional protein kinase by membrane phospholipids, *J. Biol. Chem.* **254**:3692.

Takasu, N., Yamada, T., and Shimizu, Y., 1981, An important role of prostacyclin in porcine thyroid cells in culture, *FEBS. Lett.* **129**:83.

Tam, S. W., and Dannies, P. S., 1980, Dopaminergic inhibition of ionophore A23187-stimulated release of prolactin from rat anterior pituitary cells, *J. Biol. Chem.* **255**:6595.

Taraskevich, P. S., and Douglas, W. W., 1977, Action potentials occur in cells of the normal anterior pituitary gland and are stimulated by the hypophysiotropic peptide thyrotropin-releasing hormone, *Proc. Natl. Acad. Sci.* **24**:4064.

Taraskevich, P. S., and Douglas, W. W., 1978, Catecholamines of supposed inhibitory hypophysiotrophic function suppress action potentials in prolactin cells, *Nature* **276**:832.

Taraskevich, P. S., and Douglas, W. W., 1980, Electrical behavior in a line of anterior pituitary cells (GH cells) and the influence of the hypothalamic peptide thyrotrophin releasing factor, *Neuroscience* **5**:421.

Targovnik, J. H., Rodman, J. S., and Sherwood, L. M., 1971, Regulation of parathyroid hormone secretion *in vitro*: quantitative aspects of calcium and magnesium ion control, *Endocrinology* **88**:1477.

Tashjian, A. H., Lomedico, M. E., and Maina, D., 1978, Role of calcium in the thyrotropin releasing hormone-stimulated release of prolactin from pituitary cells in culture, *Biochem. Biophys. Res. Comm.* **81**:798.

Taub, M., and Saier, Jr. M. H., 1979, Regulation of ^{22}Na transport by calcium in an established kidney epithelial cell line, *J. Biol. Chem.* **254**:11440.

Tauc, L., 1979, Are vesicles necessary for release of acetylcholine at cholinergic synapses? *Biochem. Pharmacol.* **27**:3493.

Taurog, A., 1978, Thyroid hormone: Synthesis and release, in: *The Thyroid* (S. C. Werner and S. I. Ingbar, eds.), pp. 31–61, Harper and Row, New York.

Taylor, D. L., Hellelwell, S. B., Virgin, H. W., and Heiple, J., 1978, The solation-contraction coupling hypothesis of cell movements, in: *Cell Motility: Molecules and Organization* (S. Hatano, H. Ishikawa, and H. Sato, eds.), pp. 363–377, University Park Press, Baltimore.

Theoharides, T. C., and Douglas, W. W., 1978, Secretion in mast cells induced by calcium entrapped within phospholipid vesicles, *Science* **201**:1143.

Thoa, N. B., Costa, J. L., Moss, J., and Kopin, I. J., 1974, Mechanism of release of norepinephrine from peripheral adrenergic neurones by the calcium ionophores X-537A and A23187, *Life Sci.* **14**:1705.

Thoa, N. B., Wooten, F., Axelrod, J., and Kopin, I. J., 1975, On the mechanism of release of norepinephrine from sympathetic nerves induced by depolarizing agents and sympathomimetric drugs, *Mol. Pharmacol.* **11**:10.

Thoenen, H., Huerlimann, A., and Haefely, W., 1969, Cation dependence of the noradrenaline-releasing action of tyramine, *Europ. J. Pharmacol.* **6**:29.

Thompson, M. E., Orczyk, G. P., and Hedge, G. A., 1977, *In vivo* inhibition of thyroid secretion by indomethacin, *Endocrinology* **100**:1060.

Thompson, W. J., Chang, L. K., and Rosenfeld, G. C., 1981, Histamine regulation of adenylyl cyclase of enriched rat gastric parietal cells, *Amer. J. Physiol.* **240**:G76.

Thorn, N. A., Russell, J. T., and Robinson, I. C. A. F., 1975, Factors affecting intracellular concentration of free calcium ions in neurosecretory nerve endings, in: *Calcium Transport in Contraction and Secretion* (E. Carafoli, F. Clementi, W. Drabikowski, and A. Margreth, eds.), pp. 261–275, North-Holland Publishing Company, Amsterdam.

Thorn, N. A., Russell, J. T., Torp-Pedersen, C., and Treiman, M., 1978, Calcium and Neurosecretion, *Ann. N.Y. Acad. Sci.* **307**:618.

Tidball, M. E., and Scherer, R. W., 1972, Relationship of calcium and magnesium to platelet histamine release, *Am. J. Physiol.* **222**:1303.

Tischler, A. S., Dichter, M. A., Biales, B., DeLellis, R. A., and Wolfe, H., 1976, Neural properties of cultured human endocrine tumor cells of proposed neural crest origin, *Science* **192**:902.

Tobias, J. M., 1964, A chemically specified molecular mechanism underlying excitation in nerve: A hypothesis, *Nature* **203**:13.

Topol, E., and Brodows, R. G., 1980, Effects of indomethacin on acute insulin release in man *Diabetes* **29**:379.

Torp-Pedersen, C., Saermark, T., Bundgaard, M., and Thorn, N. A., 1980, ATP-dependent Ca^{2+} accumulation by microvesicles isolated from bovine neurohypophyses, *J. Neurochem.* **35**:552.

Trifaro, J. M., 1977, Common mechanisms of hormone secretion, *Ann. Rev. Pharmacol. Toxicol.* **17**:27.

Triggle, D. J., 1980, Receptor–hormone interrelationships, in: *Membrane Structure and Function*, Vol. 3, (E. E. Bittar, ed.), pp. 1–58, John Wiley and Sons, New York.

Triggle, D. J., 1981, Calcium antagonists: Basic chemical and pharmacological aspects, in: *New Perspectives on Calcium Antagonists* (G. B. Weiss, ed.), pp. 1–18, Waverly Press, Baltimore.

Urry, D. W., 1978, Basic aspects of calcium chemistry and membrane interaction: on the messenger role of calcium, *Ann. N.Y. Acad. Sci.* **307**:3.

Usherwood, P. N. R., 1978, Amino acids as neurotransmitters, *Adv. Comp. Physiol. Biochem.* **7**:227.

Uvnäs, B., 1973, An attempt to explain nervous transmitter release as due to nerve impulse-induced cation exchange, *Acta Physiol. Scand.* **87**:168.

Uvnäs, B., and Thon, I. L., 1961, Evidence for enzymatic histamine release from isolated mast cells, *Exptl. Cell. Res.* **23**:45.

Vale, W., and Guillemin, R., 1967, Potassium induced stimulation of thyrotropin release *in vitro:* Requirement for presence of calcium and inhibition by thyroxine, *Experientia* **23**:855.

Vale, W., Rivier, C., and Brown, M., 1977, Regulatory peptides of the hypothalamus, *Ann. Rev. Physiol.* **39**:473.

Van Breemen, C., Farinas, B., Gerba, P., and McNaughton, E., 1972, Excitation contraction coupling in rabbit aorta studied by the lanthanum method for measuring cellular calcium influx, *Circ. Res.* **30**:44.

Van den Hove-Vandenbroucke, M. F., 1980, Secretion of thyroid hormones, in: *The Thyroid Gland* (M. DeVisscher, ed.), pp. 61–79, Raven Press, New York.

Vanderhoek, J. V., and Feinstein, M. B., 1979, Local anesthetics, chlorpromazine and propranolol inhibit stimulus-activation of phospholipase A_2 in human platelets, *Mol. Pharmacol.* **16**:171.

Van der Vusse, G. J., Kalkman, M. L., Van Winsen, M. P. I., and Van der Molen, H. J., 1975, On the regulation of rat testicular steroidogenesis, *Biochim. Biophys. Acta* **398**:28.

Van der Vusse, G. J., Kalkman, M. L., Van Winsen, M. P. I., and Van der Molen, H. J., 1976, Effect of Ca^{2+}, ruthenium red, and ageing on pregnenolone production by mitochondrial fractions from normal and luteinizing hormone treated rat testes, *Biochim. Biophys. Acta* **428**:420.

Van Dongen, R., and Peart, W. S., 1974, Calcium dependence of the inhibitory effect of angiotensin on renin secretion in the isolated perfused kidney of the rat, *Br. J. Pharmacol.* **50**:125.

Vila-Porcile, E., and Olivier, L., 1978, Exocytosis and related membrane events, in: *Synthesis and Release of Adenohypophyseal Hormones* (M. Jutisz and K. W. McKerns, eds.), pp. 67–103, Plenum Press, New York.

Viveros, O. H., 1975, Mechanism of secretion of catecholamines from adrenal medulla, in: *Handbook of Physiology*, Section 7, Vol. VI, (H. Blaschko, G. Sayers, and A. D. Smith, eds.), pp. 389–426, American Physiological Society, Washington, D.C.

Viveros, O. H., Diliberto, E. J., Hazum, E., and Chang, K. J., 1979, Opiate-like materials in the adrenal medulla: Evidence for storage and secretion with catecholamines, *Mol. Pharmacol.* **16**:1101.

Viveros, O. H., Diliberto, E. J., Hazum, E., and Chang, K. J., 1980, Enkephalins as possible adrenomedullary hormones: storage, secretion and regulation of synthesis, in: *Neural Peptides and Neuronal Communication* (E. Costa and M. Trabucchi, eds.), pp. 191–204, Raven Press, New York.

Vora, N. M., Williams, G. A., Hargis, G. K., Bowser, E. N., Kawahara, W., Jackson, B. L., Henderson, W. J., and Kukreja, S. C., 1978, Comparative effect of calcium and of the adrenergic system on calcitonin secretion in man, *J. Clin. Endocrinol. Metab.* **46**:567.

Vreugdenhil, A. P., and Roukema, P. A., 1975, Comparison of the secretory processes in the parotid and sublingual glands of the mouse, *Biochim. Biophys. Acta* **413**:79.

Wakui, M., and Nishiyama, A., 1980, Ionic dependence of acetylcholine equilibrium potential of acinar cells in mouse submaxillary gland, *Pflugers Arch.* **386**:261.

Walker, M. W., and Farquhar, M. G., 1980, Preferential release of newly synthesized prolactin granules is the result of functional heterogeneity among mammotrophs, *Endocrinology* **107**:1095.

Walker, A. M., and Hopkins, C. R., 1978, Stimulation of luteinizing hormone release by luteinizing hormone-releasing hormone in the porcine anterior pituitary: The role of cyclic AMP, *Mol. Cell. Endocr.* **10**:327.

Waller, M. B., and Richter, J. A., 1980, Effects of pentobarbital and Ca^{2+} on the resting and K$^+$-stimulated release of several endogenous neurotransmitters from rat midbrain slices, *Biochem. Pharmacol.* **29**:2189.

Warberg, J., Eskay, R. L., Barnea, A., Reynolds, R. C., and Porter, J. C., 1977, Release of luteinizing hormone releasing hormone and thyrotropin releasing hormone from a synaptosome enriched fraction of hypothalamic homogenates, *Endocrinology* **100**:814.

Warner, W., and Carchman, R. A., 1978, Effects of ruthenium red, A23187, and D-600 on steroidogenesis in Y-1 cells, *Biochim. Biophys. Acta* **528**:409.

Watson, E. L., Williams, J. A., and Siegel, I. A., 1979, Calcium mediation of cholinergic stimulated amylase release from mouse parotid gland, *Amer. J. Physiol.* **236**:C233.

Weakly, J. N., 1973, The action of cobalt ions on neuromuscular transmission in the frog, *J. Physiol. (London)* **234**:597.

Weber, A., 1976, Synopsis of the presentations, *Symp. Soc. Exptl. Biol.* **30**:445.

Weiner, N., 1979, Multiple factors regulating the release of norepinephrine consequent to nerve stimulation, *Fed. Proc.* **38**:2193.

Weinshilboum, R. M., 1979, Serum dopamine β-hydroxylase, *Pharmacol. Rev.* **30**:133.

Weiss, G. B., 1974, Cellular pharmacology of lanthanum, *Ann. Rev. Pharmacol.* **14**:343.

Weiss, G. B., 1981, Sites of action of calcium antagonists in vascular smooth muscle, in: *New Perspectives on Calcium Antagonists* (G. B. Weiss, ed.), pp. 83–94, Waverly Press, Baltimore.

Weissmann, G., Goldstein, I. and Hoffstein, S., 1976, Prostaglandins and the modulation by cyclic nucleotides of lysosomal enzyme release, *Adv. Prostaglandin Thromboxane Res.* **2**:803.

Weissmann, G., Anderson, P., Serhan, C., Samuelsson, E. and Goodman, E., 1980, A general method, employing arsenazo III in liposomes, for study of calcium ionophores: Results with A23187 and prostaglandins, *Proc. Natl. Acad. Sci.* **77**:1506.

Wellman, K. F., and Volk, B. W., 1977, Histology, cell types and functional correlation of islets of Langerhans, in: *The Diabetic Pancreas* (B. W. Volk and K. F. Wellmann, eds.), pp. 99–119, Plenum Press, New York.

Wells, J. N., and Hardman, J. G., 1977, Cyclic nucleotide-phosphodiesterases, *Adv. Cyclic Nucleotide Res.* **8**:119.

Westfall, T. C., 1977, Local regulation of adrenergic neurotransmission, *Physiol. Rev.* **57**:659.

White, J. G., 1972, Interaction of membrane systems in blood platelets, *Am. J. Path.* **66**:295.

White, T. D., 1978, Release of ATP from a synaptosomal preparation by elevated extracellular K^+ and by veratridine, *J. Neurochem.* **30**:329.

Whitfield, J. F., Boynton, A. L., MacManus, J. P., Sikorska, M., and Tsang, B. K., 1979, The regulation of cell proliferation by calcium and cyclic AMP, *Mol. Cell. Biochem.* **27**:155.

Whittaker, V. P., 1974, Molecular organization of the cholinergic vesicle, in: *Advances in Cytopharmacology*, Vol. 2, *Cytopharmacology of Secretion* (B. Ceccarelli, F. Clementi, and J. Meldolesi, eds.), pp. 311–317, Raven Press, New York.

Whorton, A. R., Misono, K., Hollifield, J., Frolich, J. C., Inagami, T. and Oates, J. A., 1977, Prostaglandins and renin release: I. Stimulation of renin release from rabbit renal cortical slices by PGI_2, *Prostaglandins* **14**:1095.

Wilcox, C. S., 1978, The effect of increasing the plasma magnesium concentration on renin release from the dog's kidney: Interactions with calcium and sodium, *J. Physiol. (London)* **284**:203.

Williams, G. A., Hargis, G. K., Bowser, E. N., Henderson, W. J., and Martinez, N. J., 1973, Evidence for a role of adenosine 3′-5′-monophosphate in parathyroid hormone release, *Endocrinology* **92**:687.

Williams, J. A., 1970, Effects of TSH on thyroid membrane properties, *Endocrinology* **86**:1154.

Williams, J. A., 1972a, Effects of Na^+ and K^+ on secretion *in vitro* by mouse thyroid glands, *Endocrinology* **90**:1452.

Williams, J. A., 1972b, Effects of Ca^{++} and Mg^{++} on secretion *in vitro* by mouse thyroid glands, *Endocrinology* **90**:1459.

Williams, J. A., 1980a, Regulation of pancreatic acinar cell function by intracellular calcium, *Am. J. Physiol.* **238**:G269.

Williams, J. A., 1980b, Multiple effects of Na^+ removal on pancreatic secretion *in vitro*, *Cell. Tissue Res.* **210**:295.

Williams, J. A., 1981. Electrical correlates of secretion in endocrine and exocrine cells, *Fed. Proc.* **40**:128.

Williams, J. A., and Chandler, D., 1975, Ca^{++} and pancreatic amylase release, *Amer. J. Physiol.* **228**:1729.

Williams, R. J. P., 1970, The biochemistry of sodium, potassium, magnesium, and calcium, *Quart. Rev. Chem. Soc.* **24**:331.

Williams, R. J. P., 1977, Calcium chemistry and its relation to protein binding, in: *Calcium-Binding Proteins and Calcium Function* (R. H. Wasserman et. al, eds.), pp. 3–12, North-Holland, Amsterdam.

Winkler, H., and Smith, A. D., 1975, The chromaffin granule and the storage of catecholamines, in: *Handbook of Physiology*, Vol. VI, Section 7, (H. Blaschko, G. Sayers, and A. D. Smith, eds.), pp. 321–339, American Physiological Society, Washington, D.C.

Winkler, H., Schneider, F. H., Rufener, C., Nakane, P. K., and Hortnagl, H., 1974, Membranes of adrenal medulla: Their role in exocytosis, in Advances in Cyto-pharmacology, Vol. 2, *Cytopharmacology of Secretion* (B. Ceccarelli, J. Meldolesi, and F. Clementi, eds.), pp. 127–139, Raven Press, New York.

Wolff, J., and Cook, G. H., 1973, Activation of thyroid membrane adenylate cyclase by purine nucleotides, *J. Biol. Chem.* **248**:350.

Wolff, J., and Williams, J. A., 1973, The role of microtubules and microfilaments in thyroid secretion, *Rec. Prog. Horm. Res.* **29**:229.

Wollheim, C. B., Blondel, B., Renold, A. E., and Sharp, G. W. G., 1977, Somatostatin inhibition of pancreatic glucagon release from monolayer cultures and interactions with calcium, *Endocrinology* **101**:911.

Wong, P. Y. K., Lee, W. H., and Chao, P. H-W., 1980, The role of calmodulin in prostaglandin metabolism, *Ann. N.Y. Acad. Sci.* **356**:179.

Woodin, A. M., 1968, The basis of leucocidin action, in: *The Biological Basis of Medicine*, Vol. 2 (E. E. Bittar and N. Bittar, eds.), pp. 373–396, Academic Press, New York.

Woodin, A. M., and Wieneke, A., 1970, Site of protein secretion and calcium accumulation in the polymorphonuclear leucocyte treated with leucocidin, in: *Calcium and Cellular Function* (A. W. Cuthbert, ed.), pp. 183–197, MacMillan, New York.

Wooding, F. B. P. and Morgan, G., 1978, Calcium localization in lactating rabbit mammary secretory cells, *J. Ultrast. Res.* **63**:323.

Woods, S. C., and Porte, Jr., D., 1974, Neural control of the endocrine pancreas, *Physiol. Rev.* **54**:596.

Wright, D. G., Bralove, D. A., and Gallin, J. J., 1977, The differential mobilization of human neutrophil granules, *Am. J. Path.* **87**:273.

Yoshikami, S., and Hagins, W. A., 1978, Calcium in excitation of vertebrate rods and cones, *Ann. N.Y. Acad. Sci.* **307**:545.

Zenser, T. V., and Davis, B. B., 1978, Effects of calcium on prostaglandin E_2 synthesis by rat inner medullary slices, *Am. J. Physiol.* **235**:F213.

Zenser, T. V., Herman, C. A., and Davis, B. B., 1980, Effects of calcium and A23187 on renal inner medullary prostaglandin E_2 synthesis, *Amer. J. Physiol.* **238**:E371.

Zimmermann, H., 1979, Vesicular heterogeneity and turnover of acetylcholine and ATP in cholinergic synaptic vesicles, *Prog. Brain Res.* **49**:141.

Zimmermann, H., and Whittaker, V. P., 1974, Effect of electrical stimulation on the yield and composition of synaptic vesicles from the cholinergic synapses of the electric organ of *Torpedo*, *J. Neurochem.* **22**:435.

Zusman, R. M., and Keiser, H. R., 1977, Prostaglandin E biosynthesis by rabbit renomedullary interstitial cells in tissue culture, *J. Biol. Chem.* **252**:2069.

Zwiller, J., Ciesielski-Treska, J., and Mandel, P., 1976, Effect of lysolecithin on guanylate and adenylate cyclase activities in neuroblastoma cells in culture. *FEBS Lett.* **69**:286.

Index